Nematode Diseases of Vegetable Crops and their Management

Nematode Diseases of Vegetable Crops and their Management

Dr. P. PARVATHA REDDY

Former Director
Indian Institute of Horticultural Research
Bangalore

CRC Press
Taylor & Francis Group
Boca Raton London New York

CRC Press is an imprint of the
Taylor & Francis Group, an **informa** business

-EPH-

First published 2025
by CRC Press
4 Park Square, Milton Park, Abingdon, Oxon, OX14 4RN

and by CRC Press
2385 NW Executive Center Drive, Suite 320, Boca Raton FL 33431

CRC Press is an imprint of Informa UK Limited

British Library Cataloguing-in-Publication Data
A catalogue record for this book is available from the British Library

Print edition not for sale in South Asia (India, Sri Lanka, Nepal, Bangladesh, Pakistan or Bhutan).

ISBN: 9781041032694(hbk)
ISBN: 9781041032700(pbk)
ISBN: 9781003623113(ebk)

DOI: 10.1201/9781003623113

Typeset in A Caslon Pro
by Elite Publishing House, Delhi

-EPH-

Contents

Preface

The destructive plant-parasitic nematodes are one of the major limiting factors in the production of vegetable crops throughout the world. For centuries, man has been plagued by these microscopic organisms feeding on the roots of crop plants essential for his survival and wellbeing. Roots damaged by the nematodes are not efficient in the utilization of available moisture and nutrients in the soil resulting in reduced functional metabolism. Visible symptoms of nematode attack often include reduced growth of individual plants, varying degrees of chlorosis, wilting of the foliage, and sometimes death of plants. Furthermore, roots weakened and damaged by nematodes are easy prey to many types of fungi and bacteria which invade the roots and accelerate root decay. These deleterious effects on plant growth result in reduced yields and poor quality of vegetable crops. The estimated overall average annual yield loss of the world's major crops due to damage by plant parasitic nematodes is 12.3%. In India, plant parasitic nematodes on vegetable crops cause 19.6% crop losses amounting to ₹ 14461.22 million annually. Tomato (₹ 6035.20 million), brinjal (₹ 3499.12 million), and okra (₹ 2480.86 million) among vegetable crops suffered comparatively more losses. Nematode management is therefore important for high yields and quality that are required by the high cost of modern vegetable crop production.

The information on various aspects of nematode diseases of vegetable crops and their management is very much scattered and there is no book at present which comprehensively and exclusively deals with the above aspects. The present book on "**Nematode Diseases of Vegetable Crops and their Management**" provides the detailed description of the causal organism, distribution, nature of damage and symptoms, crop losses, host range, biology and life cycle, interaction with other organisms, spread and survival, and management of nematode diseases of solanaceous, leguminous, cruciferous, cucurbitaceous, malvaceous, root, bulbous, leafy, and mushroom crops.

The book is divided into twelve sections. The First Section describes the vegetable crop area, production, and productivity; technological interventions to enhance productivity, post-harvest treatments, integrated pest management, future thrusts, and conclusions. This section also deals with different aspects of nematode diseases in vegetable crops such as economic importance, emerging nematode problems, major nematode pests, interaction with other pathogens (fungi, bacteria, and viruses), and nematode management methods (regulatory, physical, cultural, chemical, biological, host resistance and integrated methods).

The nematode diseases of solanaceous vegetable crops such as potato, tomato, brinjal, and chilli and their management are discussed in detail in Section Two. Section Three describes nematode diseases of leguminous vegetable crops such as French bean, cowpea, peas, pigeon pea, chickpea, green gram, and cluster bean and their management. The nematode diseases of cruciferous vegetable crops such as cabbage, and cauliflower and their management are discussed in Section Four. Section Five deals in detail with nematode diseases of cucurbitaceous vegetable crops such as cucumber, watermelon, musk melon, pumpkin, bottle gourd, pointed gourd, ridge gourd, bitter gourd, and sponge gourd and their management.

Section Six deals with malvaceous vegetable crop like okra and their management. The nematode diseases of bulbous vegetable crops such as onion and garlic and their management are discussed in Section Seven. Section Eight deals with root vegetable crops like carrot, beet root, parsnips, and turnip and their management. The nematode diseases of leafy vegetables such as lettuce, celery, coriander, and Amaranthus are dealt in Section Nine. Section Ten deals with nematode diseases of mushrooms like button and paddy straw mushrooms and their management.

The nematode diseases of flower crops grown under protected cultivation like tomato, bell pepper, cucumber and lettuce and their management are discussed in Section Eleven. Section Twelve deals with vision for future outlook including integrated nematode management, pillars of INM, modifications and expansions to INM for future, new building blocks of INM, and conclusion.

This book will be of immense value to the scientific community involved in teaching, research and extension activities related to vegetable crop nematodes and their management. The material can be used for teaching postgraduate courses. The book can also serve as a very useful reference to policy makers and practicing farmers. Suggestions to improve the contents of the book are most welcome (E-mail: reddypp42@gmail.com). The publisher, Elite Publishing House, New Delhi, deserves commendation for their professional contribution.

Parvatha Reddy

About the Author

Dr. P. Parvatha Reddy obtained his Ph. D. degree jointly from the University of Florida, USA, and the University of Agricultural Sciences, Bangalore.

Dr. Reddy served as the Director of the prestigious Indian Institute of Horticultural Research (IIHR) at Bangalore from 1999 to 2002 during which period the Institute was honored with "ICAR Best Institution Award". He also served as the Head, Division of Entomology and Nematology at IIHR and gave tremendous impetus and direction to research, extension and education in developing bio-intensive integrated pest management strategies in horticultural crops. These technologies are being practiced widely by the farmers across the country since they are effective, economical, eco-friendly and residue-free. Dr. Reddy has about 34 years of experience working with horticultural crops and involved in developing an F1 tomato hybrid "Arka Varadaan" resistant to root-knot nematodes.

Dr. Reddy has over 250 scientific publications to his credit, which also include 50 books. He has guided two Ph.D. students at the University of Agricultural Sciences, Bangalore.

Dr. Reddy is serving as Senior Scientific Advisor, Dr. Prem Nath Agricultural Science Foundation, Bangalore. He had also served as Chairman, Research Advisory Committee (RAC), Indian Institute of Vegetable Research, Varanasi; Member, RAC of National Centre for Integrated Pest Management, New Delhi; Member of the Expert Panel for monitoring the research program of National Initiative on Climate Resilient Agriculture (NICRA) in the theme of Horticulture including Pest Dynamics and Pollinators; Member of the RAC of the National Research Centre for Citrus, Nagpur; and the Project Directorate of Biological Control, Bangalore. He served as a Member, QRT to review the progress of the Central Tuber Crops Research Institute, Trivandrum; AICRP on Tuber Crops; AICRP on Nematodes; and AINRP on Betel vine. He is the Honorary Fellow of the Society for Plant Protection Sciences, New Delhi; Fellow of the Indian Phytopathological Society, New Delhi; and Founder President of the Association for Advancement of Pest Management in Horticultural Ecosystems (AAPMHE), Bangalore.

Dr. Reddy has been awarded with the prestigious "Association for Advancement of Pest Management in Horticultural Ecosystems Award", "Dr. G.I. D'souza Memorial Lecture Award", "Prof. H.M. Shah Memorial Award" and "Hexamar Agricultural Research and Development Foundation Award" for his unstinted efforts in developing sustainable, bio-intensive and eco-friendly integrated pest management strategies in horticultural crops.

Dr. Reddy has organized "Fourth International Workshop on Biological Control and Management of Chromolaena odorata", "National Seminar on Hi-Tech Horticulture", "First National Symposium on Pest Management in Horticultural Crops: Environmental Implications and Thrusts", and "Second National Symposium on Pest Management in Horticultural Crops: New Molecules and Biopesticides".

Chapter - 1

VEGETABLE CROPS: AN OVERVIEW

1.1. INTRODUCTION

There is a growing trend of sustainable vegetable production in organic farms, home gardens, and peri-urban systems. Thus, the technological innovations are vital in local vegetable cropping to increase yield productivity, production efficiency, food quality, and food safety. Technological trends that would benefit the modern vegetable production include rapid multiplication and production of propagules, development of modern nursery techniques, micro irrigation along with fertigation and greenhouse crop production, proper pruning, training and pollination strategies, Good Agricultural Practices (GAP) and Integrated Crop Protection Technologies (ICPT), traceability initiatives, and methods to ensure quality and safety of food. This emphasizes the need of capitalizing on intensification of cropping systems, viable seed production and quality assurance, restructuring and revitalizing of the research and extension system, etc., with a view to ensure the sustainable production of high-quality vegetables.

1.2. AREA, PRODUCTION AND YIELD

1.2.1. International scenario

The country-wise area, production, and yield of vegetable crops during 2020-21 is presented in Table 1.1.

Table 1.1. Country-wise area and production of vegetable crops (2020-21)

Country	Area (Million ha)	Production (Million tons)	Yield (tons/ha)
China	23.55	543.16	23.07
India	8.20	119.91	14.63
USA	0.99	33.12	33.62
Turkey	0.73	24.40	33.41
Iran	0.46	14.29	31.31
Egypt	0.68	16.80	24.62
Russian Federation	0.68	16.29	24.03
Mexico	0.71	14.33	20.07

Spain	0.31	13.60	44.28
Italy	0.47	12.32	26.11
Others	16.51	249.65	15.12
WORLD	57.00	1075.20	18.86

1.2.2. Indian scenario

The crop-wise area, production, and yield of vegetable crops in India during 2020-21 is presented in Table 1.2.

Table 1.2. All India area, production and yield of vegetable crops in India (2020-21)

Vegetable crop	Area ('000 ha)	Production ('000 tons)	Yield (tons/ha)
Beans	261	2595	278.141
Bitter gourd	109	1330	12.201
Bottle gourd	193	3171	186.430
Brinjal	749	12874	17.18
Cabbage	412	9560	23.203
Capsicum	37	563	15.216
Carrot	108	1885	17.453
Cauliflower	473	9225	19.503
Cucumber	117	1652	14.119
Chillies (Green)	411	4363	10.615
Elephant foot yam	38	874	23.000
Mushroom	---	243	---
Okra/Ladies' finger	531	6466	12.177
Onion	1624	26641	16.404
Parwal/Pointed gourd	62	725	11.693
Peas	567	5846	10.310
Potato	2203	56173	25.498
Radish	207	3263	15.763
Pumpkin	106	2205	20.801
Sweet potato	106	1121	10.575
Tapioca	183	6941	37.929
Tomato	845	21181	25.066
Other vegetables	1517	21550	14.205
Total vegetables	10859	200445	18.458

The major vegetable-producing states in India are Uttar Pradesh, West Bengal, Madhya Pradesh, Bihar, Gujarat, Maharashtra, Odisha, Karnataka, Haryana and Chhattisgarh (Table 1.3).

Table 1.3. Production share of leading vegetable producing states for 2017-18

State/Union Territories	Production (in '000 tons)	% Share
Uttar Pradesh	28316.45	15.4
West Bengal	27695.29	15.03
Madhya Pradesh	17545.48	9.52
Bihar	15863.21	8.61
Gujarat	12254.29	6.65
Maharashtra	12306.96	6.63
Odisha	8766.82	4.76
Karnataka	8394.15	4.55
Haryana	7151.66	3.88
Chhattisgarh	7003.25	3.80
Others	39096.96	21.21
All India Total	184394.51	100

The per capita availability of vegetables (210 g/caput/day) is still behind the recommended quantity (285 g/caput/day). To meet the per capita requirement of vegetables at the recommended level of 285 g/caput/day, the vegetable production has to be increased substantially to match the population growth rate of 1.8% per year. In order to provide balanced diet to the Indian population, the production level has to be increased to 200 million tons of vegetables by 2020 as against approximately 125 million tons of present vegetable production. Hybrid vegetable technology is one of the better options particularly at this juncture because of the fact that growing of vegetables is four times more remunerative than cereals. Hence, in next decade, hybrid vegetable technology has to go a long way in this country to meet challenges.

There is a significant gap in the national average productivity of different vegetable crops as compared to the world average productivity as well as the potential productivity for the respective vegetable crops (Table 1.4). Thus, there is a vast scope for improving the productivity of these crops at national level. The critical gap between the experimental yield and national average can be minimized by a strong extension service, training and demonstration. There is an urgent need to recast the vegetable production strategy so that vegetable production may get impetus. Further, for enhancing the vegetable production in the country, our target should be to achieve 25 tons/ha productivity by 2020, which is presently around 16 tons/ha.

Table 1.4. Vegetable productivity scenario (tons/ha)

Vegetable crop	India (Average)	World (Average)	Potential productivity	Maximum productivity
Potato	17.05	43.87	---	157.00 (Netherlands)
Tomato	14.07	26.69	60-80	70.45 (USA)
Eggplant	16.08	17.48	40-50	34.7 (Japan)
Chilli	9.18	14.4	30-40	44.5 (Spain)
Okra	9.59	6.47	15-20	17.78 (Jordan)
Peas	9.14	8.35	18-20	20 (Lithuania)
Melons	20.48	20.95	30-40	45.83 (Cyprus)
Cucurbits	9.72	12.97	25-30	41.33 (Israel)
Cucumber	6.67	16.98	40-50	67.67 (Korea)
Watermelon	12.75	27.13	30-40	40.96 (Spain)
Cabbage	21.43	21.1	30-40	42.59 (Japan)
Cauliflower	17.14	18.36	35-40	45.25 (New Zealand)
Onion	10.38	17.53	40-50	60.33 (Korea)
Garlic	4.17	12.37	15-20	23.23 (Egypt)
Cassava	31.44	31.44	---	---
Sweet potato	8.94	24.24	---	171.00 (Japan)

1.3. TECHNOLOGICAL INTERVENTIONS TO ENHANCE PRODUCTIVITY

1.3.1. Varietal improvement: High yielding varieties and F1 hybrids

Seed and planting material are critical inputs for enhancing the productivity of all horticultural crops and play a vital role in ensuring food and nutritional security. They also offer to integrate production, protection and quality enhancement technologies in a single entity, in a cost-effective way as use of quality seeds and planting material alone increases productivity by 15-20 percent highlighting its importance in olericulture. The response of all other inputs depends on the quality of seeds and planting material to a large extent, and it can be further raised up to 45 per cent with efficient management of other inputs. Therefore, any attempt to enhance vegetable productivity will largely depend on higher replacement rate of quality seeds and planting material of high yielding varieties/hybrids of vegetable crops (Chauhan *et al.*, 2013).

The objectives for varietal improvement include:

- » Improved yield
- » Better quality
- » Biotic and abiotic resistance
- » Early and uniform maturity
- » Photo- and thermo- insensitivity

» Desirable agronomic trait
» Wider adaptability

1.3.1.1. High yielding varieties: Vegetable production is still dominated by the locally available genotypes or inferior land races all over the country. It is estimated that under eggplant, approximately 32.2% area, cauliflower 46.71%, chilli 60%, gourds 77.56%, melons 70%, okra 14.62% and tomato 18.49% area are under unidentified local varieties. Thus, there exists a need to replace the local cultivars with the improved high yielding and disease resistant varieties. As a result of multi-disciplinary, multi-location testing of new research materials during the last two decades, 119 improved varieties in 16 major vegetable crops have been identified and recommended for cultivation in various agro-climatic regions of the country (Table 1.5).

Table 1.5. Improved varieties of vegetable crops

Vegetable crops	Improved varieties
Solanaceous crops	
Brinjal	Pusa Anupama, Pant Samrat, RU-IC, ARU-2C, PH-4, Pusa Kranti, Pusa Azad Kranti, H-7, Purple Cluster, Punjab K-202-9, NDB 25, Pusa Barasti, Pusa Purple Long, Jamuni Gole Baingan, Arka Navneet, BB-7, Pant Rituraj, BWR-12, H-8, Pusa Hybrid-6, T-3, Aruna, Arka Kusumakar, Annamalai, Azad B-1, CO-1, Gujarat Brinjal-6, Junagadh Oblong, Kalianpur Type 3, MDU-1, Punjab Neelam, Sada Bahar Baingan, Punjab Moti, BH-1,BH-2.
Tomato	HS-101, Pusa Early Dwarf, BT-1,CO 3, KS-2, Lanate, Pusa Gaurav, S-12, Bonita, Punjab Chauhara, Punjab Kesri, Sel-7, Arka Vikas, Pant Bahar, Pant T-3, Arka Saurabh, Pusa Ruby, Sel-120, Sioux, Pant T-2, Azad T-2, Kalianpur Angoorlata, Kalianpur-1, NTDR-1, PKM-1, Punjab Tropic, PNR-7, TH 2312, TH 802.
Chili	Chanchal, CO-1, CO-2, DH-76-6, Gujarat Chilli, K-1, MDH-1, Pant C-1, Sindhur, CH-1, Andhra Jyoti, J-218, K-2, LDC-206, Bhagyalakshmi, Musalwadi, Pusa Jwala, Sel-IX-235.
Cole crops	
Cabbage	Pusa Mukta.
Cauliflower	Early Kunwari, Pusa Early Synthetic, Pusa Deepali, 235-S, Pusa Synthetic, Improved Japanese, Pant Shubhra, Pusa Shubra, Pusa Snowball-1, Snowball-16, Pusa Snowball-2, Pusa Snowball-K-1.
Leguminous crops	
Cowpea	Pusa Komal.
Dolichos	Deepaliwal.
French Bean	VL-Boni, Arka Komal, Pant Anupma.
Pea	Arkel, Jawahar Matar-4, Early December, PM-2, Mattar Agata-6, JP-4, Jawahar Matar-1, Bonneville, Lincoln, VL-3, P-88, PRS-4, Pant Uphar.

Cucurbitaceous crops	
Muskmelon	Hara Madhu, Arka Jeet, Arka Rajahans, Pusa Sharbati, Punjab Madhu, M-3, Pusa Sunheri, Madhuras.
Pumpkin	Arka Chandan, Ambali, Arka Suryamukhi, Pusa Vishwas.
Watermelon	Arka Manik, Sugar Baby, Arka Jyoti, Durgapura Meetha.
Bulb crops	
Garlic	Jamuna Safed, Agri-Found White.
Onion	Agrifound Dark Red, Arka Kalyan, Arka Niketan, N-2-4-1, N-257-9-1, Pusa Madhavi, Punjab Selection, Punjab Red Round, Pusa Ratnar, Punjab Naroya, Pusa Red, VL-3, ALR, Pusa White Flat, Pusa White Round, PB-48.
Root crops	
Carrot	Pusa Yamdagni.
Malvaceous crops	
Orka	P-7, Parbhani Kranti, Arka Abhay, Arka Anamika.

1.3.1.2. F1 Hybrids: F1 hybrids are well known for increasing vegetable production due to their high yield potential, earliness, quality and resistance attributes. Heterosis breeding in vegetable crops in India has received serious attention only in recent years (Table 1.6). The first F1 hybrid of tomato (Karnataka Hybrid) and capsicum (Bharat) were released for commercial cultivation in 1973 by a private seed company M/s Indo-American Hybrid Seeds followed by 28 other hybrids in nine vegetable crops.

Table 1.6. Heterosis breeding for increase in yield of vegetable crops.

Vegetable crop	Heterosis for yield (%)	Inbreeding depression
Asparagus	64-140	Severe
Cabbage	12-60	High
Broccoli & cauliflower	40-90	High
Carrot	25-40	High
Onion	Up to 80	High
Cucumber	More than 100	High
Tomato	15-50	Not high
Brinjal	Up to 75	High
Chilli	10-60	High

F1 hybrids have been developed in tomato, brinjal, cabbage, cauliflower, bell pepper, chilli, cucumber, carrot, musk melon and watermelon (Table 1.7). Presently, Karnataka, Maharashtra, Gujarat, Andhra Pradesh, Uttar Pradesh, and Madhya Pradesh are leading producers of hybrid tomato. The entire cabbage production in southern parts of Maharashtra and West Bengal is under F1 hybrids.

Table 1.7. Public sector hybrids in vegetable crops

Crop	Available Hybrids	Source
Tomato	Pusa Hybrid-1, Pusa Hybrid-2, Pusa Hybrid-4, Pusa Hybrid-8, Pusa Divya (Kt-4) IARI, Delhi	IARI, New Delhi
	Arka Rakshak, Arka Ananya, Arka Samrat, Arka Shreshta, Arka Vishal, Arka Vardan, Arka Abhijit	IIHR, Bengaluru
	Kashi Abhiman	IIVR, Varanasi
	Pant Hybrid-1, Pant Hybrid-2, Pant Hybrid-10, Pant Hybrid-11	GBPUAT, Pantnagar
	Rajashree, Phule Hybrid-1	MPKV, Rahuri
Brinjal	DBHL-20, Pusa Hybrid-5 (Long), Pusa Hybrid-6 (Round), Pusa Hybrid-9, Pusa Anupama (Kt-4)	IARI, Delhi
	Arka Navneet	IIHR, Bengaluru
Chilli	CH-1, CH-3	PAU, Ludhiana
	Arka Meghana, Arka Harit, Arka Sweta	IIHR, Bengaluru
	Kashi Early, Kashi Surkh	IIVR, Varanasi
Sweet pepper	Pusa Deepti, KTCPH-3	IARI, Delhi
Cucumber	Pusa Sanyog	IARI, Delhi
Bitter gourd	Pusa Hybrid-1, Pusa Hybrid-2	IARI, Delhi
Bottle Gourd	Pusa Hybrid-3	IARI, Delhi
	Kashi Bahar	IIVR, Varanasi
	Pant Sankar Lauki 1	GBPUAT, Pantnagar
	Narendra Sankar-1	NDUAT, Faizabad
Muskmelon	Pusa Rasraj	IARI, Delhi
	Punjab Hybrid-1	PAU, Ludhiana
Pumpkin	Pusa Hybrid-1	IARI, Delhi
Summer squash	Pusa Alankar	IARI, Delhi
Watermelon	Arka Jyoti	IIHR, Bengaluru
Cauliflower	Pusa Kartik Sankar, Pusa Hybrid-2, Pusa Snowball Hybrid-1	IARI, Delhi/ R.S Katrain
Cabbage	Pusa Cabbage Hybrid-1	IARI, Delhi
Carrot	Pusa Vasudha, Pusa Nayanjyoti	IARI, Delhi/ R.S Katrain
Onion	Arka Lalima, Arka Kirtiman, Arka Bhima	IIHR, Bengaluru
Okra	Kashi Bhairav	IIVR, Varanasi
Ash gourd	Pusa Shreyali, Pusa Urmi	IARI, Delhi

However, it is estimated that presently only about 10% area of vegetables is under the hybrids of which tomato covers 36%, cabbage 30%, brinjal 18%, okra 7%, melons and gourds each 5%, cauliflower 2% and chilli 1%. It is expected that in the coming five years period there will be more than 25% area, and after ten years, hybrids will occupy more than 50% area, which will substantially contribute to enhancing the vegetable production of the country. Hybrids resistant to multiple diseases and pests are being developed in tomato, brinjal, chilli, bell pepper, okra, muskmelon, watermelon, cucumber, bitter gourd, carrot, radish, *etc.*

1.3.1.3. Disease resistant varieties: Research on breeding for disease resistance has resulted in the release of several varieties. 'Pusa Sawani' variety of okra developed as resistant to yellow-vein-mosaic virus is the first example of successful disease resistance breeding in vegetable crops in India. Okra hybrids of Indo-American Hybrid Seeds, Bangalore, such as Varsha, Vijay and Vishal are resistant to YVMV. In case of indigenous hybrids of brinjal, gourds, melons and cucumber, more area can be covered if they carry additional attributes like disease resistance or any special quality attribute. Several varieties have also been released which are resistant to abiotic stresses such as heat and moisture.

A list of varieties of different crops resistant to diseases is given in Table 1.8.

Table 1.8. Disease resistant crop varieties

Vegetable crop	Disease	Resistant varieties/rootstocks
Potato	Late blight	Kufri Sutlej, Kufri Badshah, Kufri Jawahar (in plains), Kufri Jyothi, Kufri Giriraj, Kufri Kanchan, Kufri Meghad (in hills).
Tomato	Bacterial wilt	Arka Abha, Arka Alok, Arka Shreshta, Arka Abhijit, Megha, Shakthi, Sun 7610, Sun 7611.
	Powdery mildew (PM)	Arka Ashish.
	Leaf curl virus	Avinash-2, Hisar Anmol.
Brinjal	Bacterial wilt	Arka Nidhi, Arka Keshav, Arka Neelkant, Arka Anand, Swarna Shree, Swarna Shyamali, Surya.
	Phomopsis blight	Pusa Bhairav.
	Little leaf	Pusa Purple Long, Pusa Purple Cluster (Field resistant).
Chilli	TMV, CMV, Leaf curl	Pusa Sada Bahar, Punjab Lal, Pusa Jwala.
	Powdery mildew	Arka Suphala (T).
	Mosaic (T), leaf curl (T)	Pant C-1.
	Leaf curl (T)	Jawahar 218.
	Viruses	Arka Harita, Arka Meghana.
	Bacterial blight	Ujwala.
	Dieback (T), powdery mildew (T)	Musalwadi.
French bean	Angular leaf spot, Mosaic	Pant Anupama.

	Rust, Bacterial blight	Arka Anoop.
	Rust	Arka Bold, Swarna Priya, Swarna Latha, Arka Anoop.
	Rust, *Alternaria* leaf spot	Arka Bold.
Pea	Powdery mildew	Pusa Pragati, Jawahar Matar 5, Jawahar Peas 83.
	PM, Rust	Arka Ajit, Arka Karthik, Arka Sampoorna.
	Fusarium Wilt	JP Batri Brown 3, JP Batri Brown 4.
Cowpea	Bacterial blight	Pusa Komal.
Pigeon pea	*Fusarium* wilt	Maruti.
Field bean	Viral diseases	Pusa Sem-2, Pusa Sem-3.
Cluster bean	PM, *Alternaria* leaf spot	Gomah Manjari.
Okra	YVMV	Pusa Sawani, Arka Abhay, Arka Anamika, Hisar Unnat, DVR-1, DVR-2, IIVR-10, Varsha Upkar, P-7, Pusa A-4, Parbhani Kranti (T), Punjab Kesari, Punjab Padmini.
Cucumber	PM	Swarna Poorna.
	PM, Downy mildew (DM), anthracnose, angular leaf spot	Poinsette.
Cabbage	Black rot	Pusa Mukta, K-1.
	Black leg	Pusa Drumhead.
Cauliflower	Black rot	Pusa Snowball K-1.
	Black rot & Curd blight	Pusa Shubhra.
	Curd blight	Pusa Synthetic.
	DM	Pusa Hybrid-2.
Onion	Purple blotch, Basal rot	Arka Pitamber, Arka Kirtiman, Arka Lalima.
	Purple blotch, *Alternaria porri*	Arka Kalyan.
Garlic	Purple blotch, *Stemphylium* sp.	Agri-found White.
Muskmelon	PM	Arka Rajhans, Pusa Madhuras (MR).
	PM, DM	Punjab Rasila.
	Fusarium wilt	Pusa Madhuras, Durgapura Madhu, Arka Jeet, Punjab Sunehari (MR), Harela.
Watermelon	PM, DM, Anthracnose	Arka Manik.
	Fusarium wilt, Anthracnose	Crimson Sweet, Petite Sweet, Super Sweet, All Sweet
Ridge gourd	PM, DM	Swarna Uphaar.

Bottle gourd	Blossom end rot	Arka Bahar (T).
	CMV	Punjab Komal.
Carrot	PM	Arka Suraj.
Amaranthus	White rust	Arka Arunima, Arka Suguna (MR).
Palak	*Cercospora* leaf spot	Arka Anupama.

1.3.1.4. Nutrition rich varieties: More than 800 million people worldwide are thought to be malnourished. In the developing world, around 192 million children and about 2000 million others experience malnutrition, particularly the pregnant women leading to infant mortality. British, American, and Australian researchers are all working on development of a range of vegetable varieties that the industry claims could counter malnutrition and illness. Some of the nutrient rich varieties of different vegetable crops are presented in Table 1.9.

Table 1.9. Nutrient rich vegetable crop varieties

Vegetable crop	Variety	Nutrient/s	Developed at
Tomato	Pusa Uphar, Pusa Hybrid-2	Vitamin C	IARI, New Delhi
	CO 1	Ascorbic acid – 17.50 mg/100g	TNAU, Coimbatore
	CO 2	Ascorbic acid -19.5 mg/100g	TNAU, Coimbatore
	CO 3	Vitamins (25 ml/100 ml)	TNAU, Coimbatore
	PKM 1	Vitamin C – 23.7 mg/100g fruit pulp	HC&RI, TNAU, Periyakulam
	PLR 1	Ascorbic acid – 2 mg/100 g	TNAU, Coimbatore
Chilli	Gujarat veg. chilli 121	Ascorbic acid	GAU, Anand
	PKM 1	Ascorbic acid – 306.0, 269.8 and 67.0 mg/100 g of green, ripe, and dry pods, respectively	HC&RI, TNAU, Periyakulam
Okra	CO 1	Crude fibre -14.06%, starch – 14.80%, protein – 18.08%	TNAU, Coimbatore
	CO 2	12.6% crude fibre, 15.2% starch, 19.2% protein, 2.84% fat and 50.5mg of ascorbic acid in 100 g pulp	TNAU, Coimbatore
Onion	CO 3	0.53 % sulphur	TNAU, Coimbatore
Broccoli	Palam Vichitra	Vitamins, minerals	HPKV, Palampur

French bean	Pusa Parvati	Protein	IARI, New Delhi
Pea	Arkel	Protein	IARI, New Delhi
	Ooty 1	Protein (19.94%), ascorbic acid (18.2 mg/100 g)	HRS, TNAU, Ooty
Cowpea	Sel 263	Crude protein (3.57%)	PAU, Ludhiana
Field bean	Pusa Sem-2	Protein	IARI, New Delhi
	CO 1	Protein-4.69%, vitamin C-0.12%, Fe-0.2%, calcium-0.39%, phosphorus-0.14%, crude fibre/100g edible pods-1.33%	TNAU, Coimbatore
Broad bean	Pusa Sumeet	Protein	IARI, New Delhi
Muskmelon	Arka Jeet	Vitamin C	IIHR, Bangalore
	Durgapura Madhu	Vitamin C	ARS, Durgapura
Ridge gourd	CO 1	Carbohydrates-3.5 g, proteins-0.9 g, fat-0.2 g, fibre-0.4 g, calcium-42.0 mg, phosphorous-37 mg, potassium-37 mg, iron-1.7 mg, sodium-4.2 mg, sulphur-12 mg, ascorbic acid/100 g edible fruit-5 mg	TNAU, Coimbatore
Bottle gourd	Pusa Samridhi	Vitamin C-8.07 mg, calcium-19.35 mg, phosphorus-20.39 mg, iron-1.35 mg/100 g	IARI, New Delhi
	Pusa Santushti	Vitamin C-7.12 mg, calcium-19.45 mg, phosphorus-18.15 mg, iron-1.10 mg/100 g	IARI, New Delhi
Bitter gourd	Arka Harit	Vitamin C, iron	IIHR, Bangalore
	Priya	Vitamin C, iron	KAU, Thrissur
	Pusa Hybrid-1	Vitamin C, iron, calcium	IARI, New Delhi
Snake gourd	MDU 1	Vitamin-A (44.4 mg/100g)	TNAU, Coimbatore
Pumpkin	Pusa Hybrid-1, Pusa Vikas	Vitamin A	IARI, New Delhi
	Arka Chandan	Vitamin A	IIHR, Bangalore
	CO 2	Starch-9.2%, ascorbic acid-10 mg/100 g	TNAU, Coimbatore
Carrot	Pusa Meghali	ß-carotene (Vitamin A) (11571 IU/100 g)	IARI, New Delhi
	Pusa Kesar	ß-carotene (Vitamin A) (7753 IU/100 g)	IARI, New Delhi
	Arka Suraj	Carotene (11.27 mg/100 g)	IIHR, Bangalore
	Hisar Gairic	Carotene (96.2 mg/100 g)	CCSHAU, Hisar

	Ooty 1	ß -carotene	HRS, TNAU, Ooty
Beet root	Ooty 1	Protein – 1.52%, carbohydrate – 10.25%, Vitamin C – 6 mg/100 g	HRS, TNAU, Ooty
Radish	Palam Hriday	Vitamin C	HPKV, Palampur
	CO 1	Vitamin C-10.9 mg/100g, crude fibre-1.20%	TNAU, Coimbatore
Amaranth	Pusa Kiran, Pusa Kirti, Pusa Lal Chauli	Vitamin A & C, iron, calcium	IARI, New Delhi
	CO 2	Iron - 19.0 mg, calcium – 20.0 mg/100g	TNAU, Coimbatore
	CO 3	Vitamin C-25.2 mg, crude fibre-1.74%, iron-0.8%, Ca/100g-2.48%	TNAU, Coimbatore
	CO 4 (grain type)	Protein -15.95%, lysine – 7.5 mg/ 100g, phenylalanine – 5 mg/100g, leucine – 1.2 mg, isoleucine – 1.8 mg/100g	TNAU, Coimbatore
Spinach beet (Palak)	Pusa Jyoti, Pusa Bharati	Vitamin A & C, iron, calcium	IARI, New Delhi
	Ooty 1	Vitamin A	HRS, TNAU, Ooty
Chenopodium	Pusa Bathua-1	Vitamin A, iron, calcium	IARI, New Delhi

1.3.1.5. Abiotic resistance: Several varieties have also been released which are resistant to abiotic stresses such as heat and moisture. A list of varieties of different vegetable crops resistant to abiotic stresses released so far is given in Table 1.10.

Table 1.10. Vegetable crops varieties resistant to major abiotic stresses

Vegetable crop	**Abiotic stress**	**Resistant varieties**
Tomato	Moisture	Arka Meghali
	Cold set	Pusa Sheetal
	Hot set	Pusa Hybrid-1
	Hot & cold set	Pusa Sadabahar
Chilli	Moisture	Arka Lohit
Field bean	Moisture	Arka Jay, Arka Vijay, Konkan Bushan
Cowpea	Moisture	Arka Garima
French bean	Heat tolerant	Arka Komal
Cluster bean	Moisture	Pusa Nav Bahar, Pusa Sada Bahar

Lima bean	Moisture	IIHR Sel-1, IIHR Sel-4
Round melon	Heat tolerant	Arka Tinda, Punjab Tinda
Long melon	Heat tolerant	Arka Sheetal, Punjab Long melon
Bottle gourd	Heat tolerant	Pusa Summer Prolific Long
Bitter gourd	Heat tolerant	Pusa Do Mousami, Kalyan Sona
	Warm humid climate	Arka Harit, Coimbatore Long, Konkan Tara, Priya, CO-1, MC-84, MDU-3
Cabbage	High temp. tolerant	Pusa Ageti
Cauliflower	Curd development in May in lower hills	Pusa Himajyoti
Turnip	Hot & humid climate	Pusa Sweta
Radish	High temp. tolerant	Pusa Chetki
	Low temp. tolerant	Pusa Himani
Carrot	Temperate type, bolting and seed setting under high temp.	Pusa Meghali
Palak	Not bolting in plains	Pusa Harit
Turmeric	Tolerant to drought	CO-1, BSR-1

1.3.1.6. Genetically modified crops

GM crops offer vast potential for improving the efficiency of crop production, thereby lowering the cost and increasing the quality of food. The tools of biotechnology can provide scientists with new approaches to develop higher yielding and more nutritious crop varieties, to improve resistance to pests including weeds and adverse conditions, or to reduce the need for fertilizers and other expensive agricultural chemicals.

Biotech crops represent promising technologies that can make a vital contribution, but are not a panacea, to global food, feed and fiber security. Biotech crops have also made a critically important contribution to the alleviation of poverty under the aegis of the Millennium Development Goals (MDG).

A list of commercially available genetically modified vegetable crops is presented in Table 1.11.

Table 1.11. Commercially available genetically modified vegetable crops

Crop	Trait phenotype	Target trait gene(s)	Trait designation	Originating company	Year of first commercial sale	Trade name
Squash	Resistance[1] to CMV, WMV2 and ZYMV	Coat protein genes of CMV, WMV2 and ZYMV	CZW3	Asgrow; Semini Vegetable Seeds (now Monsanto)	1998	Destiny III Conquerer III Liberator III

	Resistance[1] to WMV2 and ZYMV	Coat protein genes of WMV 2 and ZYMV	ZW-20	Asgrow; Semini Vegetable Seeds (now Monsanto)	1995	Preclude II Patriot II Declaration II Independence II

Source: Castle *et al.* (2006).

[1] CMV - Cucumber mosaic virus; WMV2 - Watermelon mosaic virus; ZYMV - Zucchini yellow mosaic virus.

i) *Bt* Brinjal: *Bt* Brinjal is the first genetically modified food crop in India that has reached the approval stage for commercialization. It was developed by inserting a gene *cry1Ac* from a soil bacterium called *Bacillus thuringiensis* through an *Agrobacterium*-mediated gene transfer. It is a genetically modified brinjal developed by the Maharashtra Hybrid Seed Company Ltd. (Mahyco), a leading Indian seed company. *Bt* Brinjal event EE1 has been developed in a Public Private Partnership mode under the aegis of the Agriculture Biotechnology Support Project (ABSP) from Cornell University where the *Bt* technology available with M/s Mahyco has been transferred (free of cost) to the Tamil Nadu Agricultural University (TNAU), Coimbatore; the University of Agricultural Sciences (UAS), Dharwad; and the Indian Institute of Vegetable Research (IIVR), Varanasi. The Event EE 1 was introgressed by plant breeding into various local varieties (Table 1.12).

Table 1.12. Development of *Bt* brinjal hybrids and open pollinated varieties in selected institutions

Mahyco's 8 *Bt* brinjal hybrids	**Public Sector's 16 *Bt* brinjal open pollinated varieties (OPVs)**		
	UAS, Dharwad (6)	**TNAU, Coimbatore (6)**	**IIVR, Varanasi (4)**
MHB-4*Bt*	Malapur local (S) *Bt*	CO2-*Bt*	Pant Rituraj
MHB-9*Bt*	Manjarigota *Bt*	MDU1-*Bt*	Uttara
MHB-10*Bt*	Rabkavi Local *Bt*	KKM1-*Bt*	Punjab Barsati
MHB-11*Bt*	Kudachi Local *Bt*	PLR1-*Bt*	VR-14
MHB-39*Bt*	Udupi Gulla *Bt*	IVBL-9	
MHB-80*Bt*	GO112 *Bt*	VR-5	
MHB-99*Bt*			
MHB-112*Bt*			

The Genetic Engineering Approval Committee (GEAC) cleared *Bt* brinjal for commercialization on 14th October 2009. Following concerns raised by some scientists, farmers and anti-GM activists; the Government of India officially announced on 9th February 2010 that it needed more time before releasing *Bt* brinjal, with Indian Environment Minister Jairam Ramesh saying that there is no overriding urgency to introduce *Bt* brinjal in India. On 17th February 2010, Jairam Ramesh reiterated that the Centre had only imposed a moratorium on the release of transgenic brinjal hybrid, and not a permanent ban, saying that "until we arrive at a political, scientific and societal consensus, this moratorium will remain".

The imposed moratorium has been criticized by some scientists as not being based on any compelling scientific evidence and potentially setting Indian biotechnology decades back (Shantharam, 2010).

ii) *Bt* brinjal in Bangladesh: The *Bt* brinjal project in Bangladesh may lay claim to be the first crop biotechnology transfer project to deliver a product to farmers. *Bt* brinjal was developed as an international public-private partnership, between an Indian seed company Mahyco generously donating technology to the Bangladesh Agricultural Research Institute (BARI), facilitated by Cornell University led project ABSP-II and funded by USAID. Bangladesh approved *Bt* brinjal for commercial cultivation on 30th October 2013 and in record time – less than 100 days – on 22 January 2014, a group of small farmers (120) planted the first four commercial *Bt* brinjal varieties (Uttara, Kajla, Nayantara, and Iswardi/ISD 006) on 12 hectares. In 2015, a total of 250 hectares of *Bt* brinjal were planted by 250 farmers and the area is expected to increase substantially in 2016. Five additional *Bt* brinjal varieties (Dohazari, Shingnath, Khatkati, Chaga, and Islampuri). – three of them received NCB approval in 2015, will be available in 2016; and the remaining two in near future.

Bt brinjal increases marketable yield by at least 30% and reduces the number of insecticide applications by 70-90%, reduced cost of production, increases marketable yield and improves fruit quality; with a net economic benefit of US$ 1,868 per hectare, equivalent to a gain of up to US$ 200 million per annum. Agronomic performance data released by BARI from 19 *Bt* brinjal demonstration plots, established in 108 farmers' fields in 19 districts showed close to zero pest infestation, increased yield up to 100% compared to non *Bt* variety. Farmers have successfully sold *Bt* brinjal fruits in the open market labelled as "BARI *Bt* Begun #, no pesticide used".

1.3.2. Micro-irrigation

1.3.2.1. Water losses under various irrigation systems: In traditional surface irrigation methods, the losses in water conveyance and application are large (Table 1.13). These losses can be considerably reduced by adopting drip and sprinkler irrigation methods. Among all the irrigation methods, drip irrigation is the most efficient and it can be practiced in a large variety of crops, especially in vegetables.

Table 1.13. Water losses under various irrigation systems

Method of irrigation	Water losses (%)
Surface	30-45
Gate pipe	15-20
Sprinkler irrigation	5-10
Drip irrigation	3-6

Micro-irrigation system is an irrigation system with high frequency application of water in and around the root zone of the plant system, which consists of a network of pipes along with a suitable emitting device. With micro-irrigation, the yields of crops are better in addition to the saving of water.

Micro irrigation can be classified as drip irrigation and sprinkler irrigation. In drip irrigation, water is applied near the plant root through emitters or drippers, on or below the soil surface, at a low rate varying from 2-20 Liters per hour. The soil moisture is kept at an optimum level with frequent irrigation. Drip irrigation results in a very high-water application efficiency of about 90-95 per cent.

Drip irrigation system is suitable for all vegetable crops. This system has also been successfully employed for close growing crops including onion and okra. The National Committee on Plasticulture Applications in Horticulture (NCPAH), Ministry of Agriculture, Government of India (GOI), has estimated a total of 27 million hectares area in the country that has the potential for drip irrigation application.

There is a need to promote increased irrigation water use efficiency. Participatory action research program initiated by the Ministry of Water Resources, Government of India, during 2007-09 in over 2000 villages all over the country to assess the impact of water saving technologies has shown that yield and income can be increased by 50 to 100% in most crops by using water saving technologies.

1.3.2.2. Advantages of micro-irrigation: The advantages of micro-irrigation are of two types – qualitative and quantitative. The qualitative advantages are production of quality horticultural produce and increase in water and fertilizer use efficiency. Excellent soil health could be maintained due to optimum soil moisture content, accelerated microbial activities and maximization of biomass decomposition converting into humus. The quantitative advantages are related to growth of micro-irrigation in terms of area coverage and overall increase in productivity. Quantity of water or fertilizer saved due to imposition of drip system, the energy conservation and reduced labor cost are the other advantages. Data on increase in crop yields and percent increase over traditional surface irrigation methods by using micro-irrigation systems are presented in Table 1.14. It shows that micro-irrigation could save water ranging from 14% to 84% with enhancement of yield from 13% and 69% under varied agro-climatic and crop conditions. The higher yield under micro-irrigation system is attributed to the fact that during crop growth favorable soil moisture (generally at ⅓ bar) is maintained and the required volume of soil mass is wetted through irrigation water, which provides optimum environment for root growth of the plant.

Table 1.14. Saving in applied water and productivity gains through drip irrigation water consumption (or application?) (mm/ha)

Vegetable crops	Water consumption (mm/ha)		Yield (ton/ha)		Water saving (%)	Yield increase (%)	Water consumption/ ton yield (mm/ton)	
	FMI	DMI	FMI	DMI			FMI	DMI
Cucumber	---	---	155.0	225.0	55.6	45.2	---	---
Ash gourd	840	740	10.84	12.03	12	12	77.49	61.51
Bottle gourd	840	740	38.01	55.79	12	47	22.09	13.26
Ridge gourd	420	172	17.13	20.00	59	17	24.52	8.60
Bitter gourd	---	---	32.0	43.0	56.6	34.4	---	---
Watermelon	800	800	29.47	88.23	Nil	179	27.15	9,07
Potato	200	200	23.57	34.42	Nil	46	8.49	5.81
Tomato	498	107	6.18	8.87	79	43	80.58	12.06
Brinjal	900	420	28.00	32.00	53	14	32.14	13.13
Chilli	1097	417	4.23	6.09	62	44	259.34	68.47
Cabbage	660	267	19.58	20.00	60	2	33.71	13.35
Broccoli	---	---	140.0	195.0	14.3	39.3	---	---
Cauliflower	389	255	8.33	11.59	34	39	46.67	22.00
Okra	535	86	10.00	11.31	84	13	53.50	7.60
Onion	602	451	9.30	12.20	25	31	64.73	36.97

Radish	464	108	1.05	1.19	77	13	441.90	90.76
Beet root	857	177	4.57	4.89	79	7	187.53	36.20
Sweet potato	631	252	4.24	5.89	61	40	148.82	42.78

Source: INCID (1994) and NCPA (1990) as cited in Narayanamoorthy (2004): pp 122

Note: FMI - Flood method of irrigation. DMI - Drip method of irrigation.

Water use efficiency (WUE) can be enhanced through an appropriate choice of crop/cultivars and implementation of strategies that identify the best match between crop type and time of cultivation for a specific environment. WUE of bell pepper with respect to growing season (spring and fall), soil type (loamy sand soil and sandy soil), and irrigation regime (irrigation at soil water tensions of -25, -50, or -75 kPa) is presented (Smittle *et al.*, 1994). The impact of water saving technologies has shown that yield and income can be increased by 50 to 100% in most crops.

The new research initiatives will have to be taken to improve the water productivity focusing on development of water efficient cultivars through plant genomic research along with water application technology based on precision farming.

Appropriate scheduling of irrigation results in economy of water use and increase in yield per mm of water applied. A differential irrigation scheduling based on the phonological stages of crop growth not only reduces the water requirement but also avoids water stress (both deficit and excess) during critical stages and thereby enhancing yield. The irrigation schedule and crop management practices have been standardized for several vegetable crops to economize the water (Table 1.15).

Table 1.15. Economizing the water use in fruits and vegetable crops by irrigation scheduling

Vegetable Crop	Recommended irrigation schedule	Irrigation schedules proposed at non-critical stages	Non-critical growth stage for reduced irrigation	Possible water saving
Brinjal	50% ASM	30% ASM	Vegetative stage* (30-45 days)	100 mm
Chilli	70% ER	50% ER	After establishment* (30 days)	120 mm
Okra	40% ER	20% ER	20 days after sowing* (25 days)	70 mm
Cauliflower	100% ER	50% ER	15 days after planting* (20 days)	90 mm
Onion	60% ER	35% ER	Vegetative stage* (30 days)	100 mm
Carrot	100% ER	50% ER	25 days after sowing* (25 days)	90 mm
Tomato	70% ER	30% ER	25 days after sowing* (20 days)	56 mm
Capsicum	70% ER	35% ER	15 days after planting (25 days)	90 mm

Watermelon/ Muskmelon	50% ER	30% ER	Vegetative state (15 days)	76 mm
Garden pea	70% ER	39% LER	20 days after sowing (15 days)	36 mm

*The stress induced need not be at a stretch but could also be in intervals.

ASM-Available Soil Moisture; ER – Evaporation Requirement.

Besides irrigation scheduling for different crops for different regions, critical stages and water requirements have been worked out and given in Table 1.16.

Table 1.16. Critical stages and water requirement of vegetable crops

Crop	Optimum soil moisture regimes	Critical stages	Yield t/ha	No. of irrigations	Water requirement (mm)	WUE kg/ha. mm water
Brinjal (June-Oct)	60-80% ASM	Flowering and fruit enlargement	27.12	7-8	486	43.2
Chilli (July-Nov)	Irrigate if dry spell exceeds one week	Flowering and fruit enlargement	12.42	3-5	640	19.4
Onion (July-Nov)	0.65 bar (15 cm depth)	Bulb development, bulb enlargement	38.6	3-5	500	77.2
Okra (July-Oct.)	40 mm CPE	Flowering, pod enlargement	17.28	8-9	345	50.1
Cauliflower (Oct.-Jan)	30 mm at 100 % Epan	Seedling, curd formation to harvest	16.77	20	541	31.3
Carrot (Winter)	40 mm at 100 % Epan	Seedling, root enlargement	24.2	14	14	59.1
Tomato (Oct.-Feb)	Furrow 0.7 Epan Drip	Seedling, fruit set and fruit development	56.5 70.31	12 Daily	336 336	165.6 209.3
Capsicum	Furrow 0.18 bar (drip)	Flowering, fruit set and development	12.3 16.7	16 Daily	480 330	25.6 50.6
Garden Pea (Oct – Feb)	IW /CPE of 1.00	Flowering and pod development	9.91	08	240	41.2
Watermelon (Dec-April)	0.50 Epan (drip) 1.00 Epan (furrow)	Flowering, fruit development, fruit	27.2 32.1	Daily 14	506 469	53.6 68.4
Muskmelon Summer	1.0 Epan Drip	Flowering, fruit development and fruit ripening	6.58 7.65	15 Daily	348 230	18.9 33.3

Cucumber Summer	IW/CPE 1.2 Drip	Seedling and Reproductive phase	37.7 44.4	16 Daily	436 310	86.5 143.2

Crop varieties continue to be released through crop breeding programs, with yield improvement in water shortage environments (Morison *et al.*, 2008). Several workers reported that about half of this yield increase is due to the crop improvement and half due to improved agronomy and management (Passioura, 2002; Richards, 2004; Turner, 2004). Beyond a threshold, further nitrogen application leads to little improvement in WPET.

As a result of the controlled and targeted application of water, soil moisture can be maintained at optimum levels. This leads to an increase in the productivity of the crops. Across various studies, it has been found that the adoption of micro irrigation systems helped boost the yield of vegetable crops. An average increase of 52.8 percent increase in the productivity of vegetable crops was seen. This in turn helps increase incomes of the farmers.

Micro-irrigation helps in increasing benefit: cost ratio to the extent of 1.83 to 2.74 in different states.

Enhancing the water use efficiency for increasing the yield per unit of irrigation water applied is called deficit irrigation. The relative ET deficit imposed at certain growth stages of vegetable crops such as common bean and potato showed comparatively less yield reduction (Table 1.17).

Table 1.17. Effect of planned evapotranspiration deficit of 25 percent on expected relative yield and relative water use efficiency in different crops.

Crop	Stage when ET deficit occurred	Expected relative yield	Relative water use efficiency	Reference
Common bean	Vegetative. yield formation	0.86 0.78	1.14 1.04	Calvache & Reichardt (1999)
Potato	Whole season. vegetative	0.79 0.90	1.06 1.20	Kovacs *et al.* (1999) Iqbal *et al.* (1999)

Supplemental irrigation (SI) which involves the application of small amounts of water during critical crop growth stages can significantly enhance water productivity (ensuring more sustainable use of groundwater) and enhance yield. The application of relatively small amounts of water at the time of water shortage during the most critical stages of crop growth (Table 1.18) that assure the survival of an economically valuable crop, which otherwise can lead to death of plants and significant reduction in yields.

Table 1.18. Critical soil moisture periods for different horticultural crops

Vegetable crop	Available critical soil moisture limit (%)	Critical soil moisture stage of crops
Chilli	50	Tenth leaf to flowering and fruit development
Potato	65	Stolon formation, tuberization, and tuber enlargement
Onion	60	Bulb formation and bulb enlargement
Tomato	60	Flowering and fruit development
Peas	40	Flowering and pod development

Cabbage	60	Head formation and enlargement
Cauliflower	70	Curd formation and enlargement
Eggplant	50	Flowering and fruit development
Cucumber	50	Flowering and fruit development
Okra	45	Flowering and fruit development

In arid and semiarid regions, supplementary irrigation will be desirable because it is mainly used at the critical periods of the crop growth cycle, to stabilize and improve yield quantity and quality.

1.3.3. Soilless culture

Soilless culture is one of the best techniques to overcome local water shortages, while also producing high quality produce, even in areas with poor soil structure and problematic conditions. Soilless culture is a method of growing plants without soil. The application of these systems is likely to increase close to existing cities as well as in mega-cities worldwide soon.

To meet the growing demand for soilless culture technology, ICAR-Indian Inst. of Hort. Res., Bangalore, has initiated a project on development and standardization of soilless cultivation of vegetables on Arka Fermented Cocopeat under open and protected conditions during the year 2015. Under this project, the production technology including nutrient formulations for open and polyhouse soilless cultivation of most consumed vegetables, viz. tomato, chilli, cabbage, cucumber, brinjal, cowpea, Dolichos, French bean, garden peas, ridge gourd and leafy vegetables and few exotic vegetables like zucchini, broccoli and color cabbage using Arka Fermented Coco peat as substrate has been standardized. The results of most of the experiments conducted with different vegetable crops in grow bags under open-field and polyhouse soilless culture indicated that the plants grown in soilless culture recorded higher yield and better quality particularly in mineral nutrient content compared to soil grown plants.

1.3.4. Fertigation

Among the various factors that are responsible for increasing the productivity of horticultural crops, a regular and optimum supply of water and fertilizers at appropriate times is very crucial as it enables the crops to attain higher growth rate and give increased yields. As both water and fertilizers are costly inputs and are becoming scarce in recent years, there is an immediate need for adoption of techniques with which these inputs can be utilized more efficiently. Fertigation is one such technique of applying fertilizers to the crops along with water through micro irrigation systems (drip/sprinkler) on a continual basis in controlled manner so as to allow steady uptake of nutrients by the plants and to save the costly inputs including water and fertilizers.

1.3.4.1. Advantages: The fertigation allows applying the nutrients exactly and uniformly only to the wetted root volume, where the active roots are concentrated. This remarkably increases the efficiency in the application of the fertilizer, which allows reducing the amount of applied fertilizer. This not only reduces the production costs but also lessens the potential of ground-water pollution caused by the fertilizer leaching. Fertigation allows adapting the amount and concentration of the applied nutrients to meet the actual nutritional requirement of the crop throughout the growing season. To make a correct planning of the nutrients supply to the crop according to its physiological stage, we must know the optimal daily nutrient consumption rate during the growing cycle that results in maximum yield, quality and production. These functions are specific for each crop and climate, and were determined in different experiments for the main crops in Israel like tomatoes, cucumbers, melons, *etc.* The optimal curve of consumption of nutrients defines the minimal application rate of a certain nutrient that is required to maintain a constant nutrient concentration in the soil solution. These data constitute the basis for the recommendations given by the Israeli Soil Extension Service for the farmers regarding the fertigation regime for different crops.

There are three beneficial effects of fertigation:

i) Maximizing water and fertilizer use efficiency

- Delivers water and nutrients to the root zone.
- Increases the availability and uptake of nutrients.
- Decrease nutrient losses from leaching.
- Accurate and uniform application of water and nutrients.
- Maximum saving of water and nutrients.

ii) Maximizing crop and soil productivity

- Higher yields and high-quality produce.
- Steep-sloped and marginal lands can be made cultivable.

iii) Minimizing production cost

- Economy in labor, water and nutrients.

1.3.4.2. Nutrient use efficiency: The crops do not efficiently utilize fertilizers applied under traditional methods of irrigation. The efficiency of different nutrients and methods of application is presented in Table 1.19.

Table 1.19. Efficiency of different nutrients and methods of application

Element	Method of application	Efficiency (%)
N	Fertigation	90-95
	Surface Flow	50-70
	Broadcasting (Solid fertilizers)	30-50
P	Fertigation	40-45
	Surface Flow	10-20
K	Fertigation	75-80
	Surface Flow	< 50

1.3.4.3. Optimum leaf nutrient status: At optimum leaf nutrient status, the growth, yield, and quality of the crop are satisfactory and there is no need to make any changes in the schedule of manures and fertilizers. Changes in nutrient concentration in the specified plant part do not increase or decrease growth or production. Leaf is the principal site of plant metabolism and changes in the nutrient supply are reflected in composition of the leaf. The concentrations of the nutrient in the leaf at a specific growth stage are related to the performance of the crop.

1.3.4.4. N fertigation: Urea is well suited for injection in the micro-irrigation system. It is highly soluble and dissolves in non-ionic form, so that it does not react with other substances in the water.

Also, urea does not cause precipitation problems. Urea, ammonium nitrate, ammonium sulphate, calcium ammonium sulfate, and calcium ammonium nitrate are used as nitrogenous fertilizers in drip fertigation. Measured N demands in vegetable crops are given in Table 1.20.

Table 1.20. Nitrogen uptake by various vegetable crops with respect to relative time of growth (Kafkafi and Kant, 2005).

Crop	Plants (no./ha)	Total uptake (g/plant)*	Expected yield (t/ha)**
Tomato	20,000	11.25 (83)	100
Eggplant	20,000	9.75 (54)	40
Sweet pepper	50,000	4.60 (46)	55
Potato	60,000	2.83 (28)	50
Muskmelon	25,000	6.00 (60)	50
Watermelon	12,000	15.00 (150)	75
Cabbage	50,000	3.60 (56)	29
Cauliflower	50,000	4.20 (66)	9

*The numbers in parentheses are the daily amounts of N (mg N plant^{-1} day^{-1}) to be added with drip fertigation.

1.3.4.5. P fertigation: P fertilizers used in fertigation include Phosphoric acid (75%) (H_3PO_4), Urea phosphate, Mono potassium phosphate (MKP) [(NH_2)2CO·H_3PO_4], Acidulated mono potassium phosphate (MKP) (KH_2PO_4), and Mono ammonium phosphate (MAP) ($NH_4H_2PO_4$). Measured phosphorus demand in field grown crops is given in Table 1.21.

Table 1.21. Phosphorus uptake by various vegetable crops with respect to relative time of growth (Kafkafi and Kant, 2005).

Crop	Plants (no./ha)	Total uptake (g/plant)*	Expected yield (t/ha)**
Tomato	20,000	0.95 (7.0)	100
Eggplant	20,000	1.10 (6.0)	40
Sweet pepper	50,000	0.45 (4.5)	55
Potato	60,000	0.39 (3.9)	50
Muskmelon	25,000	0.82 (8.2)	50

*The numbers in parentheses are the daily amounts of phosphorus (mg P_2O_5 plant^{-1} day^{-1}) to be added with drip fertigation.

1.3.4.6. K fertigation: Potassium nitrate (KNO_3), Potassium chloride (KCl) or Muriate of potash (MOP), Potassium sulphate (SOP), and Mono potassium phosphate (MKP) are used in drip fertigation.

Table 1.22 shows K uptakes by different crops, converting the absolute number of growing days of each crop to a relative value as a % of the total time needed from seeding to harvest.

Table 1.22. Potassium uptake by various vegetable crops with respect to relative time of growth between seeding (0%) and harvest (100%) (Kafkafi and Kant, 2005).

Crop	Plants (no./ha)	Total uptake (g/plant)*	Expected yield (t/ha)**
Tomato	20,000	16.50 (121)	100
Eggplant	20,000	11.50 (64)	40
Sweet pepper	50,000	5.70 (57)	55
Potato	60,000	4.70 (47)	50
Muskmelon	25,000	12.00 (120)	50

*The numbers in parentheses are the daily amounts of K (mg K_2O plant-1 day-1) to be added with drip fertigation.

The nutrient recommendations for different crops through proportional dosification approach are presented in Table 1.23.

Table 1.23. Nutrient recommendations for various crops (proportional approach)

Crop	Concentration in mg/L (ppm)				
	N	P	K	Ca	Mg
Tomato	190	40	310	150	45
Cucumber	200	40	280	140	40
Pepper	190	45	285	130	40
Melon	200	45	285	115	30

Fertigation requirements and responses of important horticultural crops are summarized in Table 1.24.

Table 1.24. Fertigation requirement and its response in major fruits, vegetables, and plantation crops.

Vegetable crop	Fertigation requirement	Crop response	Reference
Tomato, *Solanum lycopersicum*	75% RDF (262.2 kg N, 40.4 kg P_2O_5 and 90.5 kg K_2O/ha), using MAP Multi K and urea in surface drip irrigation	Improved fruit yield, higher leaf area index. Root growth and NPK uptake was higher with fertigation.	Kalanjiyam & Manickam (2015)
Potato, *Solanum tuberosum*	100% RDF (50% nitrogen mineral, 10 t/fed compost + humic substances, 10 t/fed compost + effective microorganism 4L hr^{---1})	Highest potato tuber quality and quantity in sandy soil conditions	Mohamed *et al.* (2014)
Brinjal, *Solanum melongena*	80% RDF (150 kg N: 50 kg P_2O_5: 50 kg K_2O ha^{-1}), IW/CPE ratio of 1.0, Irrigation level – 0.6 PE.	Significantly higher value of total N, P and K uptake, maximum fertilizer use efficiency of NPK and yield	Ugade *et al.* (2013)

Chilli, *Capsicum annuum*	RDF: 100 kg N – 50 kg P_2O_5 50 kg K_2O/ha at irrigation of 1.0 PE– (40.18 days)	Increase in number of leaves, plant height and yield response	Chaurasiya & Sahu (2016)
Onion seed crop, *Allium cepa*	Drip irrigation at 100% nutrients 240 kg ha^{-1} (120 kg N; 60 kg P_2O_5; and 60 kg K_2O) - F3 ETC carried out the experiment in semi-arid climate.	Significantly increase in seeds/ umbel, seed yield/umbel/plant as well as soil nutrient uptake	Dingre *et al.* (2016)
Cabbage, Brassica *oleracea*	100% RDF (254.4 kg N, 19.1 kg P_2O_5 and 324 kg K_2O ha^{-1}) along with surface irrigation at IW/CPE ratio of 1.0	Increase in head diameter. TSS and ascorbic acid content was highest at 125% RDF but B:C ratio of 3.03 was recorded in daily fertigation with 100% RDF.	Vasu & Reddy (2013)
Cauliflower, Brassica *oleracea*	RDF: 225 kg N - 75 kg P_2O_5 - 75 kg K_2O/ha with irrigation @ 2.0 L hr^{-1} flow rate at 100 kPa pressure (KCP 0.75-1.25)	Improved yield and significant difference in curd weight and diameter, yield, AGB, number of leaves per crop and curd weight/ AGB ratio.	Bozkurt *et al.* (2011)
Spinach, *Spinach oleracea*	RDF (170 kg N – 85 kg P_2O_5 – 85 kg K_2O/ha) drip irrigation twice a week	Leaf chlorophyll SPAD nitrogen balance index values, growth response, increase in nitrogen and WUE	Zhang *et al.* (2014)
Coriander, Coriandrum *sativum*	75% RDF {N (50, 60 and 70 kg/ha), P_2O_5 (50 and 60 kg/ha) + FYM (20 t/ ha), K_2O (40 kg/ ha)} and foliar spray of soluble fertilizer (13: 0: 45) with water soluble fertilizer through drip irrigation	Improved quality and growth response	Rajaraman & Paramaguru (2011)
Curry leaf, *Murraya koenigii*	RDF: 150g N: 25g P_2O_5: 50g K_2O/plant/year with drip irrigation twice in a week	Significant effect on physiological parameters, improved plant growth, increased secondary branches and leaves	Sharon *et al.* (2012)

Water saving, yield and profit under drip and drip-fertigation systems on tomato are presented in Table 1.25.

Table 1.25. Water saving, yield and profit under drip and drip-fertigation systems.

Crops	Water Saving (%)	Yield (t/ha)			Profit (Rs/ha)		
		Conventional	Drip	Drip + Fertigation	Conventional	Drip	Drip + Fertigation
Tomato	32	45	56	65	56000	77000	95000

The percentage yield increase and fertilizer saving through fertigation has been established in many crops and is summarized in Table 1.26.

Table 1.26. Recommended dose of major nutrients and yield under fertigation

Crop	N (kg/ha)	P_2O_5 (kg/ha)	K_2O (kg/ha)	Yield (t/ha)
Cabbage	370	85	480	70
Carrot	125	55	200	30
Cauliflower	250	100	350	50
Cucumber	70	50	120	40
Egg plant	175	40	300	60
Okra	60	25	90	20
Onion and garlic	120	50	160	35
Spinach	120	45	200	25
Tomato	140	55	190	50

Source: Biswas (2010).

1.3.5. Integrated nutrient management

Integrated nutrient management refers to maintaining the soil fertility and plant nutrient supply to an optimum level for sustaining the desired crop productivity through optimization of the benefits from all possible sources of plant nutrients in an integrated manner. It is a holistic approach, to know what exactly is required by the plant for an optimum level of production, in what different forms these nutrients should be applied in soil and at what different timings in the best possible method, and how best these forms should be integrated to obtain the highest productive efficiency on the economically acceptable limits in an environment friendly manner.

The encouraging results obtained by conjunctive use of inorganic fertilizers, organic manures, green manures, biofertilizers and fertigation can provide leads to the future strategies for the rational use of fertilizers and manures for enhancing the productivity of vegetable crops without detrimental effect to the environment.

1.3.5.1. Organic manures: The manures are waste organic materials of plants and animals (crop residues, FYM, compost, vermicompost, oil cakes, animal slurry, sewage sludge, fish meal, bone meal, flesh meal, dried blood, *etc.*). Green manure crops can also be used as organic manures. The manuring is done to replenish the soil with nutrients resulting in stronger plants and better crops. The benefits of manuring include improvement in soil texture and moisture retention capacity of the soil, enhancement of nutrients and soil biota (beneficial microbes) in soil and increase soil porosity to facilitate exchange of gases.

1.3.5.2. Biofertilizers: There are 6 types of biofertilizers:

» Biological N-fixing microorganisms (*Rhizobium* spp., *Azotobacter* spp., *Azospirillum* spp., *Bradyrhizobium* spp., and *Beijerinckia* spp.).

» Phosphate solubilizing and mobilizing microorganisms (*Bacillus megaterium, B. arculans, B. subtilis, B. polymyxa, Pseudomonas striata, P. fluorescens, Aspergillus awamori, A. flavus, A. niger, Pencillium bilaii,* and *P. digitatum*).

» Potash solubilizing microorganisms (*Frateuria aurantia*).
» Sulfur mobilizing microorganisms (*Acetobacter pasteurianus*).
» Arbuscular mycorrhizal fungi (*Glomus mosseae, G. fasciculatum, Gigaspora* spp., *Acaulospora* spp., and *Sclerocystis* spp.).
» Growth promoting substance excreting microorganisms (PGPR).

Biofertilizers improve soil physical properties and sustain soil fertility by providing aeration, biomass and nutrients. They not only make atmospheric N available but also solubilize K, mobilize S, and soil fixed P and improve P uptake in plants. They enhance plant growth due to release of hormones, vitamins, auxins, *etc.* They also help for proliferation of beneficial microorganisms which in turn suppress soil-borne pathogens.

1.3.5.3. Inorganic fertilizers: The commonly used nitrogen, phosphatic, and potassic fertilizers are presented in Table 1.27.

Table 1.27. Commonly used nitrogen, phosphatic and potassic fertilizers.

Nitrogen fertilizers	Phosphorus fertilizers	Potassium fertilizers
Ammonium sulphate, Calcium ammonium nitrate Urea	Single superphosphate Rock phosphate	Potassium sulphate, Potassium nitrate, Muriate of potash (Potassium chloride)

The basis for manure and fertilizer application is based on:

i) Soil testing: The first step in maintaining soil fertility is to know current nutrient levels. This is accomplished by soil testing, which is an important soil management tool and effective basis for nutrient and lime recommendations.

ii) Plant analysis: Plant analysis is a nutrient management tool most effectively used in conjunction with a regular soil-testing program. The crop integrates effects of soil fertility and other growth factors, and balanced plant nutrition is the goal of crop nutrient management, so it makes sense to directly analyze plants.

The integrated nutrient management (INM) can be achieved by adopting the following principles:

» Integrated and optimum usage (dose) of organic manures, inorganic fertilizers, and biofertilizers in a correct proportion.
» To obtain full potential of all the resources.
» To maintain soil fertility and plant nutrient supply at an optimum level for sustaining soil health and desired crop productivity.

1.3.6. Vertical farming

Vertical farming is the practice of growing crops on a smaller land area, by the utilization of vertical space, which is usually left unutilized in traditional agriculture. Though vertical farming has been in vogue since ancient times as evinced by the Hanging Gardens of Babylon, the modern concept of vertical farming involves the union of plant biology and suitable engineering know-how. It can be considered as an extension of indoor farming that evolved in the 1700s with the advent of greenhouses, with the primary objective of harnessing the off-season crop cultivation potential during unfavorable seasons. This involves the stacking of crops growing platforms in a vertical fashion and providing the necessary nutrition and lighting by artificial means to cultivate crops all-round the year. Most commercial vertical farm ventures in developed countries operate from existing warehouses or abandoned factories/buildings that have been suitably converted to hydroponic/aeroponic facilities with LED based illumination systems. Further, utilizing vertical space is highly energy intensive since artificial illumination needs to be provided for crop production in multi-tiered structures. Therefore, uninterrupted power supply would be a limiting factor for vertical farming under Indian conditions. Apart from this, the high initial costs of the infrastructure and the operating costs may act as deterrents for large scale vertical farming. Therefore, there is a need to develop country specific infrastructure and technology before vertical farming can be adopted on a large scale in India. Other limitations of the vertical farming include the cultivation of crops without sufficient scientific validation, lack of varieties/hybrids that have been exclusively bred for the purpose of vertical farming, lack of good agricultural practices for vertical farming situations and the design of existing high-rise structures that are not amenable for vertical farming. The above limitations could be taken up as researchable issues to promote vertical farming in India.

1.3.7. Quality planting material production

Availability of planting material of good quality is one of the most important elements of successful horticultural production. Planting material available to small-scale farmers in different areas is often of insufficient quality, which undermines potential yield and performance of crop production. There has been a significant increase in the production of vegetable crops in recent years in India.

1.3.7.1. Seedling production: To ensure high productivity and high quality of the produce, raising of high-quality seedlings through use of good quality seeds at right time and at an appropriate place is one of the cheapest but most important strategies. A large number of virus free healthy seedlings of different vegetables can be raised in a small area of greenhouse in plastic pro-trays by using soil-less media for growing vegetables either for main season or for their off-season cultivation. These trays help in proper germination, provide independent area for each seed to germinate, reduce the mortality rate, maintain uniform and healthy growth of seedlings, are easy in handling and storing, reliable and economical in transportation. With the use of this technology, it is now almost possible to raise healthy vigorous seedlings of different cucurbits, otherwise it was not possible in the traditional system of nursery raising.

Hardening of vegetable seedlings before transplanting in the main field is very essential for reducing transplanting shock, and also to have better crop stand. Plants should be gradually hardened or toughened by acclimatizing them to anticipated growing conditions of fields, at least a week before planting them in the main field. This is done by slowing their rate of growth to prepare them to withstand such as chilling, drying winds, shortage of water or high temperature. Generally, seedlings of all vegetables become ready for transplanting 28-30 days after sowing in plug tray.

The seedlings of various cucurbits can be grown for seed production during January in north India, which otherwise is not possible to grow early seed crop in open field conditions. Similarly, the production of seedlings of brinjal free from shoot borer and leaf miner could be organized during the onset of rainy season by growing in net houses. The virus free seedlings of tomato and chilli can be produced over the beds during rainy and autumn season by covering the nets.

1.3.8. Protected cultivation

Among the productivity enhancing technologies, protected cultivation has a tremendous potential to increase the yield of vegetables crops by several folds.

Protected cultivation technology, using different kinds of structures, offers several advantages to produce vegetables of high quality with high yields and minimum risks due to uncertainty of weather. It also ensures chances for efficient use of land and other natural resources. The main purpose of protected cultivation is to create a favorable environment for the sustained growth of plant so as to realize its maximum potential even under adverse climatic conditions. There is good opportunity for greenhouse cultivation of vegetables by economic methods of water and fertilizer use such as fertigation. Low-cost greenhouses along with micro-irrigation and fertigation techniques may be popularized in areas where evaporation exceeds precipitation during several months in a year. This becomes relevant to the growers in various parts of the country who have small holdings, say less than 2 ha. They would be interested in a technology, which helps them to produce more crops each year from their land, particularly during off-season when prices are higher in nearby markets. Vegetable production under location-specific and cost-effective structures may give substantial benefits even to small growers.

1.3.8.1. Increase in productivity: The yield under polyhouse cultivation can be achieved to the level of 5-8 times as compared to the open crop cultivation (Tables 1.28 to 1.31). Various trials conducted at Agricultural Research Centres in northern India indicate that bell pepper (planted in mid-September), cucumber (planting in mid-October) and tomato (November planting) under polyhouse produced 106, 146, and 153 tons per ha. The duration of these crops were 4-9 months and more than 90% of total yield was obtained during off-season (during winter before the start of summer) which fetches significantly higher market price (2-4 times than normal season). Further, the crop duration can be extended up to July–August with the application of micro-irrigation and fertigation and yield can be achieved to the level of 20-25 kg/m^2. Therefore, it is possible to harvest a single crop round the year with minimum additional inputs and higher income can be generated.

Table 1.28. Comparative performance of Solanaceous vegetable crops in greenhouse

Crop	Fruit yield (kg/m^2)		
	Fiberglass reinforced panel	Trench	Open
Bell pepper	1.82	0.83	0.54
Brinjal cv. PH-5	1.25	1.09	0.63
Chilli cv. BSS 344	1.35	1.24	0.13
Tomato cv. BSS 347	12.69	---	7.25

Table 1.29. Effect of greenhouse in achieving earliness in some Solanaceous vegetables

Crop	Days for first picking		Harvesting period (days)	
	Fiberglass reinforced panel	Open	Polycarbonate	Open
Tomato	70	105	100	50
Bell pepper	90	127	90	45
Brinjal	92	128	---	---

Table 1.30. Yield of vegetable crops under greenhouse at different locations in India

Location	Bell pepper (t/ha)	Tomato (t/ha)	Cucumber (t/ha)
Pune	203	124	---
Coimbatore	148	186	---

Bangalore	147	152	---
Solan	79	95	---
Hissar	302	211	151

Table 1.31. Yield of major vegetables grown in greenhouses across the world

Country	**Yield (tons /ha)**			
	Tomato	**Sweet pepper**	**Cucumber**	**Lettuce**
USA	484	100	-	-
The Netherlands	460	262	690	-
Canada	463	258	530	334
United Kingdom	413	248	480	36
Finland	337	138	396	126
Russia	300	100	-	-
Mexico	153	-	-	-
Syria	141	-	-	-
Spain	150	70	95	25
Israel	-	100	-	-
Turkey	106	104	105	-
Nicaragua	-	49	-	-

Source: Hickman 2011.

Depending upon different varieties, season and crop stage, fertigation schedule through the drip irrigation system for different vegetable crop varies (Table 1.32).

Table 1.32. Fertigation scheduling in vegetables under protected cultivation

Crop	**Spacing (m)**		**Plants number/ 1000 m²**	**Total dose (kg/1000 m²)**			**Yield (t/ha)**
	Plant to plant	**Row to Row**		**N**	**P**	**K**	
Tomato	0.45	0.4	2773	35	12	39	150-200
Cherry tomato	0.45	0.4	2773	35	12	39	30-40
Bell pepper	0.3	0.4	4160	32	15	36	46-50
Cucumber	0.3	0.4	4160	10-16	4-7	11-18	30-40

*Dosages vary depending upon varieties/ season/ crop stage.

Source: Hasan *et al.*, 2010

1.3.9. Precision horticulture

Precision horticulture can be defined as a comprehensive system designed to optimize agricultural production through the application of crop information, advanced technology, and management practices. To be a truly comprehensive system, it must begin during the planning stages of the crop or commodity and continue through the post-harvest processing phase of production. Information, technology, and management are the keys to success in this production system.

An information and technology-based farm management system identifies, analyses and manages variability in fields by conducting crop production practices at the right place and time and in the right way, for optimum profitability, sustainability and protection of the land resource. Precision horticulture technologies system approach helps to re-organize the total system of agriculture towards low-input, high-efficiency, and sustainable agriculture. Precision farming is an approach where inputs are utilized in precise amounts to get increased average yields, compared to traditional cultivation techniques.

Sustainable precision horticulture is this century's most valuable innovation in farm management that is based on using Information and Communication Technologies (ICTs). This is the most recent innovation technology based on sustainable agriculture and healthy food production and it consists of profitability and increasing production, economic efficiency and the reduction of side effects on the environment.

Precision horticulture is a comprehensive system designed to optimize production. The key elements of information, technology, and management, precision agriculture can be used to increase production efficiency, improve product quality, improve the efficiency of crop chemical use, conserve energy, and protect the environment. Technology and management practices such as field scouting, field mapping, variable rate control, yield mapping, and post-harvest processing can be readily adapted to horticultural crop production. Much of this technology is still in its infancy. More research will be necessary to allow the systems to reach maturity. While technically feasible, further research is also needed to clarify the economic and environmental benefits of many elements of precision agriculture.

Some of the areas in olericulture where precision farming is taking hold with implications for the economics of farming are listed below.

1.3.9.1. Precision seeding: Seeding (or sowing) is a critical step in crop growing. For a successful seeding process, two challenges need to be overcome:

i) Correct depth: If sown too deep into the soil, roots will not be able to breath. If sown on the surface, birds may damage the seeds.

ii) Proper distance: If plants are overcrowded, they will not get enough water, nutrients and sunlight, resulting in yield loss. If they are planted too far from each other, valuable land is left unused.

Hand-seeding used to be a laborious, burdensome task with a lot of inaccuracies involved. Today, seed drills and planters do the job of sowing with the help of tractors. Modern precision seeding equipment manages to place the seeds uniformly at proper distances and depths for optimal access to moisture and sunlight. The equipment ensures that more seeds are planted in soil with favorable growing conditions, while less seeds are planted in poor soils to avoid loss of crop production or void areas. The fast and uniform germination ensures that the crop can compete with weeds and promotes even maturity across the whole field for easier harvesting and greater yields. The system can be combined with geo mapping (a map showing soil density, quality *etc.*). The latest precision technologies are helping farmers to refine the seeding process even further and achieve a higher yield with less seed.

1.3.9.2. Soil fertility management: This involves dividing a field into several small and equal divisions using the sub-inch accuracy of GPS. To do this, the tractor is fitted with a dish antenna to receive signals from satellites, which are recorded on a tractor-mounted computer. Soil samples are mechanically taken from each sub-division and this process is technically known as "Grid Sampling".

1.3.9.3. Pest management: The role of homogeneous vegetation in pest outbreaks, which can also be brought about by the uniform application of pesticides over large areas, has long been recognized by applied crop protection scientists. The simplification of pest communities, their impoverishment of many natural enemies and reduction of their clustered distribution are also brought about by pesticide applications (Johnson and Tabashnik, 1999). A judicious application of pesticides (insecticides, fungicides, nematicides, and herbicides) at a scale smaller than a field was not possible until recently, due to mechanized large scale crop production systems. The implementation of precision agriculture has made it possible for fine-tuning of inputs in the field, through the new technologies that have become available in the last decade or so. The site-specific pest management systems combined with an understanding of pest population in time and space is suggested as one of the strategies to reduce the negative impact of pesticide use and to avoid the traditional applications based on average pest density (Park *et al.*, 2007).

1.3.9.4. Post-harvest management: The post-harvest management begins as soon as the crop is harvested. Improper handling of the crop during this stage can be detrimental to the quality. Horticulture precision farming applications of post-harvest process management uses sensors to monitor conditions in curing or storage to achieve the optimum parameters and preserve the quality. Automatic controls are used to regulate temperature, humidity, and fresh air-delivery. By continuously monitoring the curing or handling conditions, adjustments can be made that would not be possible with the conventional method of manual control. As in other facets of horticulture, the feedback control loop is critical element in precision farming. By continuously monitoring the state of the crop in storage or in curing, and analyzing the data in real time, adjustments can be made in storage or curing parameters to preserve or enhance quality.

1.3.9.5. Irrigation management: Recent developments are being released for commercial use in sprinkler irrigation by controlling the irrigation machines motion with GPS based controllers. In addition to motion control, wireless communication, and sensor technologies are being developed to monitor soil and ambient conditions, along with operation parameters of the irrigation machines (i.e., flow and pressure) to achieve higher water application efficiency and utilization by the crop. These technologies show remarkable potential. The GPS-guided grid system helps to apply variable rates of irrigation tailored to the requirement of each grid area improving water use efficiency.

1.3.9.6. Yield monitoring: At harvest, crop yield information is recorded on a grid section basis. Solutions for differences of yield between grid sections are sought through computer analysis of all variables controlling yield of crops that are stored in the computer. Based on this, the farmer fine-tunes variable rates of application of fertilizers and other impacting parameters for use in future cropping programs.

1.3.9.7. Time management: One other great advantage of the GPS system of farming involves the ability of the farmer to achieve greater efficiency in time control of his farm operations. This is because the GPS system enables him to operate his equipment round the clock irrespective of factors restricting visibility such as fog, darkness, or even showers. The sub-inch accuracy of GPS-based operations provides the farmer maximum efficiency with equipment operations.

1.3.9.8. Success story: The study conducted in the Dharmapuri district in Tamil Nadu, India, has revealed that adoption of precision farming has led to 80 per cent increase in yield in tomato and

34 per cent in eggplant production. Increase in gross margin has been found as 165 and 67 per cent, respectively in tomato and eggplant farming. The contribution of technology for higher yield in precision farming has been 33.71 per cent and 20.48 per cent, respectively in tomato and eggplant production. The elasticity of 0.39 for the adoption in tomato and 0.28 in eggplant has indicated that as the probability of adoption increases by 10 per cent, net return increases by 39 per cent and 28 per cent in tomato and eggplant cultivation, respectively.

1.3.10. Nanotechnology

In the last decade, nanotechnology has emerged as a recent evolving sector in the field of science and technology. Nanoparticles due to their small size show exceptional physical, chemical, optical, and magnetic properties. These unique properties of nanoparticles make them useful in olericulture for obtaining sustainability in production through their application in production of quality planting material, nano-fertilizers for improving plant nutrition, nano-sensors for monitoring plant health and quality of soil and water, nano-pesticides for disease and pest management and improved packaging for enhancing shelf-life. Nanoparticles can be used for improving germination percentage of seeds and controlling contamination rate in *in vitro* propagation. They can also improve stress tolerance in plants against various biotic and abiotic stresses. Nanotechnology provides the possibilities of exploring nanoscale materials in the form of nano-fertilizers or nano-pesticides which can improve the nutrient use efficiency, enhance the efficacy and specificity of pesticides and at the same time minimize the risk of pollution to ecosystem. Smart packaging developed using nanomaterials can reduce the occurrence of food-borne diseases and improve food safety by ensuring fresh and tasty purchase to the consumers. Nano biosensors can not only be used for monitoring soil and plant health but also can help in predicting the outbreak of pest or diseases, thus reducing the economic loss. Better understanding about the interaction of nanomaterials in the olericultural field, however, is required for eliminating the adverse effects of nanoparticles in the vegetable crops.

Various kinds of nanomaterials have shown high potential for enhancing shelf-life, reducing post-harvest damage and improving the quality of vegetable crops. Antimicrobial nanomaterials as nanofilm on harvested products and/or on packaging materials are suitable for the storage and transportation of vegetables. The nanotechnology could be used to generate cutting-edge techniques towards promoting productivity and quality of vegetable crops to ensure food and nutritional security of ever-increasing population of the world.

1.3.10.1. Nano fertilizers: The application of fertilizers is indispensable for the higher productivity of crops in modern agriculture. One of the big bottlenecks of fertilization in crops is that substantial part of the used fertilizers is lost in various ways and ultimately pollute the environment and increase the production cost. A notable recent progress in reducing the loss of applied fertilizers to the environment is the application of nano fertilizers. Use of Nanotechnology for developing nano-fertilizers could improve the nutrient use efficiencies, reduce nutrient loss, replenish soil fertility, increase crop yield, maintain ecosystem, soil health and curtail serious soil problems. Nano-fertilizers are formulated to deliver and emit nutrient tardily and deliberately. Regular release of nutrients by nano-fertilizers helps in augmenting nutrient use efficiency beyond several related adverse outcome. Nano-fertilizers may increase the fruit yield and quality of various vegetable crops which will in turn increase the nutritive content of products as well as increase the global food security.

1.3.10.2. Nano-plant growth stimulators: The application of NMs in olericulture is remarkably increased in the last few decades (Tarafdar *et al.*, 2014; Vance *et al.*, 2015). The form and constituents of the NMs are used in pest and disease management to enhance the growth and productivity of plants (Servin and White, 2016). Nevertheless, both beneficial and detrimental effects of application of nanomaterials on plants have also been reported.

1.3.10.3. Nutrient uptake and subsequent translocation in plant: Nanoparticles are easily absorbed to plant surfaces and uptake by plants via nano- to micrometre-scale natural openings of plants. Nanoparticles (NPs) uptake into the plant body can use different pathways. Uptake rates depend on the surface properties and size of the NPs. Very small-sized NPs can be penetrated via the cuticle. Large size NPs can enter via non-cuticle areas e.g., hydathodes, stomata, and the stigma of flowers. Nanoparticles must traverse the cell wall to enter into the protoplast of the plant cell. Several lines of evidence suggest that NPs less than 5 nm in diameter are efficiently traversed the wall of the undamaged plant cell (Tarafdar, 2015).

1.3.10.4. Nano pesticides: Pests are the most important limiting factors for crop yield and need to be efficiently controlled. Pest control by traditional means involves the use of large quantities of chemical pesticides, which results in environmental problems and increases the cost of production (Sharon *et al.*, 2010). Pesticide dilution with nano-treated water could greatly improve their efficiency and reduce the quantity of chemicals used. Nano pesticides are more efficient than conventional pesticides in controlling pests. There use also reduces the cost by half compared to conventional pesticides (Huang *et al.*, 2014).

1.3.10.5. Post-harvest management: Increasing production efficiency and decreasing post-harvest wastage with using the findings of novel scientific research such as biotechnology and nanotechnology in products, could be counted as the best solution to this problem. Nanotechnology using particular characteristics of nanoparticles can be a very useful technology in all science and industry branches. Nowadays, a lot of usages of nanotechnology in agricultural sciences have been established. In relation with extension of vegetable products shelf life, nanotechnology can help us in some grounds, e.g., controlling growth and development of microorganisms, introducing a new generation of packaging coverage's (films) and controlling influence of gases and the harmful rays (UV), increasing strength, quality and packaging beauty, and using the multiple chips (nano biosensors) for labelling products that considered as fundamental step to automated control of storages.

1.3.10.6. Nano sensors: A nano sensor can be defined as any device that is capable of conveying data and evidence about the behavior and characteristics of NPs at the nano scale level to the macroscopic level (Scoville, 2013). Nano sensors are necessary for facilitating real-time tracking of crop, crop growth, and pest and disease incidence. Nanosized materials which can be used for sensor manufacturing are metal nano tubes, nano wires, nano fibers, nano composites, nano rods, and nano structured polymers, different allotropes of carbon including carbon nanotubes, graphene, and fullerenes (Márquez and Morant, 2015). Real-time monitoring can minimize the excess use of pesticides and fertilizers in crop production, which is helpful in the reduction of environmental pollution and production costs. Application of nano sensors changes conventional agriculture into smart agriculture, which is more energy-efficient and eco-friendly for sustainable agricultural practices.

1.3.10.7. Seed science: Nanotechnology can be used to harness the full potential of seed. Use of bio-nano sensors specific to contaminating pollen can help alert the possible contamination and thus reduce contamination. The same method can also be used to prevent pollen from genetically modified crops from contaminating field crops. Novel genes are being incorporated into seeds and sold in the market. Tracking of sold seeds could be done with the help of nano barcodes (Nicewarner Pena *et al.*, 2001) that are encodable, machine-readable, durable and sub-micron sized taggants. Nano-coating of seeds using elemental forms of Zn, Mn, Pa, Pt, Au, and Ag will not only protect seeds from seed-borne pathogens but used in far less quantities than done today. Su and Li (2004) developed a technique known as quantum dots (QDs) as a fluorescence marker coupled with immuno-magnetic separation for *E coli* 0157: H7, which will be useful to separate unviable and infected seeds.

1.3.10.8. Water sustainability: The small size and unique properties of engineered nanomaterials (ENMs) are particularly promising for addressing the pressing technical challenges related to water quality and quantity. For example, the increased surface area and reactivity of ENMs can be exploited to create precious-metal-free catalysts for water purification, and the enhanced strength-to-weight properties of nanocomposites can be used to make stronger, lighter, and more durable piping systems and components. The goal of the Water Sustainability through Nanotechnology Signature Initiative (the "Water NSI") is to take advantage of the unique properties of engineered nanomaterials to generate significant breakthroughs in addressing our Nation's water challenges. This initiative is designed to aid in the development of technological solutions that can alleviate current stresses on the water supply and provide methods to sustainably utilize water resources in the future.

i) Solanaceous vegetables: Productivity of potato *cv.* Arizona, fertilizer use efficiency and agronomic use efficiency can be increased by fertigation with nano NPK fertilizers (Hayyawi *et al.*, 2019).

Tomato (*Solanum lycopersicum*) yielded highest number of fruits per plant, fruit weight, fruit diameter by application of 300 kg/ha K nano fertilizer, and the highest plant height and stem diameter is observed under application of 400 kg/ha K nano fertilizer (Ajirloo *et al.*, 2015).

ii) Crucifers: Wang *et al.* (2013) experimented on effects of nano-preparation on growth and nitrogen fertilizer use efficiency in cabbage. It was observed that addition of nano-hydroquinone and nano-thiophenols in nitrogen fertilizer improved absorption of N, P, K and production of chlorophyll in cabbage. There is increase in N fertilizer use efficiency and production of cabbage by 134.1 and 44.3%, respectively. This proves that nano preparation improves fertilizer use efficiency and boost crop yields.

iii) Cucurbits: Application of Ferbanat nanotechnology liquid fertilizers @ 3 L/ha to cucumber crop gives highest fruit diameter (Melek Ekinci *et al.*, 2014). Application of Nanonat at 3.0 L/ha to cucumber crop gave highest TSS (Melek Ekinci *et al.*, 2014).

The nano fertilizer has a positive effect on all stages of plant growth and development. *Cucurbita pepo* cv. White Bush Marrow when treated with nano SiO_2 increased growth and germination of plant, enhanced photosynthetic activity, reduced degradation, improved water use efficiency thereby improving plant defense mechanism of plant and stress (Siddiqui *et al.*, 2014).

iv) Leafy vegetables: Ladan Moghadam *et al.* (2012) investigated the effect of nano iron chelate on growth and yield of 2 spinach varieties 'Varamin 88' and 'Viroflay'. The research findings shows that Iron chelate Nano fertilizers improves wet weight by 58% and 47% maximum leaf area index.

Drumstick, *Moringa oleifera* was treated with nano chelated iron at 0, 1, 2, 3 and 4 mg/L, GA_3 at 0, 200 and 400 mg/L and organic fertilizer Acadian at 0 and 1 mg/L. Nano fertilizer at GA_3 at lower concentration showed positive response on the production of α-tocopherols, stigmasterols and campesterol (Kadim, 2018).

v) Legumes: Red bean, *Phaseolus vulgaris* treated with N bio-fertilizers showed increase in yield and yield components. It was also observed that K- chelate Nano fertilizers could replace chemical fertilizers (Farina and Ghorbani, 2014). Carbon nanotubes NPK and chitosan nanoparticles NPK fertilizer applied as foliar spray on French bean proved beneficial at low dose, it improved the uptake and absorption of nutrients and enhanced their overall growth and development (Hasaneen *et al.*, 2016).

vi) Okra: Foliar organic nano NPK fertilization in okra at 0.4% recorded higher nutrient status on post-harvest soil (Nibin *et al.*, 2019).

1.4. POST-HARVEST MANAGEMENT

There is a considerable gap between the production and net availability of the horticultural produce due to post-harvest losses due to lack of elaborate harvesting equipment, collection centers in major producing areas, suitable containers, commercial storage plants and lack of cold chain in entire post-harvest operation (Table 1.33). Considerable losses occur during transport. Minimizing these losses has, therefore, been recognized as an effective means of increasing vegetable availability without bringing additional areas under these crops. In fact, the reduction in post-harvest losses is a complementary means of increasing the productivity.

Table 1.33. Post-harvest losses scenario in vegetables

Vegetables	Post-harvest losses (% of production)
Beans and peas	7-12
Brinjal	10-13
Cabbage	7-15
Cauliflower	10-15
Garlic	1-3
Onion	15-30
Potato	15-20
Tomato	10-20

1.4.1. Pre-harvest treatments

» Pre-harvest application of maleic hydrazide (MH) reduces sprouting of onion and potato during storage. In onions, application of 1500 to 2000 ppm MH at 75 to 90 days after transplanting reduced sprouting during 4 to 5 months storage in ventilated structures.

» Post-harvest diseases of tomato and onion can be controlled by 3 pre-harvest sprays of 0.2% Difolatan at an interval of 10 days before harvest.

» Pre-harvest application of N-benzyladenine (growth promoter) at 10 to 20 ppm prolongs shelf-life of leafy vegetables.

» Stoppage of irrigation 10 to 15 days before harvesting onion results in enhanced storage life of bulbs.

1.4.2. Post-harvest treatments

1.4.2.1. Use of chemicals: Post-harvest losses can be reduced by use of chemicals, fungicides, waxing etc. Protective skin coating with wax increases storage life of fruits and vegetables at ambient temperatures. This treatment protects fruits and vegetables against excessive moisture loss and higher rate of respiration. To protect against microbial spoilage, fungicides can be added to the wax emulsion.

1.4.2.2. Irradiation: Sprouting of onion can be checked by gamma irradiation at a dose of 0.06 to 0.1 KGY. In potato, gamma irradiation at 0.1 KGY can inhibit sprouting completely and also inhibits the light induced synthesis of chlorophyll and the toxic alkaloid solanin. The irradiated potatoes could be stored successfully for 6 months at 15 °C with 10% weight loss.

1.4.2.3. Cold storage: Storage at low temperatures immediately after harvest reduces the rate of respiration resulting in reduction in build-up of heat, thermal decomposition, microbial spoilage and, also helps in retention of color and quality and freshness for a longer period. At present, there is a cold storage capacity of 30 million cubic meters in India of which 90% capacity is utilized for storage of potatoes. Fruits and vegetables together utilize only about 8.1% of total cold storage capacity.

Different methods employed in cold storage include Controlled atmospheric (CA) storage and Zero energy cool chamber.

1.4.2.4. Value addition: Processing of vegetables for value addition is an indispensable part of horticultural industry. In countries like USA and Brazil, nearly 60 to 70% of fruits and vegetables are used for processing as against 1 to 2% in our country. Steps should be initiated for achieving at least 15 % level of processing by 2025 which would generate an investment of Rs. 1 lakh crores and create almost half a million direct employment and another 2 million indirect employments. Hence, attempts to promote processing industries in this country is very essential. In onion, our share is nearly 8% of the total world production and hence a great scope exists for development of newer products from this commodity. Dehydrated onion and tomato products have great potential as export commodities.

1.4.3. Reduction of post-harvest losses

- » Follow optimum maturity standards for harvest to reduce loss at the farm level.
  - Firm breaker and semi ripe stage harvest to reduce transit loss (Tomato)
- » Pre-harvest practices to reduce PHL.
  - Use of harvester can reduce loss considerably.
  - IPM – tomato, cabbage
  - Pre-harvest sprays to reduce post-harvest losses.
- » Hot water treatment
- » Packaging - CFB cartons, plastic crates to reduce loss, individual shrink wrapping.
- » Specialized transport vehicles (cushioning and cooling) to reduce transit losses.
- » Sorting immediately after unloading at processing unit
- » Conversion of vegetables into processed products would reduce glut in the market and reduce the losses, besides enhancing income prospects of the producers – tomato into puree.
- » Establishing processing units in the production area – reduce number of handlings, infection, backward and forward linkages, minimize price fluctuations.
- » Efficiency of the marketing system to be studied using PHL as an item of cost.
- » Strengthening marketing infrastructure at the local and distant market levels would help reduce the losses and also increase the returns to the producers.

1.5. INTEGRATED PEST MANAGEMENT

1.5.1. Weeds

Integrated weed management system (IWMS) involves co-ordinated efforts towards development of a long lasting and environment friendly practice of weed management. IWMS includes all preventive, crop husbandry and direct control measures aimed at reducing and maintaining weed population at sub-economic level. In this, all weed control methods like preventive, cultural, mechanical, and biological are included in different permutation and combinations are tried to achieve maximum weed control which is not possible in any one single method. IWMS is well suited for adoption in different cropping systems.

1.5.2. Insect pests, diseases pathogens and nematodes

Integrated Pest Management (IPM) is an important principle on which sustainable crop protection can be based. IPM allows farmers to manage pests in a cost effective, environmentally sound, and socially acceptable way. According to FAO, IPM is defined as "A pest management system that in the context of the associated environment and the population dynamics of the pest species, utilizes all suitable techniques and methods, in a compatible manner as possible and maintains the pest populations at levels below those causing economic injury". IPM strategies are based on 3 main components/tools like prevention, monitoring/ observation/regulation, and intervention.

Biointensive integrated pest management is a systems approach to pest management based on an understanding of pest ecology. It begins with steps to accurately diagnose the nature and source of pest problems, and then relies on a range of preventive tactics and biological controls to keep pest populations within acceptable limits. Reduced-risk pesticides are used if other tactics have not been adequately effective, as a last resort, and with care to minimize risks.

An important difference between conventional and biointensive IPM is that the emphasis of the latter is on proactive measures to redesign the agricultural ecosystem to the disadvantage of a pest and to the advantage of its parasite and predator complex.

BIPM options may be considered as proactive or reactive.

1.5.2.1. Proactive options: Proactive options, such as crop rotations and creation of habitat for beneficial organisms, permanently lower the carrying capacity of the farm for the pest. The carrying capacity is determined by the factors like food, shelter, natural enemy complex and weather, which affect the reproduction and survival of a pest species. Cultural control practices are generally considered to be proactive strategies. Proactive practices include crop rotation, resistant crop cultivars including transgenic plants, disease-free seed and plants, crop sanitation, spacing of plants, altering planting dates, mulches, *etc.*

The proactive strategies (cultural controls) include:

- » Healthy, biologically active soils (increasing below-ground diversity).
- » Habitat for beneficial organisms (increasing above-ground diversity).
- » Appropriate plant cultivars.

1.5.2.2. Reactive options: The reactive options mean that the grower responds to a situation, such as an economically damaging population of pests, with some type of short-term suppressive action. Reactive methods generally include inundative releases of biological control agents, mechanical and physical controls, botanical pesticides, and chemical controls.

1.6. FUTURE THRUSTS AND CONCLUSION

The olericulture sector in India possess unique advantage because of availability of varied climatic conditions for production of various vegetable crops and several ICAR institutes and SAUs are working for the benefit of farmers and other stakeholders. Besides, there is a huge varietal diversity in most of the cultivated vegetable crops, which increase the harvesting season and more revenues to the farmers. The production and supply of quality planting materials, implementation of INM and IPM, multi-cropping system, integrated farming system (IFS), pollinators, pre- and post-harvest management, forward and backward linkages in marketing are very important for successful cultivation of vegetable crops. In spite of higher production, the quantity of processed products from the raw materials are poor and very low in produce quantity, which is due to the lack of GAP for most of the horticultural crops and poor post-harvest handling. Several prediction models have suggested that the area under vegetable crops is going to increase, and cereal crops will be decreasing, hence, it is time to breed the climate resilient varieties to withstand the climatic aberrations. Hence, a comprehensive plan covering R and D aspects for all vegetable crops is a must and convergence of scientists, policy makers and public are the need of the hour for marching ahead and to take the olericulture to newer heights. However, there are certain cause of concerns like-olericulture does not enjoy a safety net like the Minimum Support Price (MSP) for food grains; lack of good cold chain storage and transport networks to extend the life of perishable products; very less or limited input by machinery and equipment so it is tough to minimize the time restraints; higher input costs than food grains make it a difficult set up, especially when there is no support from the local governments to the smaller farmers; and limited availability of market intelligence, mainly for exports–make it a tougher option to choose.

Development of olericulture in India needs some critical management inputs particularly that of supply chain management - collaboration among various stakeholders along with efficient vertical and horizontal integration. The vegetable sector in particular has to prioritize development of research in the issues of genetics, biotechnology, integrated and sustainable production systems, post-harvest handling, storage, marketing and consumer education. Diversification offers an attractive option and a major source of pushing up growth of horticultural sector. While technological up-gradation and associated institutional changes are identified as thrust areas for future development of the olericulture sector, exports are considered to be most important for the growth of the sector. India can look forward to emerging as a major producer of vegetable products and thus secure reasonable market access for its agro exports, which are largely dependent on the competitive technologies that will help in enhancing export potential. This development will also help in overall growth of the economy through generation of extra foreign exchange, creating employment opportunities and also upliftment of the small and marginal farmers, with definite positive implications on income and employment. The government should create a positive environment that will ensure a mutually beneficial relationship between farmers and organized sector.

Vegetable crop diversification should be encouraged by intercropping with non-horticultural crops. This will yield more food, more income and better soil health. To increase the production and productivity of vegetables, introduction of vegetables in the crop rotation and adoption of recommended practices is very important. The use of vegetables in intercropping also helps in increasing the incomes of farmers during the period when the fruit orchard has not become commercially viable. The diversification plan of the horticulture sector needs to identify potential crops area-wise and the area under low yielding vegetables should be shifted to more productive and profitable one. There is a strong need to strengthen the research on vegetable crops to develop demand-driven technology by improved variety, pest management, *etc.* The horticultural development requires a minimum set of basic production factors, an optimal crop management infrastructure, post-harvest infrastructure, entrepreneurial management and horticultural expertise, logistical infrastructure and supporting financial infrastructure.

Development of vegetable sector should be accompanied by the growth of the agro- processing industry. The opportunity exists to promote the industry by intensifying production of a required, appropriate variety of tomatoes, cucumbers *etc.*, for the products like ketchup, sauce, juice and pickles. Thus, the production strategy should target not only meeting the domestic and export demand of fresh products but also of the processed products. There is the need to improve post-harvest operations related to handling, storage and marketing of fresh and processed produce. Volumes saved in post-harvest losses are actually the surpluses generated, without additional cost.

The vegetable sector has an immense potential for generating additional employment by developing the horticulture based agro-processing units. Empirical analytical evidence has shown that the vegetable export has increased the opportunity for higher earnings for smallholders and that the much higher land sizes owned by horticultural smallholders are indeed a cause or consequence for their participation in the sector. This sector needs to be developed as an organized industry and has to be run collectively by all the stakeholders with farmers as entrepreneurs and private sectors.

Our future line of action should focus on strengthening of the integrated cold chains at farmgate pack houses; expanding the harvesting window by mainstreaming of vegetable crops even during off-season to increase availability; and last but not the least, developing nutrient dense biofortified varieties for nutritional security. It goes without saying that our last mile, more than 700 Krishi Vigyan Kendras (KVKs), need to be aggressively plugged into farmgate supply and value chains to advise on increasing productivity, yield, and market reach of vegetable crops. India can achieve the goal of bringing to the plate vegetable crops to the poorest of the poor, not as a special food but as a daily necessity.

Chapter - 2

NEMATODE DISEASES OF VEGETABLE CROPS: AN OVERVIEW

2.1. INTRODUCTION

Intensive cultivation of vegetable crops is gaining importance because of their increasing value in exports and domestic markets. Monocropping of these crops increased the nematode populations to an alarming dimension and hence plant parasitic nematodes have become one of the major limiting factors in vegetable crop production systems. Vegetable production in every part of the world is reduced to a greater or lesser extent by nematode pests. Bulbs, rhizomes, roots, tubers, or other subterranean portions of plants are infected by nematodes and infected tissue makes such products undesirable as marketable food items. Roots damaged by the nematodes are not efficient in the utilization of available moisture and nutrients in the soil resulting in reduced functional metabolism. Visible symptoms of nematode attack often include reduced growth of individual plants, varying degrees of chlorosis, wilting of the foliage and sometimes death of plants. Furthermore, roots damaged by nematodes are easy prey to many types of fungi and bacteria which invade the roots and accelerate root decay. These deleterious effects on plant growth resulted in reduced yields and poor quality of vegetable crops.

Root-knot nematodes (*Meloidogyne* spp.) are serious pests of vegetable crops (tomato, eggplant, okra, onion, carrot, cowpea, French beans, peas, amaranths, cabbage, cauliflower, and cucurbits). Reniform nematode (*Rotylenchulus reniformis*) is another important nematode inflicting considerable losses in the yield of several vegetables. Other nematodes such as *Pratylenchus* spp., *Heliotylenchus* spp., *Tylenchornchus* spp. and *Hoplolaimus* spp. do cause considerable damage to the vegetable crops. Further the nematode problems are widespread, almost in all the production systems followed by all the categories of farmers.

Nematode management is, therefore, important for high yields and quality that are required by the prohibitive cost of modern vegetable production. It was observed that no single method of control of nematodes was effective in the management of nematodes in vegetable crops on a sustainable basis. Hence, the possibilities of effective use of different components for the integrated management of the nematodes in vegetable crops on a sustainable basis needs to be investigated.

2.2. ECONOMIC IMPORTANCE

2.2.1. International scenario

Crop losses due to plant parasitic nematodes in different vegetable crops are presented in Table 2.1.

Table 2.1. Estimated annual yield losses in vegetable crops due to damage by plant parasitic nematodes–World basis (Sasser and Freckman, 1987)

Vegetable crops	Loss (%)
Potato	12.2
Tomato	20.6
Eggplant	16.9
Bell pepper	12.2
Cowpea	15.1
Pigeon pea	13.2
Field bean	10.9
Okra	20.4
Melons	13.8
Beet root	10.9
Sweet potato	10.2
Cassava	8.4
Yam	17.7

In the United Kingdom, the potato cyst nematode (PCN) problem (*Globodera rostochiensis* and *G. pallida*) accounts for an estimated ~$US 70 million or 9% of UK potato production (Nicol *et al.*, 2011). In the USA alone, vegetable crops suffer 11 per cent annual loss due to nematodes.

2.2.2. Indian scenario

Based on data generated through AICRP on Nematodes over the years, a critical analysis was made on losses in different vegetable crops. Tomato (₹ 6035.2 million), brinjal (₹ 3499.12 million), and okra (₹ 2480.86 million) among the vegetable crops suffered comparatively more losses. The details of crop losses in different vegetable crops are provided in Table 2.2 (Walia and Chakraborty, 2018).

Table 2.2. Estimated losses due to economically important plant parasitic nematodes to vegetable crops in India (2014-15) (Walia and Chakraborty, 2018).

Crop	Nematode	Production (million tons)	Yield loss (%)	Price per metric ton (₹)	Monetary loss (₹ in million)
Vegetable crops					
Bitter gourd	*M. incognita*	00.08 (00.77)	13.5	23410	252.82
Bottle gourd	*M. incognita*	00.19 (01.82)	22	9660	403.78

Brinjal	*Meloidogyne* spp.	01.25 (12.58)	21	13330	3499.12
Capsicum	*Meloidogyne* spp.	00.02 (00.18)	10	26460	52.92
Carrot	*Meloidogyne* spp.	00.10 (00.96)	34	22180	754.12
Chilli	*Meloidogyne* spp.	00.20 (01.99)	15	24830	744.90
Cucumber	*Meloidogyne* spp.	00.07 (00.67)	12	13150	110.46
Okra	*Meloidogyne* spp.	00.57 (05.70)	19.5	22320	2480.86
Potato*	*Globodera* spp.	0.032 (0.032)	26	15270	127.04
Tomato	*Meloidogyne* spp.	01.64 (16.38)	23	16000	6035.20
Mean yield loss in vegetable crops -19.6%		**Total monetary loss in vegetable crops – ₹ 14461.22 million**			

Figures on production are from Anonymous (2016) and AGMARKETNET Portal (2014).
Loss estimations are based on work done at various Centers of AICRP on Nematodes.
Only 10% of total area under crop production infested with phytonematodes is used for estimation of losses.
Figures in parentheses show the total production of the crops in the country.
**Globodera* spp. is widespread only in Nilgiris Hills (Tamil Nadu), Hence, the area and production of Nilgiris Hills alone is used for estimation of losses.

Various researchers reported a crop loss of 70 per cent in chillies, brinjal, tomato and okra. In Mahendragarh district of Haryana, in a field naturally infested with *M. incognita* @ 2800-3460 1arvae per kg soil, losses in yields of okra, tomato and brinjal were 90.9, 46.2 and 27.3 percent, respectively. The yield of tomato was reduced by 39.8 percent in a field at population level of 20 1arvae per g of soil. In peas, the avoidable loss in yield was estimated to be 19-20 per cent.

In Tamil Nadu, losses in yields of Capsicum and tomato were 19.7 and 61.0 per cent respectively, due to *M. incognita*. In Haryana, in a field having initial *M. javanica* population of 51 and 296 per 250 g soil, the avoidable losses in okra yield ranged from 20.2 to 41.2 per cent, respectively. Percent avoidable losses in yield of okra, brinjal, French bean and cowpea due to *M. incognita* under field conditions at Bengaluru were 28.08, 33.68, 43.48 and 28.60, respectively. Yield losses in tomato, brinjal and bitter gourd due to *M. incognita* race 3 under field conditions in Maharashtra were 46.92, 32.73 and 36.72 per cent, respectively.

Scientists investigated the effect of *R. reniformis* on yield potential of a few vegetables and yields from fumigated plots of okra, tomato, cowpea, and lettuce were 19, 20, 15 and 52 per cent higher, respectively, than the non-fumigated plots.

2.3. EMERGING NEMATODE PROBLEMS

The changing cropping patterns, introduction of new crops, crop diversification, and agronomic practices *etc.* change the spectrum of plant parasitic nematodes. For example, diversification towards horticultural crops, particularly protected cultivation systems; and widespread and unchecked movement of planting materials from horticultural nurseries, have led to emergence of new nematode problems in newer areas and intensification of existing nematode problems. A number of new nematode problems have emerged in the intensive cropping systems.

Numerous challenges in the discipline of Nematology have opened up new opportunities to prove nematodes as important limiting factors in vegetable production: root-knot nematode problem on vegetables grown under protected cultivation; cyst nematodes problem on potato; impact of climate change; threat of introduction of new nematode pests under globalization of agricultural produce; lack of management strategy in standing crops, *etc.* are some of the challenges in near future.

2.3.1. Cyst nematodes on potato

Potato cyst nematodes (*Globodera rostochiensis* and *G. pallida*), which were hither to restricted to South Indian states such as Tamil Nadu (Nilgiris Hills) and Kerala (Kodai Hills), have now been detected in North Indian state like Himachal Pradesh (Shimla, Mandi, Kullu, Chamba and Sirmaur districts), which is a serious concern with ramification on export and quarantine issues (Ganguly *et al.*, 2010). Recently, potato cyst nematodes have also been recorded from Jammu and Kashmir and Uttarakhand states.

2.3.2. Nematode problems on polyhouse vegetable crops

Vegetable crops (tomato, bell pepper, cucumber, and okra) are being grown throughout India under protected cultivation (in polyhouses/greenhouses/shade nets) which are seriously infested with nematodes such as *Meloidogyne incognita*, *M. javanica* (root-knot nematodes) and *Rotylenchulus reniformis* (reniform nematode). Nematode problems on all these crops under protected conditions have assumed alarming proportions leading to huge losses (up to 80%) in select crops. The nematode infestations exacerbate severity of fungal diseases leading to complete crop losses. *M. incognita* infection makes the plants highly susceptible to *Fusarium oxysporum* f. sp. *dianthi* attack in bell pepper.

2.4. MAJOR NEMATODE PESTS

2.4.1. Root-knot nematodes, *Meloidogyne* spp.

The root-knot nematodes are by far the most important pests of vegetable crops. At least four species viz., *M. incognita, M. javanica, M. hapla* and *M. arenaria* have been reported and all form conspicuous root galls. *M. incognita* and *M. javanica* are most widespread in distribution and have a wide host range among vegetables, whereas *M. hapla* attacks potato and *M. arenaria* infects chillies. Four races of *M. incognita* have been reported from India.

Physiological races/biotypes are known to be prevalent in a few species of root-knot nematodes. The international *Meloidogyne* project has identified four races in *Meloidogyne incognita* and two in *M. arenaria*. In India, researchers reported two races of *M. incognita*. In Haryana and Karnataka, the existence of three races of *M. incognita*, i.e., race 1, 2 and 4 has been reported. The prevalence of race 4, which is very rare (2 percent of 472 world population) has been reported for the first time from India.

2.4.2. The reniform nematode, *Rotylenchulus reniformis*

The reniform nematode infects tomato, brinjal, okra, cowpea, Dolichos, French beans, pointed gourd (parwal) and other vegetables. Two races of this nematode have been reported from India. The emergence of cowpea seedlings was delayed by 7 to 9 days and seedling stand was reduced to the tune of 6 to 11 per cent due to *R. reniformis* at one nematode per one g of soil.

Comprehensive work done by various researchers has established the existence of two biotypes (A and B) in *R. reniformis*. Biotype A completes its life cycle on all the three differential hosts (cowpea, castor, and cotton) while biotype B would not complete its life cycle on castor or cotton. The two races interbreed with each other, and the hybrids were fertile and more polyphagous than the parents.

2.4.3. Root lesion nematodes, *Pratylenchus* spp.

P. delattrei reduces the vigor of chillies and induces downward cupping and crinkling of leaves. *P. delattrei* was highly pathogenic to okra.

2.4.4. Stunt nematodes, *Tylenchorhynchus* spp.

Although widely distributed both in temperate and tropical zones, only a few species are known to be pathogenic to various vegetable crops. In Southeastern USA, *T. claytoni* has been responsible for stunting of pea and *T. martini* in sweet potato. *T. brassicae* damages cabbage and cauliflower, *T. dubius* parasitizes cauliflower, pea, radish and turnip in The Netherlands. Besides cauliflower and cabbage, tomato, radish, sugar beet and lettuce are also good hosts.

The stunt nematode, *Tylenchorhynchus brassicae* reduce the yields of cabbage and cauliflower in Uttar Pradesh.

2.4.5. Spiral nematodes, *Helicotylenchus* spp.

H. dihystera causes perceptible reduction in root growth of chillies. Okra, tomato, brinjal and onion were good hosts for this nematode.

2.4.6. Cyst nematodes, *Heterodera* spp., *Globodera* spp.

Potato cyst nematodes (*G. rostochiensis* and *G. pallida*) are the most important pests of potatoes. This nematode is reported from Nilgiris and Kodai Hills of Tamil Nadu and Munnar Hills of Kerala. Out of 9,000 ha under potato; 3,000 ha are infected by this nematode in Nilgiris. In Kodai Hills, about 200 ha are infected. Tomato and brinjal are also hosts of this nematode. Total failure of the crop took place under severe infestation conditions. Recently, potato cyst nematodes have been recorded causing damage in Himachal Pradesh, Uttarakhand, and Jammu states.

The other cyst nematode, *Heterodera cajani* is known to infect cowpea, guar and *Dolichos lablab*.

2.4.7. The lance nematode, *Hoplolaimus indicus*

H. indicus causes growth reduction in tomato.

2.4.8. The stubby root nematode, *Trichodorus allius*

T. allius infects onion.

2.4.9. The stem and bulb nematode, *Ditylenchus destructor*

The stem and bulb nematode, *D. destructor* cause dry rot of potato tubers in Shillong; while *D. dipsaci* is reported on onion and garlic.

2.4.10. The burrowing nematode, *Radopholus similis*

R. similis has been reported on beans, cabbage, beet, cantaloupe, carrot, cowpea, okra, pea, sweet potato, pumpkin, radish, squash, and tomato.

2.5. INTERACTION WITH OTHER MICROORGANISMS

The root-knot nematode, *M. incognita*, present along with *Ralstonia solanacearum*, enhance the incidence of bacterial wilt of brinjal. Pani and Das (1972) have reported the association of root-knot nematode with bacterial wilt of tomato. The potato cyst nematode, *Globodera rostochiensis* in association with the wilt bacterium, *R. solanacearum* causes wilt complex in potatoes. The potato cyst nematode, *Globodera rostochiensis* in association with the fungus *Rhizoctonia solani* caused root rot complex in potatoes.

In French bean, the root-knot nematode, *M. incognita* reduced the number of *Rhizobium* nodules in the root system. Nodules containing root-knot nematodes deteriorate more rapidly than those free of nematodes. Thus, the concurrent nematode infection reduces the total benefit from the nitrogen-fixing bacteria. *M. incognita* and *Heterodera cajani* adversely affected the root nodulation and nitrogen content of cowpea (Sharma and Sethi, 1976b). The root-knot nematode also reduced the leghaemoglobin content of the cowpea root nodules (Sharma and Sethi, 1975).

M. incognita and *Rhizocotnia bataticola* or *Sclerotium rolfsii* when inoculated simultaneously in soil, reduce the germination of seeds in okra, brinjal and tomato (Chabbra and Sharma, 1981, Shukla and Swarup, 1970). *M. javanica* increased the extent of damage by pre- and post-emergence phases of damping off caused initially by *R. solani* and *Pythium debaryanum* in tomato (Nath *et al.*, 1984). *M. incognita* and *Ozonium texanum* var. *parasiticum* together acted synergistically by reducing the germination of brinjal to 28 per cent (Nath *et al.*, 1976).

M. incognita in association with soil-borne fungi such as *R. solani* and *F. solani* was responsible for root-rot/wilt disease complex in French beans. In presence of the nematode, the root rot due to *R. solani* and wilt due to *F. solani* increased considerably (Parvatha Reddy *et al.*, 1979, Singh *et al.*, 1981).

The incidence and severity of root rot of brinjal, tomato and okra, caused by the soil-borne fungi such as *R. solani, R. bataticola* and *Phompsis vexans* was increased in the presence of *M. incognita* (Hazarika and Roy, 1974, Khan *et al.*, 1971, Chalal and Chhabra, 1984, Chhabra *et al.*, 1977, Sharma *et al.*, 1980).

Okra plants succumbed to wilt at an early stage when the reniform nematode, *R. reniformis* and *R. solani* were present (Kumar and Sivakumar, 1981).

Tylenchorhynchus brassicae alone did not affect the percentage of emergence of cauliflower seedlings. *R. solani* was highly destructive by itself, but the adverse effect on the emergence of seedlings was greater when nematode infestation occurred along with fungal infection (Khan *et al.*, 1971).

The growth of tomato plants was significantly reduced in simultaneous inoculations with *M. incognita* and tobacco mosaic virus or when the nematode preceded the virus inoculation (Goswami and Chenulu, 1974). In the case of brinjal and bottle gourd, root-knot development has increased in the presence of mosaic virus (Mahmood *et al.*, 1974).

The combined effect of *M. incognita* and mycoplasma-like organisms (little leaf) on growth of brinjal was more than their individual effects (Dhawan and Sethi, 1977).

2.6. MANAGEMENT STRATEGIES

2.6.1. Regulatory methods

From the imported germplasm material, economically important nematodes like *Heterodera goettingiana, Meloidogyne incognita, Ditylenchus dipsaci, Radopholus similis* (Sethi *et al.*, 1972) and *Globodera rostochiensis* (Renjhen, 1973) have been intercepted during quarantine.

In India, there is a quarantine Act against cyst nematodes of potato (*Globodera rostochiensis, G. pallida*) in Nilgiris. Infected seed potatoes are not allowed to be transported from Nilgiris to other parts of India for seed purpose.

Potato seed pieces free of the cyst nematode can be produced commercially for seed certification.

2.6.2. Physical methods

The most important nematode pests controlled by greenhouse steaming are cyst nematode of potato attacking tomatoes and root-knot nematodes on tomatoes, cucumbers, and lettuces.

Incubation of potato tubers at 45 ^{0}C for 48 hours has been shown to kill about 98.9 per cent *M. incognita* without affecting tuber viability (Table 2.3) (Nirula and Bassi, 1965).

Table 2.3. Time and temperature recommendations for hot-water treatment for denematizing planting stock.

Nematode	Planting stock	Temp. C	Time min.
Meloidogyne spp.	Potato tubers	46.0-47.5	120
	Sweet potatoes	46.7	65
	Yam tubers	51.0	30

2.6.3. Cultural methods

Crops such as cereals (sorghum, millet, maize, wheat, rice), cruciferous (cabbage, cauliflower, kohl rabi, mustard), onion, garlic, groundnut, cotton, roselle (*Hibiscus sabdariffa*) and pigeon pea reduce root-knot populations in the soil.

A number of crops and other plants are reported resistant to the reniform nematode, *Rotylenchulus reniformis* which include sugarcane, *Coffea robusta*, jack bean (*Canavalia ensiformis*), velvet bean (*Stizolobium deeringianum*), chillies, maize, sorghum, *Leucaena glauca*, finger millet (*Eleusine coracana*), *Crotolaria, Capsicum frutescens*, carrot, *Anethum graveoleus*, coriander, *Foeniculum vulgare, Spinacea oleracea, Beta vulgaris* and *Raphanus sativus* (Khan and Khan, 1973). Two successive crops of maize or sorghum preceding susceptible crops are effective in controlling *R. reniformis*. Rotations with sugarcane and pangola gross are recommended for control of the reniform nematode.

2.6.3.1. Nematode-free planting material: The potato cyst nematode (*Globodera rostochiensis*) can be eliminated by selecting nematode-free planting material.

2.6.3.2. Time of planting: Planting potatoes during the third or fourth week of March in Shimla Hills would reduce the damage due to *M. incognita* (Prasad *et al.*, 1983). The yields were maximum which was concomitant with the lowest tuber infestation and lowest larval population in the soil at harvest.

2.6.3.3. Crop rotation: Continuous cropping with potatoes increases the cyst population of *Globodera rostochiensis*. More than three-year rotations with wheat, strawberry, cabbage, cauliflower, peas, maize, and beans reduce the nematode population to a safe level. Likewise, decrease in root-knot nematode population on tomato, brinjal, okra, chillies and sponge gourd occur following marigold, spinach, and bottle gourd (Khan *et al.*,1975).

There was a sudden decline in population of *Tylenchorhynchus brassicae* when cabbage and cauliflower were rotated with wheat (Siddiqi *et al.*,1973). The mustard (rabi), radish (summer), sesamum (kharif) sequence reduced the population of *Tylenchorhynchus* spp. (Haque and Gaur, 1985). Considerable reduction in population of *T. brassicae* was noted in potato and tomato, of *Hoplolaimus indicus* in cabbage, chilli and brinjal (Alam *et al.*, 1980).

2.6.3.4. Intercropping: The root-knot development on tomato and okra was low when interplanted with *Tagetes erecta*. The populations of *Tylenchorhynchus, Helicotylenchus, Hoplolaimus, Rotylenchulus,* and *Pratylenchus* were also markedly reduced (Alam *et al.*,1977, Khan *et al.*, 1971). Alpha-terthienyl is the active principle in *Tagetes* spp. which is toxic to these nematodes.

2.6.3.5. Trap crops: Cowpea causes the root-knot nematode eggs to hatch, the larvae enter the roots and develop to immobile stage. Then the crop is uprooted before the nematodes mature.

Crotolaria is highly susceptible to invasion by the root-knot nematode but is resistant to the development of larvae into adults.

2.6.3.6. Enemy plants: Potatoes grown with white mustard in a pot of infested soil were less infected by nematodes than potatoes growing alone. Potato root diffusate was ineffective in the presence of leaching's from the roots of mustard seedlings. Mustard oils increased the yield of potatoes by reducing the severity of nematode attack. The active principle involved in mustard is allyl isothiocyanate which is toxic to the nematodes.

Asparagus officinalis would not support populations of *Trichodorus chiristiei* for more than 40 to 50 days. Tomato, normally a good host of this nematode, supported only a low population when asparagus was growing in the same pots. Glycoside (asparagusic acid) is the active principle involved which is toxic to *T. christiei* and other nematode species.

Atwal and Mangar (1969) showed that root exudates from sesame (*Sesamum orientale*) have nematicidal properties against *M. incognita*. When okra was grown along with sesame in *M. incognita* infested soil it was only slightly attacked and there were fewer nematodes compared with when okra was grown in the absence of sesame.

2.6.3.7. Fallowing: Decrease in *M. incognita* population on tomato, brinjal, okra, chillies and sponge gourd occurs when the field was left fallow (Khan *et al.*, 1975).

2.6.3.8. Flooding: Flooding for three months is effective against root-knot and stunt nematodes on vegetables.

2.6.3.9. Organic amendments: Singh and Sitaramaiah (1966) applied finely divided oil cakes to root-knot infested soil and noted a reduction in disease incidence in okra and tomato. Amendment of soil with oil cakes of neem, groundnut, mustard, and castor was effective in reducing the population of *Tylenchorhynchus brassicae* around the roots of cabbage and cauliflower (Siddiqi *et al.*, 1976). The above cakes were also effective in reducing parasitic nematode population (*Hoplolaimus, Tylenchorhynchus, Meloidogyne* and *Helicotylenchus* species) and in increasing yields of tomato, potato, carrot and turnip. Neem and karanj oil cakes at 2 tons per ha were most effective in reducing *Rotylenchulus reniformis* infecting French beans.

Best results in respect of root-knot reduction due to *M. javanica* and increase in yield of okra and tomato were obtained by amending the soil with sawdust at 2.5 tons per ha 3 weeks before planting and then applying N through urea at 120 kg per ha (Singh and Sitaramaiah, 1971). Rice hull ash at 2.5 tons per ha increased the yield of tomato by 133 to 317 percent and reduced root-knot incidence by 46 to 100 percent (Sen and Dasgupta, 1981).

2.6.3.10. Fertilizer application: Increased levels of potash have significantly reduced the number of galls by *M. javanica* in tomato (Gupta and Mukhopadhyaya, 1971). Application of potash in combination with phosphorus or nitrogen or potash alone checks the reniform nematode multiplication on okra to a great extent (Sivakumar and Meerazainuddin, 1974).

2.6.4. Chemical methods

2.6.4.1. Nursery bed treatment: Nursery bed treatment with carbofuran at 2 g a.i. per m^2 was effective in increasing seedling growth and reducing root-knot nematode population on tomato, brinjal and chillies (Jain and Bhatti, 1983, Ramakrishnan and Balasubramanian, 1981). In the main field, the above treatments were also effective and increased the fruit yields.

2.6.4.2. Seed treatment: Fenamiphos and carbofuran each at 1 per cent concentration were effective in controlling the root-knot nematode (*M. incognita*) infecting cowpea, French bean and peas and in increasing their pod yields (Parvatha Reddy, 1984). Seed treatment with carbofuran at 6 and 12 per cent was effective against the root-knot and reniform nematodes infecting okra (Sivakumar *et al.*, 1976). Carbofuran at 3 per cent was also effective against root-knot nematodes infecting chilli and bottle gourd (Kandasamy and Sivakumar, 1981). Jain and Bhatti (1981) reported effective control of *M. javanica* on okra by seed treatment with fenamiphos at 2, 4 and 6 per cent.

2.6.4.3. Bare-root dip treatment: Bare-root dip treatment with carbofuran and oxamyl at 1,000 ppm for 15 to 30 min. (Parvatha Reddy and Singh, 1975); fenamiphos at 250 to 750 ppm for 30 min. (Thakur and Patel, 1985); dimethoate at 500 ppm for 6 hr, phosphamidon and dichlofenthion both at 1000 ppm for 8 hr (Jain and Bhatti, 1978) and thionazin at 500 ppm for 15 min (Reddy and Seshadri, 1975) gave effective control of root-knot nematodes on tomato. Alam *et al.* (1973) reported that oxamyl was effective for the control of root-knot nematodes infecting brinjal and okra. VC-13 was found effective against *M. incognita* in chillies, brinjal, and tomato (Alam *et al.*, 1973, Saxena *et al.*, 1974).

Bare-root dip treatment of brinjal seedlings with carbofuran and turbofos at 500 to 1000 ppm were effective in reducing the reniform nematode population (Prasad and Krishnappa, 1981).

2.6.4.4. Main field treatment: Carbofuran, ethoprophos and fenamiphos each at 1 to 2 kg a.i. per ha were found effective in reducing the root-knot nematode population and in increasing fruit yields of tomato, cauliflower (Ahuja, 1983), brinjal (119 to 145 per cent); okra (56 to 87 per cent), French bean (100 to 112 per cent); cowpea (40 to 50 per cent), peas (58 to 85 per cent) (Parvatha Reddy, 1985 a, Rao and Singh, 1978, Singh *et al.*, 1978) and chillies (Handa and Mathur, 1981).

Carbofuran at 1.5 kg a.i. per ha was effective in reducing *Tylenchorhynchus brassicae* and *T. dubius* population and in increasing yields of cabbage (Varma *et al.*, 1978).

Reddy and Seshadri (1972) reported that thionazin at 4 kg a.i. per ha proved effective against *R. reniformis* infecting tomato.

2.6.5. Biological methods

2.6.5.1. Antagonistic bacteria: *Pasteuria penetrans* was found effective against root-knot nematodes on various vegetable crops (Rao *et al.*, 1997e).

2.6.5.2. Antagonistic fungi: *Purpureocillium lilacinum, Trichoderma harzianum, Pochonia chlamydosporia* and *Verticillium lecanii* are effective parasites of *Meloidogyne* eggs. Egg parasites are more dramatic in reducing the root-knot nematode populations on various vegetable crops (Rao *et al.*, 1997a, b, c, d; 1998a, b, c, d). Nematode eggs of the group Heteroderidae and those deposited in a gelatinous matrix are more vulnerable to attack by these organisms than are those of migratory parasites. Once in contact with the cysts or egg masses, the fungus grows rapidly and eventually parasitizes all the eggs that are in the early embryonic developmental stages. Protocols for mass production of these bio-agents and their effective use were developed (Rao *et al.*, 1993b,1996a, 1997b, 1998a, c, d; Rao and Reddy, 1994).

2.6.6. Host resistance

Sources of resistance were identified in certain vegetable crops. Nematode-resistant varieties of a very few vegetable crops are developed against one or two nematodes out of many that exist (Table 2.4).

Table 2.4. Nematode-resistant cultivars of vegetable crops.

Vegetable crop	Nematode	Resistant cultivars	Reference(s)
Tomato	*Meloidogyne* spp.	Nematox, SL-120, NTR-1, SL-12, Patriot, VFN-8, VFN Bush, Piersol, Radiant, Nemared, Ronita, Anahu, Bresch, Helani, Campbell-25, Hisar Lalit, Punuui, Arka Vardan, Pelican, Hawaii-7746, Hawaii-7747.	Rao and Tikoo, 1973, Patel *et al.*, 1979, Jain *et al.*, 1983, Kanwar and Bhatti, 1990
Brinjal	*M. incognita*	Giant of Banaras, Black Beauty, Gola	Alam *et al.*, 1973b
Chilli	*M. javanica*	579, CAP-63, Pusa Jwala	Jain *et al.*, 1983
Potato	*M. incognita, Globodera rostochiensis*	Kufri Dewa, Kufri Swarna	Raj and Gill, 1983
Cowpea	*M. incognita*	Barsati Mutant, Iron, New Selection, G-152, 92-1-B, IC 9642-B, TVU 2430-P	Sharma and Sethi, 1976a, Darekar and Patil, 1981, Singh and Parvatha Reddy, 1982a
French bean	*M. incognita*	Banat, Blue Lake Stringless, Kenya-3, Bountiful Flat, Brown Beauty, W5-114, Cambridge Countess, Gallaroy, Pinto Seafarer, Suttan's Masterpiece	Singh *et al.*, 1981
Muskmelon	M. incognita M. javanica	Scarsol S-445	Khan *et al.*, 1971b
Watermelon	*M. incognita*	Shehajanpuri	Khan *et al.*, 1971b
Ridge gourd	*M. incognita*	Panipati, Meerut Special	Khan *et al.*, 1971b
Ash gourd	*M. incognita*	Jaipuri, Agra	Khan *et al.*, 1971b
Pumpkin	*M. incognita*	Jaipuri, Dasna	Khan *et al.*, 1971b

Among four species of *Solanum* tested for their reaction to *M. incognita, S. torvum* and *S.*

seaforthianum gave resistant reaction which was reflected in the reduction in number of galls, egg masses and fecundity of females (Shetty and Reddy, 1985 b).

Nature of resistance to *M. incognita* in cowpea selections IC 9642-B and TVU 2430-P was investigated by Singh and Parvatha Reddy (1985). The resistance was associated with reduced larval invasion, root galling, egg mass production and fecundity, delayed development of larvae to adult females coupled with high concentration of magnesium.

Effects of manifestation of Mi gene in a tomato hybrid and its parental lines on the penetration and development of *M. incognita* was studied by Rao *et al.* (1998e).

Histological studies revealed that there was a direct correlation between the number of cork layers and resistance to *M. incognita* in cowpea. Cortical sclerids were noticed in resistant selections while they were absent in the susceptible cultivars (S-488). There were more numbers of starch grains in the cortex of susceptible cultivars of cowpea. Death of cells (hypersensitive reaction) around infecting *M. incognita* larvae occurred in the roots of resistant selections (Singh *et al.*, 1985).

The formation of phytoalexin 'Kievitone' was observed in root-knot resistant cowpea cv. C-152 in response to *M. incognita* infection, while it was absent in susceptible cv. S-488. Higher concentrations of phenolic and Ortho dihydroxy phenolic compounds and more active IAA-oxidase and peroxidase systems have been observed in resistant tomato varieties (Bajaj and Mahajan, 1977, Ganguly and Dasgupta, 1981, Masood and Hussain, 1976).

2.6.7. Integrated methods

To alleviate nematode problems, integrated nematode management (INM) procedures are adopted based on the principles of prevention, population reduction and tolerance. INM seeks to stabilize populations of target nematodes at acceptable levels resulting in favorable long term socio-economic and environmental consequences. In practice, integration of different methods of nematode control is the rule than the exception. In most cases, practical control of a nematode disease involves integration of diverse control measures. In this context, biological control agents form ideal components in the INM along with botanicals and to a limited extent chemicals.

2.6.7.1. Bio-agents and botanicals

i) Tomato and brinjal: Integration of neem cake with *Trichoderma harzianum* gave effective control of *M. incognita* infecting tomato and increased root colonization with biocontrol agent (Rao *et al.*, 1997d).

Incorporation of neem cake/castor cake/neem leaf/castor leaf/pongamia leaf in nursery beds at 400 g/m^2 and application of spore suspension of *P. lilacinum* or *P. chlamydosporia* in tomato and brinjal nursery beds amended with the above plant products facilitated the effective management of root-knot nematode, *M. incognita* and reniform nematode, *R. reniformis* in the nursery beds and also protected the crop from these nematodes after transplanting in the main field. These treatments in general increased plant growth parameters and significantly increased the yields under field conditions (Rao and Parvatha Reddy, 1992, Rao et al., 1997c, d; 1998a).

Root-dip treatment of tomato seedlings in 5% aqueous suspension of neem leaf/cake mixed with the spores of *P. lilacinum* or *P. chlamydosporia* for 20-30 minutes before transplanting was effective in the management of root-knot and reniform nematodes in the main field.

Transplanting of tomato seedlings raised in nursery beds treated with neem cake and *P. penetrans* in pits incorporated with *P. lilacinum* gave effective management of *M. incognita* and increased tomato fruit yield (Parvatha Reddy *et al.*, 1997).

Application of extracts of neem cake/neem leaf with spores of *P. lilacinum/P. chlamydosporia* in the nursery beds and subsequent root-dip treatment in the above botanicals with the spores of bio-agents protected brinjal in the main field from the attack of root-knot and reniform nematodes. All these treatments significantly increased the yield under field conditions (Rao *et al.*, 1993c).

Zaki and Bhatti (1991) also found that integration of *P. lilacinum* and castor leaves was effective in increasing the growth of tomato and reducing the infestation by root-knot nematodes. Management of *M. incognita* and *R. reniformis* in nursery beds to get healthy seedlings was attempted by integrating soil solarization and oil cake (Mahua cake) incorporation.

ii) Okra: Application of 5 percent inoculum of *Arthrobotrys conoides* to the pot soil amended with FYM, effectively reduced the larval penetration of *M. incognita* and reduced root galling by 30-40 per cent in okra.

Neem cake suspension (5%) mixed with the spores of *P. lilacinum* is effective for the treatment of okra seeds for the management of *M. incognita* under field conditions (Rao and Parvatha Reddy, 1992a, b; 1993; 1994; 1996; Rao *et al.*, 1994; 1997b, c).

Seed treatment of okra with 5% aqueous extracts of neem leaf/neem cake containing spores of *P. lilacinum/P. chlamydosporia* was effective for the management of root-knot nematode under field conditions and these treatments have increased the okra fruit yield (Rao *et al.*, 1997a).

2.6.7.2. Endomycorrhizae and botanicals

i) Tomato: Inoculation of *Glomus mosseae* or *G. fasciculatum* in the nursery beds amended with neem cake/castor cake/neem leaf/calotropis leaf reduced the infestation of root-knot and reniform nematodes to the maximum extent on tomato. Amendment of botanicals in the nursery beds increased the multiplication of endomycorrhizae and colonization of tomato root which in turn could protect the crop from these nematodes in the main field resulting in the increased yields (Rao *et al.*, 1995b; 1996; 1997e).

ii) Brinjal: Similarly, inoculation of *G. mosseae/G. fasciculatum* in the brinjal nursery beds and subsequent application of 5% aqueous extracts of neem cake/castor cake/neem leaf/calotropis leaf in the nursery beds resulted in the effective management of root-knot and reniform nematodes and yielded healthy eggplant seedlings which would withstand the attack of these nematodes after transplanting in the main field (Rao *et al.*, 1993a, c).

iii) Okra: Spot treatment or inoculation with endomycorrhizae, *G. mosseae/G. fasciculatum* and subsequent application of neem cake/castor cake extracts (5%) facilitated the management of root-knot nematode and increased the yield of okra (Rao *et al.*, 1997b).

2.6.7.3. Bio-agents and endomycorrhizae

i) Tomato: Interestingly, integration of bio-agent (*P. lilacinum*) and endomycorrhizae *(G. mosseae/G. fasciculatum)* have culminated into the successful management of *M. incognita* infecting tomato. This phenomenon facilitated standardization of a strategy wherein inoculation of mycorrhizae and bio-agent in the nursery beds protected the seedlings of tomato from the attack of *M. incognita*. Further, these mycorrhizal seedlings (colonized either with *G. mosseae* or *G. fasciculatum*) can be given a root-dip treatment with spore suspension of *P. lilacinus* for 5-10 minutes for the effective management of these nematodes in the main field after transplanting (Rao *et al.*, 1993b, d).

2.6.7.4. Bio-agents, endomycorrhizae and botanicals

i) Brinjal: Inoculation of *G. fasciculatum* in the castor cake amended nursery beds followed by the root-dip treatment of mycorrhizal seedlings of brinjal in spore suspension of *P. lilacinum* for the management of *R. reniformis* (Rao *et al.*, 1993b).

ii) Tomato: Inoculation of *G. mosseae* in neem leaf/neem cake amended nursery beds followed by the root-dip treatment of mycorrhizal seedlings of tomato in spore suspension of *P. lilacinum* was effective for the management of *M. incognita* under field conditions (Rao *et al.*, 1995a).

2.6.7.5. Bio-agents and chemicals

i) Tomato and brinjal: Integration of a bio-agent, *P. lilacinum* with carbofuran at 1 kg a.i. /ha was effective for the management of reniform nematode, *R. reniformis* infecting tomato and eggplant (Parvatha Reddy and Khan, 1988, 1989).

Maheshwari *et al.* (1987) reported that application of *P. penetrans* in combination with carbofuran, aldicarb, miral, sebufos and phorate significantly improved plant growth of tomato by reducing galling due to *M. javanica*.

ii) Okra: An integrated management of root-knot nematodes, *M. incognita* infecting okra using neem or karanj oil cake at 0.5 ton per ha along with carbofuran at 1 kg a.i. per ha was achieved. The above treatments gave maximum reduction in root galling with consequent increase in okra fruit yields (Parvatha Reddy and Khan, 1991).

2.6.7.6. Two bioagents

i) Tomato: Integration of two bio-agents, *P. lilacinum* and *P. chlamydosopria* resulted in combined and complimentary effects for the successful management of *M. incognita* infecting tomato (Rao and Parvatha Reddy, 1992a). Integration of *G. deserticola* with *P. chlamydosopria* gave effective management of *M. incognita* on tomato seedlings and increased the egg parasitisation by the bioagent (Rao *et al.*, 1997a).

Maheshwari and Mani (1988) observed that the population of *M. incognita* and *M. javanica* were effectively suppressed and tomato yields were greater when *P. penetrans* and *P. lilacinus* were applied together.

2.6.7.7. Botanicals and chemicals

i) Tomato: In tomato, application of carbofuran at 1 kg a.i. per ha in combination with neem cake and urea each at 10 kg N per ha, at transplanting, produced maximum yield with lowest gall index (2.5) and nematode population, 90 days after planting (Routaray and Sahoo, 1985).

ii) Brinjal: Integration of *P. lilacinum* and carbofuran at 2 kg a.i. per ha has proved to be effective in the management of reniform nematode, *R. reniformis* on brinjal.

iii) Okra: An integrated management of *M. incognita* infecting okra using neem or karanj oil cake at 0.5 t/ha along with carbofuran at 1 kg a.i. /ha gave maximum reduction in root galling with consequent increase in okra fruit yield (Parvatha Reddy and Khan, 1991).

2.6.7.8. Two cultural methods

i) Okra: Deep plowing (20 cm) followed by fallowing for one month or fallowing for one month after weeding, resulted in the control of the reniform nematode (*Rotylenchulus reniformis*) and better yield of okra (Lakshmanan and Sivakumar, 1981).

2.6.7.9. Cultural and chemicals

i) Okra: Deep plowing (20 cm) together with carbofuran seed treatment resulted in the control of the reniform nematode (*Rotylenchulus reniformis*) and better yield of okra (Lakshmanan and Sivakumar, 1981).

2.7. CONCLUSION

Although most of the research conducted under greenhouse conditions indicated that bio-control agents have the potential to nullify the damage due to phytonematodes, intensive investigations need to be carried out under field conditions to assess their practical utilization. The biological control agents with the highest potential for the management of plant nematodes are likely to be those that are endemic in agricultural soils and can be induced to proliferate by methods compatible with conventional farming practices.

Among bacteria, *P. penetrans* provides ample scope in exploiting its potential as a biocontrol agent against root-knot and lesion nematodes. The importance of endomycorrhizal fungi such as *G. mosseae* and *G. fasciculatum* in the management of phytonematodes such as root-knot, burrowing, and reniform nematodes is well documented on major vegetable crops. Since *P. penetrans* and VAM fungi are obligatory parasites, these cannot be cultured on a large scale which hinders their effective utilization in integrated nematode management programs.

The opportunistic fungi which infect eggs and females belonging to the genera, *Purpureocellium, Trichoderma* and *Pochonia* seem to be most promising. These fungi spread rapidly, are ubiquitous and are present abundantly in the rhizosophere. They are well adapted to the environment, being successful competitors and can survive host-free periods. Besides, they also have the added advantage of ease of multiplication in axenic culture and introduced into soil. In France, two commercial agents, 'Royal 300' and 'Royal 350', are available for control of *Ditylenchus myceliophagus* on mushrooms and *Meloidogyne* spp. on vegetables. More recently, the parasitic fungus, *P. lilacinum* was produced in the Philippines as 'Biocon' which can be used for the control of several nematode species. In view of the above advantages, opportunistic fungi form ideal biological control components in integrated nematode management programs under field conditions.

Chapter - 3

POTATO, *Solanum tuberosum*

The cyst nematodes (*Globodera rostochiensis* and *G. pallida*) and the root-knot nematodes *(Meloidogyne incognita, M. javanica* and *M. hapla*) are the major problems on potato in India.

3.1. CYST NEMATODES, *Globodera rostochiensis, G. pallida*

The potato cyst nematode (PCN) is the most important quarantine pest that is causing concern to potato production and trade throughout the world. The ability of this nematode to build up to damageable levels in a short span of 5-6 years, substantial yield reductions in the crop, lack of inexpensive nematicides for soil treatment capable of providing adequate level of control under field conditions, the relative ease with which the cysts are dispersed with soil adhering to the seed tubers and the long persistence of eggs within the cyst in the absence of the host makes this nematode as probably the most important pest problem on potatoes.

In India, the potato cyst nematode, *Globodera rostochiensis* was first reported by Jones (1961) who detected the cysts from the roots of potato plants which exhibited the symptoms of yellowing of leaves from Nilgiris Hills of Tamil Nadu state. This report initiated a great awareness amongst administrators in the country resulting in establishment of Nematology Unit in Tamil Nadu as well as initiating an Indo-German Collaborative Program of intensive chemical application (Dasanit) in the infested area to avoid further nematode spread. The Government of Tamil Nadu amended the Destructive Insect Pest Act 1919 during 1971 to prevent the potential damage due to potato cyst nematodes by ensuring strict prohibition of movement of seed potatoes from infested areas to any other parts of the state or the country.

3.1.1. Distribution

The earlier surveys have indicated the presence of both species of potato cyst nematode (*Globodera rostochiensis* and *G. pallida*) restricted to potato growing regions of Nilgiris, Palani and Kodaikanal Hills of Tamil Nadu (Krishna Prasad, 1986) and Kerala (Ramana and Mohandas, 1988).

Ganguly *et al.* (2010) reported the presence of potato cyst nematodes from the Potato Breeding Farms of Central Potato Research Institute located in Kufri, Shimla District of Himachal Pradesh. Nematode infestation to the extent of 58–548 cysts per 100 g soil were recorded. The European and Mediterranean Plant Protection Organization (EPPO) had also reported potato cyst nematode, *G.*

pallida from Himachal Pradesh earlier to Ganguly *et al.* (2010). Recently potato cyst nematodes were reported from the states of Uttarakhand and Jammu and Kashmir. These reports have a grave concern with ramifications on export and quarantine issues.

3.1.2. Crop losses

Potato cyst nematodes inflict huge quantitative and qualitative losses in potato. While 26% of losses are normal, the damage may be as high as 53-66% in badly infested fields. In severely infested fields, the yields may be even less than the tubers used for planting.

The avoidable yield loss in susceptible potato due to cyst nematodes was 99.5 to 99.8% in summer and autumn crops. Total failure of the crop has been reported under severe infestation conditions. An average loss of about 9% of global potato is accounted for the cyst nematodes amounting to about forty-five million tons (Prasad, 1989).

Potato cyst nematodes are responsible for 26% annual yield loss amounting to Rs. 127.04 million in Nilgiris Hills, Tamil Nadu, India (Walia and Chakraborty, 2018).

3.1.3. Symptoms

Under conditions of severe infestations, plant growth is stunted, and wilting occurs during hot parts of the day. The disease usually occurs in patches. Plants show tufting of leaves on the top as the outer leaves turn yellow and die. Root system is drastically reduced, secondary roots are induced at the collar region and the plants can be easily pulled out. Tubers formed are less in number and reduced in size. In extreme cases, the tuber formation is arrested. Outside the root surface, exceedingly small (about the size of a pin head - 0.5 mm) white females which turn to yellow or brown cysts after death can be observed at flowering or later. The total photosynthesis per plant is also significantly reduced because of reduced leaf area and this is reflected in reduced potato yield.

3.1.4. Host range

All known varieties of potato, tomato and brinjal are severely attacked by the cyst nematode. A number of other Solanaceae are recorded as hosts, but the majority of hosts are species and varieties of tuberous *Solanum*. The only susceptible plant outside the family is snapdragon, *Antirrhinum majus*.

3.1.5. Biology and life cycle

Cysts contain 200-500 eggs that remain viable for several years. Hatching commences with the planting of host crop; potato root diffusates strongly stimulate the emergence of juveniles from cysts. Only 30-40% of eggs hatch in a season, the remaining eggs hatch over a period of several years. Under Nilgiris Hills conditions, the nematode completes its life cycle in about 5-7 weeks.

3.1.6. Ecology

Well-drained, aerated sands, silts and peat soils with a moisture content of 50-75% water capacity are suitable for survival and movement of the free-living stages of the potato cyst nematode. A minimum soil temperature of 10 °C is necessary for larval activity.

3.1.7. Spread and survival

The cysts can be spread by the movement of soil through seed material for an exceedingly long distance. They can also be spread through field tools, implements, seedlings grown in infested soil for transplanting elsewhere, laborers' feet, *etc.* from one field to another. The cysts can remain viable in the soil without a host for 30 years.

3.1.8. Interaction with other pathogens

The potato cyst nematode and the black-scurf fungus, *Rhizoctonia solani* together caused much greater reduction in yield than did the organisms separately. The severity of the wilt, *Verticillium dahlia*, also increased in the presence of potato cyst nematodes. The leaf symptoms appeared earlier, and the tuber yields were depressed in the presence of the nematode.

The potato cyst nematodes, *Globodera rostochiensis* and *G. pallida* in association with the wilt bacterium, *Ralstonia solanacearum* causes wilt disease complex in potatoes.

3.1.9. Physiological races

At present, five pathotypes in *G. rostochiensis* and four pathotypes in *G. pallida* are differentiated using the following differentials (Table 3.1):

Table 3.1. Pathotypes of *Globodera rostochiensis* and *G. pallida*

Differential host	Nematode pathotypes*				
	1	2	3	4	5
	Globodera rostochiensis				
Solanum tuberosum	+	+	+	+	+
ex *andigena* CPC 1673	-	+	+	-	+
ex *kurtzianum* 60.21.19	-	-	+	+	+
ex *vernei* 58.1642/4	-	-	-	+	+
ex *vernei* 62.33/3	-	-	-	-	+
ex *vernei* 65.346/19	-	-	-	-	-
	Globodera pallida				
Solanum tuberosum	+	+	+	+	
ex *multidissectum* P 55/7	-	+	+	+	
ex *vernei* 62.33/3	-	-	+	+	
ex *andigena* x *multidissectum* D 47/1	-	-	-	+	
ex *andigena* x *multidissectum* D 47/2	-	-	-	-	

* (+) indicates ability to produce females, (-) indicates inability to produce females

In Nilgiri and Palani hills in South India, 5 pathotypes – Ro 1, Ro 2 of *G. rostochiensis*, and Pa 1, Pa 2 and Pa 3 of *G. pallida* have been detected (Amalraj and Rao, 1992).

3.1.10. Management

3.1.10.1. Regulatory methods: The infested area should be quarantined and cultivation of host plants to be discontinued indefinitely. Volunteer seedlings in the infested area should be destroyed. The seed material should not be allowed to move out of the infested area. The infested area can be put to non-agricultural use like construction of buildings, *etc.*

Potato cyst nematodes mainly spread through infected planting stock. Hence, all seed producers and growers selling seeds of commercial crops should be registered under certified schemes in all states to produce nematode-free seed tubers.

3.1.10.2. Physical methods: Fassuliotis and Sparrow (1955) demonstrated that irradiation of potato tubers with x-rays inhibited sprouting and development of *G. rostochiensis*. Cysts of this nematode exposed to 20,000 r contained only brown and dead eggs, while at 40,000 r the eggs lost their contents completely.

Careful washing of soil adhering to potato tubers can reduce the risk of spreading cyst nematodes. Commercial potato tuber washing apparatus has been developed.

3.1.10.3. Cultural methods

i) Crop rotation: It is the most practical method of control. It is recommended to grow the host crop once in every 3 years if the infestation is confined to a few patches, but only once in 4 or 5 years if the infestation is extensive. Grow non-host crops like cabbage, cauliflower, beet root, radish, carrot, beans, strawberry, and turnip in rotation. In Nilgiris Hills, potato is rotated with wheat and strawberry to control cyst nematodes. Growing of resistant potato cultivar, Kufri Swarna in rotation is beneficial.

ii) Trap cropping: Trap cropping should only be considered where PCN infestations are very large and need to be reduced quickly. The trap crop, *Solanum sisymbriifolium* generates high mortality in PCN but requires a full growing cycle to be effective. The technique is costly (at least as costly as soil fumigation) and needs careful management.

iii) Intercropping: Cyst nematode in Tamil Nadu hills is managed by intercropping potato with beans or wheat.

iv) Disease escape: Early varieties of potatoes like Irish Cobbler, usually suffer less than do late varieties. The potato crop should be planted during spring when the soil temperatures are very low for nematode activity, and the crop can mature before the nematode can multiply to sufficient level to cause damage to the crop.

v) Botanicals: Incorporation of eucalyptus litter at 10 t/ha at the time of preparation of land reduces the population of the potato cyst nematodes. Application of eucalyptus distillation waste at 2.5 t/ha helps to reduce the cyst nematode population in soil.

vi) Biofumigation: Black Mustard, Caliente Mustard, Marigolds and Sudan Grass have the potential to reduce PCN by biofumigation. Chopped and incorporated into soil, the green manure releases chemicals that 'fumigate' the soil killing the cysts as well as fungal spores and weed seeds.

Among the 6 plants extracts (broccoli, cauliflower, collards of *Brassica rapa*, collards of kale, Portuguese cabbage and watercress) evaluated at two concentrations (0.2 and 0.05 μmoles 100 g^{-1} dry weight), least reproductive factor (Rf) was observed for watercress, followed by cauliflower and *Brassica rapa*, based on the levels of 2-phenthyl and 2- propenyl glucosinolates. The results suggest also that the incorporation of plant extracts, simultaneously to the potato cultivation could be an effective strategy to suppress the potato cyst nematode population (Aires *et al.*, 2006).

3.1.10.3. Chemical methods: The infested area can be treated with an eradication dosage of Fesnulfothion or Carbofuran each at 3 kg a.i. per ha in two doses at an interval of 15 to 20 days. Dazomet, which liberates methyl isothiocyanate, is more nematicidal in pearls and clay soils and often gives large yield increases.

3.1.10.4. Biological methods

i) Antagonistic bacteria: Application of *Pseudomonas fluorescens* at 10 g/m^2 gives good nematode control (Mani *et al.*, 1998).

Application of talc-based formulation of *P. fluorescens* (15 x 10^8 cfu/g) at 10 kg/ha reduced the penetration of roots by the juveniles of potato cyst nematodes at 45 days after planting by 63.4%. A significant reduction in cyst count and egg production was observed with the above bioagent. The above treatment also increased potato tuber yield significantly (76.2% over control) (Table 3.2) (Seenivasan *et al.*, 2007).

Table 3.2. Effect of *Pseudomonas fluorescens* on cyst nematode and tuber yield of potato

Treatment	No. of J_2/g root at 54 DAP	No. of cysts /100g soil	No. of eggs/g soil	Tuber yield (tons/ha)
P. fluorescens – 10 kg/ha	60	130	58	14.89
P. fluorescens - 5 kg/ha	114	179	72	11.43
P. fluorescens – 2.5 kg/ha	138	195	85	9.46
Control	164	241	102	8.42
CD (P = 0.05)	11.24	18.47	5.42	1.12

Pseudomonas fluorescens at 10 kg/ha gave maximum reduction in cyst nematode multiplication (Reproduction factor-1.12) and improvement in tuber yield (19.40 t/ha) (Umamaheswari *et al.*, 2010).

Plant health promoting rhizobacteria namely *Agrobacterium radiobacter* and *Bacillus sphaericus* have been found to reduce toxic metabolites affecting penetration of *G. pallida* by 41% in potato roots and improving yield (Racke and Sikora, 1992). The root penetration of *G. pallida* is reduced in the presence of *B. sphaericus* or *A. radiobacter* which may be due to resistance induced by bacteria (Hallman *et al.*, 1998).

Reitz *et al.* (2000) showed that the lipopolysaccharides of *Rhizobium etli* G12 induce the systemic resistance to infection by the cyst nematode *Globodera pallida* in potato roots.

ii) Antagonistic fungi: Experiments on control of *G. rostochiensis* using *Purpureocillium lilacinum* revealed that the fungus-treated plots had higher yield than did the control plots and those treated with carbofuran (Table 3.3). Similarly, there were significantly lower numbers of cysts in the *P. lilacinum*-treated plots than in nematicide-treated or control plots (Jatala, 1985).

Table 3.3. Biological control of *Globodera rostochiensis* in potato by *Purpureocillium lilacinum*

Treatment	Tuber yield (MT/ha)		No. of cysts/ 50 ml soil
	Marketable yield	Total yield	
P. lilacinum	20.58	26.64	58.88
Carbofuran at 3 kg a.i./ha	19.68	24.90	67.33
Control	15.72	20.52	70.35
P = 0.05	N.S.	N.S.	5.73

In greenhouse tests, *P. lilacinum* was also found effective against *G. pallida*. Results showed that a significantly higher percentage of *G. pallida* eggs hatched in the control (non-inoculated) cysts as compared to those inoculated with *P. lilacinum*. Approximately 50% of the healthy *G. pallida* eggs were either infected or damaged by *P. lilacinus* during the experimental period (Jatala, 1985).

Divide and Zorilla (1995) found that tuber-dip and soil mix method of application of *P. lilacinum* were equally effective against *G. rostochiensis* on potato and their combined application enhanced nematode control efficacy.

Sikora (1981) found that *G. rostochiensis* population was reduced in potatoes colonized by *Glomus fasciculatum* which may be due to either composition of habitat's space or nutrients.

Catenaria auxillaris parasitized females of *G. rostochiensis*.

iii) Predatory nematodes: Small (1979) found a significant reduction in the population of *G. rostochiensis* in the presence of predatory nematode, *Prionchulus punctatus*.

3.1.10.5. Host resistance: *Solanum vernei* (multiple gene resistance) and *S. tuberosum* s. sp. *andigena* (single gene resistance) were found to have marked resistance to the cyst nematode of potato. *S. andigena* roots produce a hatching factor and are readily invaded by the larvae, which, however, fail to mature and reproduce. The failure is because the root tissues are not induced to produce giant cells, but instead the cells near the nematode head become necrotic and there is inadequate nutrition for the female to develop.

In the United States, both the USDA and Cornell University scientists have evolved two commercial potato varieties *viz.*, 'Peconic and `Wauseon' which are resistant to the cyst nematode. `Peconic' derives its resistance from the subspecies andigena. 'Wauseon' is a selection from a cross between B 4159-8 and Katahdin.

In India, Potato variety Kufri Swarna developed by the Central Potato Research Institute Substation at Ooty is resistant to both the species of cyst nematode and the late blight disease. The variety is also draught resistant and has shorter dormancy which is advantageous to Nilgiris region where potato is taken up as a rain fed crop almost throughout the year and thus covers about 40% of potato area (Anon,1985).

3.1.10.6. Integrated methods: The integrated management of cyst nematodes can be achieved by combining practices like planting early maturing cultivars (nematode escape) and soil application nematicides. Other practices like use of certified seed, crop hygiene, along with rotation of potato with non-host crops can also be used for the nematode management in seed production regions. Use of a resistant variety followed by a nematicide would kill 99 per cent of the nematode population.

i) Bioagents, botanicals and physical methods: Sitaramaiah and Naidu (2003) found that soil solarization during peak summer along with soil incorporation with *Pasteuria penetrans* and karanj cake/ farmyard manure gave good control of *G. rostochiensis*.

ii) Bioagents and botanicals: Application of Neem cake at 1 t/ha or FYM at 20 t/ha or compost at 10 t/ha enriched with *Trichoderma harzianum/ Purpureocillium lilacinum* gave effective control of cyst nematodes.

Umamaheswari *et al.* (2012) reported that reduction in nematode reproduction (Reproduction factor-1.09) and increase in tuber yield (23.14 t/ha) was achieved by integrating soil incorporation of Neem cake at 5 t/ha along with *Trichoderma viride* at 0.5 kg/ha.

iii) Cultural, chemical and host resistance: The most effective measure combines crop rotation, use of nematicides and resistant varieties to keep the nematodes at an economically acceptable level (Table 3.4) (Jones, 1969).

Table 3.4. Integrated management of the potato cyst nematode.

Management method (s)	Resulting kill population (% initial population)	Kill (%)	Population after growing and harvesting a susceptible cultivar, calculated at 2 assumed nematode multiplication rates (% initial population)	
			30 x	70 x
4 years without potato	3	97	90	‹ 100
1 year with resistant Potato	20	80	› 100	› 100
Nematicide treatment (Fumigant)	25	75	› 100	› 100
1 and 2	0.6	99.4	18	42
1 and 3	0.75	99.25	22.5	52.5
2 and 3	5	95	› 100	› 100
All 3 methods	0.15	99.85	4.5	10.5

An integrated system utilizing resistant cultivars, non-host crops and susceptible cultivars with chemical treatment (Carbofuran) will successfully manage densities of *Globodera rostochiensis* at desired levels.

A well-known integration of methods can control *G. rostochiensis* in potato in a 4-year rotation combining soil fumigation (causes 30% reduction in nematode population), cultivation of non-host (50% reduction), cultivation of resistant potato variety (30% reduction), cultivation of non-host (50% reduction) and then cultivation of susceptible variety of potato (Oostenbrink, 1972).

3.2. INTERACTION OF CYST NEMATODES WITH *VERTICILLIUM* WILT

3.2.1. Interaction

The association of fungus like *Verticillium dahliae* with *Globodera rostochiensis* has been reported on potato indicating that the cyst nematode participates in the disease complex.

Storey and Evans (1987) found that the pathology of the wilt fungus *V. dahliae* is dependent on the timing of invasion by the cyst nematode *G. pallida*, and on the potato cultivar. The fungus was found to enter and use the invasion channels of juvenile *G. pallida* if it was introduced at the same time as the nematode on cvs. Pentland Javelin, Maris Peer and Maris Anchor. On both Maris Peer and Maris Anchor, lignified tissue developed following the invasion of the nematodes. If the fungus was introduced 8 days after the nematode, wilt symptoms caused by *V. dahliae* were less severe than on plants treated with the fungus alone. When sections of root tissue were examined, it could be seen that the lignified tissue had sealed the nematode penetration sites. In fact, the living tissue was partially protected by areas of woody tissue where this response had taken place. This response was not produced by cv. Pentland Javelin after nematode invasion, and *V. dahliae* was still able to colonize roots to a larger extent than on plants where it was introduced alone. In addition to the cavities caused during nematode invasion, nematodes produce other forms of mechanical damage to plant roots that are open to exploitation by soil borne fungi.

G. rostochiensis interacts with *V. dahliae* enhancing the severity of wilt caused by the fungus. The leaf symptoms appeared earlier, and the tuber yields were depressed in the presence of the nematode (Jones, 1968).

3.2.2. Management

3.2.2.1. Chemical methods: Soil fumigation with methyl bromide resulted in yield increases of 68% for potato fields infested with *Globodera* spp. and *V. dahliae* (Hide and Corbett, 1974).

In *V. dahliae-G. rostochiensis* disease complex on potato, a mixture of methyl bromide and chloropicrin was found to have comparatively more fungicidal and nematicidal efficacy (Hide and Corbett, 1974).

3.3. INTERACTION OF CYST NEMATODES WITH ROOT ROT

3.3.1. Interaction

Potato cyst nematodes, *G. rostochiensis* and *G. pallida* in association with the fungus *R. solani* caused root rot complex in potatoes. In the glasshouse experiment, the combined effects of *G. pallida* with *R. solani* or *G. rostochiensis* with *R. solani* showed greater *R. solani* diseases of potatoes compared with only *R. solani*. Grainger and Clark (1963) found that potato cyst nematode and black-scurf fungus *R. solani* together caused much greater reduction in yield than did the organisms separately.

The field experiment revealed a clear positive relationship between initial population densities of *G. pallida* and the incidence of stolons infected by *R. solani*, subsequent stolon pruning and stem canker. In both experiments a clear positive relationship was found between densities of nematodes within the potato roots and the incidence of infected stolons, stolon pruning and stem canker.

3.3.2. Management

Nematicide applied together with *R. solani* in the field experiment reduced the effect of *G. pallida* on the incidence and severity of diseases caused by *R. solani*.

3.4. ROOT-KNOT NEMATODES, *MELOIDOGYNE* SPP.

Thirumalachar (1951) noticed root-knot galls on potato tubers from Shimla during 1950 for the first time and since then it has been recorded from all the potato growing regions of the country. Root-knot nematodes, *Meloidogyne* spp., are by far the most important nematode parasites of potatoes in India.

3.4.1. Crop losses

An initial level of 2 root-knot larvae per gram of soil resulted in 100% tuber infection with an overall yield reduction of 42.5% (Prasad, 1989).

Severe infestation of root-knot nematodes leading to crop failure has been noticed in Mahasu district of Himachal Pradesh and Hassan in Karnataka where farmers did not get even as much as planted by them.

3.4.2. Symptoms

Plants affected by root-knot nematodes are generally stunted, slightly yellow and may wilt during hot weather. The diagnostic symptoms of a *Meloidogyne* attack on potatoes are the galls found on both roots and tubers. The tuber surface becomes uneven and warty because of numerous blisters like galls. When an infested tuber is transversely cut, the white glistening swollen females of the size of pinhead can be easily seen embedded in the potato tissue. Tubers infected with blisters are responsible for significant crop losses due to their exceedingly small size, quickly rot, and not marketable.

3.4.3. Host range

Species of *Meloidogyne* usually have a wide host range amongst crops and weeds in many botanical families.

3.4.4. Life cycle

Species of *Meloidogyne* are sedentary root and tuber endoparasites, the life cycle being like that of *G. rostochiensis*, with some important differences. The white, swollen females do not develop into persistent, brown cysts, but deposit all their eggs in an external egg mass. Hatching occurs when the physical conditions are suitable and is not dependent on a hatching factor from host roots. Root-knot nematodes reproduce parthenogenetically and males even though present are not required for reproduction.

3.4.5. Interaction with other pathogens

Meloidogyne species are associated with the fungi *Rhizoctonia solani* and *Verticillium* sp. and the bacterium *Ralstonia solanacearum* in causing disease complexes.

3.4.6. Ecology

Root-knot damage to potatoes is usually associated with light soil. Spread of nematodes through irrigation water enhances the nematode attack. Soil temperature requirements of these nematodes vary with species. *M. hapla* require cool climate, whereas *M. incognita* and *M. javanica* require higher temperatures for their development and reproduction.

3.4.7. Physiological races

In India, all 4 races of *M. incognita* have been reported and the race 1 is predominant (Sharma and Gill, 1992).

3.4.8. Survival and spread

Root-knot nematodes survive on several weeds and volunteer crop plants present in the field after harvest of the main crop. They can also survive in the form of eggs and anhydrobiotic second stage larvae. Infected tubers, plant parts and planting material, as well as movement of infested soil by farm machinery and irrigation water are the main avenues of disseminating *Meloidogyne* spp.

3.4.9. Management

3.4.9.1. Physical methods: *Meloidogyne* spp. in seed tubers can be controlled by hot water treatment at 46-47.5 °C for 2 hr. and by dry heat treatment at 45 °C for 48 hr. (Nirula and Bassi, 1965).

In small holdings, burning of trash in the field before taking up tuber planting not only helps in reducing the nematode population but also enriches the soil.

3.4.9.2. Cultural methods

i) Field sanitation: Tubers from *Meloidogyne* infested crops should not be used as seed, as an even outwardly clean seed may be infested. The movement of soil and water from the infested fields should be avoided. The field should be kept free from weeds, since root-knot nematodes have a wide host range and most of the weeds help in the buildup of nematodes. Thus, clean cultivation (free from weeds which act as hosts) reduces nematode population to a great extent.

ii) Deep plowing: Deep plowing and drying of soil in the summer months facilitates the drying and death of infective larvae thereby reducing initial inoculum in the soil. Dry fallow with 2 or 3 deep plowings during the hot summer months gives excellent control of root-knot nematodes.

iii) Early planting: In long season, warm climate regions, potatoes may be planted during the winter months and harvested before injury occurs in the spring without visible infection. Under Shimla Hill conditions, both root and tuber infestation of *M. incognita* can be prevented by early planting of potato during March third or fourth week without affecting the yield (Ahuja, 1983; Prasad *et al.,* 1983). Late planting of autumn crop and early planting of spring crop in Northwestern Plains; while in hills, early planting of summer crop in fourth week of March is ideal. At Jalandhar, early planting of spring crop in the first week of January and late planting of autumn crop in second and third week of October reduced the root-knot infection in potato.

iv) Crop rotation: Crop rotation with maize, wheat and barley effectively check the buildup of *Meloidogyne* population. Follow crop rotation with vegetables (radish, garlic, cruciferous crops) and green manuring with lupin.

In Northwestern Hills, rotation of potato with French beans, leafy vegetables, maize and wheat helped in minimizing the nematode damage and increasing the productivity of the crop (Raj and Nirula, 1969). Crop rotation with non-host crops like maize, wheat, barley, beans, etc. reduces nematode population. Similarly, growing *Cassia* sp. also helped in reducing tuber infestations. Tomato crop can be followed by peanuts without any damage to peanuts and nematode populations will be reduced.

v) Trap crops: Growing of trap crop like French marigold, *Tagetes patula* in alternate rows with potato reduced the root-knot nematode larval population in soil, root galling and tuber infestation while the yields increased up to 123% over control. Growing nematode antagonistic crops such as crotalaria, asparagus, and mustard as cover crops prior to potato planting, significantly brought down soil population of plant parasitic nematodes by 41.6 %.

vi) Intercropping: Potatoes grown with white mustard in a pot of infested soil were lightly infested by root-knot nematodes than potatoes growing alone. Potato root diffusate was ineffective in the presence of leaching's from the roots of mustard seedlings. Mustard oil increased the yield of potatoes by reducing the severity of nematode attack. In mustard, the nematode toxic substance involved is allyl isothiocyanate. The intercropping of marigold with potato has resulted in a 7% increase in tuber yield.

Intercropping of onion and maize reduce galling by *M. incognita* on potato roots. Potatoes intercropped with maize have been found to be extremely vigorous with well- developed root systems.

vii) Organic amendments: Application of organic amendments such as FYM, compost, green manures and crop residues suppress the nematode population. This may promote fungal and other natural enemies of nematodes or release toxic compounds.

The oil cakes (neem, groundnut, mustard, and castor) were effective in reducing parasitic nematode population (*Hoplolaimus*, *Tylenchorhynchus*, *Meloidogyne* and *Helicotylenchus* species) and in increasing yields of potato. Application of eucalyptus leaf waste from oil distillation plants at 2.5 t/ha helps in reducing the nematode population. Soil incorporation of saw dust at 2.6 t/ha helped in reducing tuber infestations.

viii) Biofumigation: On-station trials using shredded crucifers applied at a rate of 50 t/ha or 5 kg/ m^2 as biofumigants against root-knot nematodes (RKN) showed that the RKN population count was significantly reduced in all shredded treatments 90 days after incorporation. Shredded mixed crucifers showed the highest percentage reduction in RKN population of 83.6% followed by radish

(79.2%), broccoli (74.2%), and cabbage (68.4%). In untreated plots, however, a 24.1% increase in RKN population was noted.

On-farm trials utilizing shredded crucifers as biofumigants against RKN showed that the nematode population decreased significantly with mixed crucifers by 86.7% followed by radish (85.4%), broccoli (82.2%), cauliflower (71.3%), and cabbage (70.9%). However, a significant increase in nematode population of 17.2% was observed in untreated plots.

The use of mixed crucifers was recommended because it had the highest percentage reduction in nematode counts both on-station (83.57%) and on-farm (86.7%) trials and highest marketable yield (11.44 t/ha), and ROI of 192%.

3.4.9.3. Chemical methods: Treatment of the tubers with Carbosulfan is beneficial. Application of carbofuran at 3 kg a.i. /ha in heavily infested fields and 2 kg a.i. /ha in moderately infested fields in two equal splits, first at planting and second at the time of earthing up, is recommended. Application of carbofuran at 3 kg a.i. /ha or ethoprop at 10 kg a.i. /ha in two equal splits, first at planting and second at the time of earthing up, increased the yields by 18 to 23% while accounting for 80% reduction in tuber infestation.

Application of Carbofuran 3G or Phorate 10G (1 kg a.i. /ha) in furrows while seeding was found effective.

3.4.9.4. Biological methods

i) Antagonistic bacteria: The bacterium, *Pasteuria penetrans* has offered possibilities of biocontrol.

ii) Antagonistic Fungi: *Purpureocillium lilacinum* significantly reduced the tuber and root galling due to root-knot nematodes on potatoes (Jatala, 1985). Potato plants grown in plots inoculated with the fungus had a significantly lower root galling index than did those grown in the nematicide-treated plots. Similarly, 86% of the egg masses collected from the roots of fungus-treated plots were infected with *P. lilacinum*, and over 54% of the eggs within these egg masses were destroyed. Yield of marketable potatoes (non-infected tubers) was significantly higher in the fungus-treated plots than in those of the control and the nematicide-treated plots (Table 3.5).

Table 3.5. Effect of *Purpureocillium lilacinum* and nematicides on *Meloidogyne incognita* infecting potato.

Treatment	Root gall index	% Egg masses infected	% Eggs destroyed	Marketable yield/400 m^2
Purpureocillium lilacinum	2.93	86.0	54.7	302.59
Fenamiphos at 2.5 kg a.i./ha	3.44	---	---	---
Carbofuran at 2.5 kg a.i./ha	3.57	---	---	---
Organic matter at 10 t/ha	3.62	---	---	---
Control	3.60	---	---	122.69
P = 0.05	0.14	---	---	85.10

Results of field experiments conducted in three consecutive years indicated consistent reduction of *M. incognita* population as expressed by reduction in root and tuber galling. Additionally, the fungus spread rapidly within the experimental plots, infesting the adjacent control plots. This is a desirable characteristic of a successful and competitive biocontrol agent. It is noteworthy to mention that in the field where this fungus was originally isolated, after 3 years of consecutive cropping of *M. incognita*-susceptible hosts, no nematode could be found infecting the root system. Artificial inoculation of this field with *M. incognita* was not successful in establishing the nematode after this period. The continuous presence of *P. lilacinum* in the field was confirmed by its isolation from the soil (Jatala, 1985).

iii) Arbuscular mycorrhizal fungi: Arbuscular mycorrhizal fungi such as Glomus fasciculatum and G. mosseae have shown promise in reducing the root-knot nematode infestation.

3.4.9.5. Host resistance: Potato cv. Kufri Dewa was found resistant to root-knot nematodes.

3.4.9.6. Integrated methods

i) Two cultural methods: Following 2 years crop rotational sequence of maize-wheat- potato-wheat coupled with summer fallow after 2 or 3 deep ploughings in Northwestern Plains reduced root-knot damage significantly.

ii) Bioagents and botanicals: Application of Neem cake/FYM/compost enriched with *Trichoderma harzianum/Purpureocillium lilacinum* gave effective control of root-knot nematodes.

iii) Physical and botanicals: Integration of soil solarization with soil application of fresh chicken manure gave effective management of root-knot nematodes (Table 3.6).

Table 3.6. Effect of integration of physical and organic amendment on yield of potato tubers

Treatment	Potato yield (kg/m^2)	Relative yield
Soil solarization + fresh chicken manure	2.45 ab	109
Soil solarization alone	2.40 ab	107
Control (no disinfestation)	1.81 c	81

3.5. INTERACTION OF ROOT-KNOT NEMATODE WITH BACTERIAL WILT

3.5.1. Interaction

The influence of root-knot nematodes on the development and severity of bacterial wilt has been elucidated (Akiew *et al.*, 1991). Plants infected with both *Meloidogyne* spp. and *R. solanacearum* have been found to express wilt symptoms earlier with greater severity than those infected with *R. solanacearum* alone (Johnson and Powell, 1969; Akiew *et al.*, 1991). Sitaramaiah and Sinha (1985) have suggested that the nematode induced stress is more important as a wilt-triggering factor than as a wounding agent. Infection by *Meloidogyne* spp. is known to predispose potato plants to *R. solanacearum* (Ateka *et al.*, 2001).

Differences in bacterial wilt severity were significant ($P \leq 0.05$) among the treatments for all the three cultivars used in this test (Table 3.7). Bacterial wilt severity was higher in plants inoculated with *R. solanacearum* and *M. incognita* than those inoculated with *R. solanacearum* alone (Ateka *et al.*, 2001).

Table 3.7. Wilt severity in three potato cultivars with varying levels of resistance seven weeks after inoculation with *Ralstonia solanacearum* alone and together with *Meloidogyne incognita* (Mean of 18 replicates)

Inoculum	**Wilt severity in potato cultivars**		
	Asnate (susceptible)	**B 53 (intermediate)**	**Kenya Dhamana (resistant)**
R. solanacearum	1.7	1.5	1.0
R. solanacearum + Root wounding	3.6	1.8	1.4
R. solanacearum + *M. incognita* + Root wounding	4.5	3.7	2.0
R. solanacearum + *M. incognita*	4.1	3.5	1.9
Mean	3.4	2.6	1.6
CD (P = 0.05)	0.84	0.86	0.62

Analysis of variance revealed significant ($P \leq 0.05$) differences in bacterial wilt severity among the three cultivars. Plants whose roots were severed and inoculated with the two pathogens had the highest (3.4) overall bacterial wilt severity. Wilt severity was lower (3.2) but not significantly ($P>0.05$) different in plants inoculated with both pathogens without root severing. Wilt severity in plants inoculated with *R. solanacearum* without root-wounding was significantly ($P=0.05$) lower (1.1) than in plants inoculated with *R. solanacearum* and whose roots were severed (2.6). Disease progress in the three potato cultivars is depicted in Table 3.8. For all the cultivars, significant ($P=0.05$) differences in bacterial wilt severity were observed among potato plants inoculated with *R. solanacearum* alone without root wounding and those inoculated with *R. solanacearum* and *M. incognita* throughout the seven weeks. Bacterial wilt severity was consistently higher in plants whose roots were severed. When plants inoculated with *R. solanacearum* alone were compared with those inoculated with *R. solanacearum* in combination with *M. incognita*, significant ($P \leq 0.05$), differences in wilt severity were observed in each of the three cultivars throughout the assessment period. Bacterial wilt development was faster and the severity consistently higher in plants inoculated with the two pathogens. Similarly, comparisons between plants inoculated with the two pathogens simultaneously with severed roots and those treated the same way but with unsevered roots indicated non-significant ($P=0.05$) differences in each cultivar and for every week (Ateka *et al.*, 2001).

Table 3.8. Severity of bacterial wilt in three potato cultivars from second to seventh week after inoculation with *Ralstonia solanacearum* alone and together with *Meloidogyne incognita*

Cultivar/Inoculum	**Weeks after inoculation**					
	2	**3**	**4**	**5**	**6**	**7**
Asnate (Highly susceptible)						
R. solanacearum	0.4	1.5	2.0	2.1	2.1	2.2
R. solanacearum + Root wounding	1.4	3.2	3.7	4.2	4.3	4.4
R. solanacearum + *M. incognita* + Root wounding	3.3	3.6	4.6	4.7	4.7	5.0
R. solanacearum + *M. incognita*	2.7	4.2	4.1	4.2	4.8	4.0

B 53 (Intermediate)						
R. solanacearum	0.0	0.3	1.0	2.2	2.4	2.6
R. solanacearum + Root wounding	0.0	0.4	2.1	2.4	2.7	2.8
R. solanacearum + *M. incognita* + Root wounding	1.4	2.9	3.6	4.2	4.7	4.9
R. solanacearum + *M. incognita*	1.2	2.4	3.7	4.4	4.5	4.6
Kenya Dhamana (Resistant)						
R. solanacearum	0.0	0.1	0.3	1.3	1.9	2.6
R. solanacearum + Root wounding	0.0	1.0	1.2	1.3	2.3	2.6
R. solanacearum + *M. incognita* + Root wounding	0.0	0.8	2.0	2.6	3.1	3.1
R. solanacearum + *M. incognita*	0.0	0.6	1.6	2.6	3.0	3.4
CD (P = 0.05)	0.58	0.77	1.01	0.99	1.02	1.07

Greenhouse studies revealed that infection by *R. solanacearum* in the presence of root-knot nematodes resulted in a higher severity and a faster development of bacterial wilt. Similar findings were reported by Ravichandra *et al.* (1990) and Akiew *et al.* (1991). The fact that bacterial wilt development was faster and severity consistently higher in plants inoculated with *R. solanacearum* in combination with *M. incognita* than in plants inoculated with *R. solanacearum* alone suggests that the resistance mechanism(s) of potato plants may have been weakened in the presence of *Meloidogyne* (Ateka *et al.*, 2001).

Nagesh *et al.* (1997) suggested a breakdown of resistance in potato plants inoculated with *R. solanacearum* together with *M. incognita*. Bacterial wilt severity was lower in potato cv. B53 plants inoculated with *R. solanacearum* with root-wounding than in plants inoculated with *R. solanacearum* and *M. incognita* without root wounding. This shows that the role of *M. incognita* in bacterial wilt development is more than just providing avenues for the entry of bacterium.

According to Napiere and Quimio (1980) and Samuel and Mathew (1983), the synergism between the two pathogens was mostly associated with wounds caused during penetration of larvae into roots. However, findings by Sitaramaiah and Sinha (1985) suggested that nematode induced stress was more important as a wilt-triggering factor than the wounding. Biochemical and physiological changes resulting from infection by root-knot nematodes may be responsible for the enhancement of wilt. According to Trudgill (1991), the biochemistry of a plant is altered following nematode infection. These changes may either weaken the chemical resistance mechanisms of the plant or modify conditions in the infected potato tissue making the plant more suitable for bacterial colonization (Khan, 1993). This is in harmony with the suggestion by Jatala (1975) that infection by one pathogen may alter the response of a host to subsequent infection by another.

3.5.2. Management

3.5.2.1. Host resistance

Nematode infection caused reduction of plant height in all the potato cultivars screened (Table 3.9). Cultivar Rutuku CIP-720097 recorded the lowest (12.4%) reduction while the highest (50.6%) reduction was observed in cultivar Dutch Robjin. Differences in plant height between infected and uninfected plants were significant (P=0.05) in cultivars Desiree, K. Pink, K. Dhamana, Mauritius, Tigoni and Nyayo (Ateka *et al.*, 2001).

Table 3.9. Galling and egg mass indices and plant heights of sixteen potato cultivars 60 days after soil infestation with 6000 eggs/juveniles of *Meloidogyne incognita* (Mean of 8 replicates)

Cultivar	Gall index	Egg mass index	Plant height		
			Infected	Uninfected	% change
Tomato Cal J	**7.6**	**5.0**			
KP9373926	5.5	5.0	30.6	37.2	17.7
BS3	5.4	5.0	24.2	36.4	33.5
Dasrtee CIP-800048	5.1	5 0	22.0	38.4	42.7
KP92	5.0	5.0	4.4	40.4	14.9
K Pink	3.9	5.0	32.4	46.0	29.6
Kenya Dhamana	3.6	4.9	38.0	58.4	34.9
Cnua 720118	3.8	4.8	44.0	52.0	15.4
MaontiuS	3.3	4.5	28.6	36.4	21.4
Dutch Robjin	3.0	4.5	32.2	65.2	50.6
Asanie CIP38138120	3.0	4.8	39.0	59.0	33.9
H Tana	2.6	4.9	32.8	44.8	26.8
Rutuku OP-720097	2.6	4.5	50.0	57.0	12.3
Furaha	2.5	4.4	27.2	32.6	16.6
Tigcni	2.5	4.4	42.8	59.6	28.2
Nyayo	1.8	4.5	47.0	58.4	19.5
CD (P = 0.05)	0.9	0.4	---	---	---

The result of this study reinforces the need to control root-knot in fields heavily infested with *R. solanacearum* to gain maximum benefit from the use of wilt resistant potato cultivars. Use of potato cultivars resistant to *Meloidogyne* are recommended as a component of integrated bacterial wilt disease management.

3.6. POTATO ROT NEMATODE, *Ditylenchus destructor*

D. destructor is the most important pest of potato tubers and is responsible for dry rot of tubers. High yield losses occur in the areas where climatological conditions favor establishment of the potato rot nematodes. The effect of nematodes will manifest itself at harvest or storage when infected tubers will rot.

3.6.1. Symptoms

D. destructor enters potato tubers through lenticels and initially causes small white mealy spots just below the surface that are only visible if the skin is removed. Infested areas enlarge and coalesce and light brown lesions, consisting of dry granular tissue, may be visible beneath the skin. As the infestation progresses, the tissues dry and shrink and the skin becomes cracked and papery. Internal tissues gradually darken and there are often secondary invasions of fungi, bacteria, mites, *etc.* If stored in moist conditions, a general rot may ensue and spread to neighboring tubers.

3.6.2. Life cycle

The life cycle is remarkably like *D. dipsaci*. *D. destructor* does not seem to have a resistant stage, as it does not form 'nematode wool' or readily withstand desiccation but survives in soil on alternative weed and fungal hosts. Thorne (1961) suggests that it may overwinter as eggs.

3.6.3. Survival and spread

D. destructor has a wide host range, can survive on weeds and on a wide range of soil-inhabiting fungi. It can also survive on infected tubers left in the field. Spread occurs by introduction of infected tubers and in soil adhering to seed pieces. Irrigation water and cultivation by infested farm tools and machinery are other sources of inoculum dissemination.

3.6.4. Management

3.6.4.1. Cultural methods: The use of healthy seeds, destruction of infected plant parts in the field, control of weed hosts, late planting when soil moisture is inadequate for optimum invasion and early harvest reduces losses. Two to three years rotation of potato with beans, corn and fallow kept free of weeds gave a very marked reduction in nematode populations. Satisfactory control of *D. destructor* can also be obtained by crop rotation with small grains, vetch and lupine provided potatoes are grown once in 3-4 years. Rotation of potatoes with sugar beet and other non-host crops can reduce nematode populations.

The application of ammonium nitrate and ammonium sulphate (at 144 and 118 kg/ha, respectively) while planting potato reduced infection from 7% to 0.2-0.5% and raised the yield from 10.6 t/ha to 16.3 and 25.2 t/ha, respectively.

3.6.4.2. Integrated methods: The control of potato rot nematode was achieved by the combination of disease escape, hygiene, in the form of seed certification and crop rotation as follows (Winslow and Willis, 1972):

- Healthy seed, planted late, harvested early, and stored as cool and dry as possible.
- Proper rotation of potatoes with non-host crops and growing of potatoes not more frequently than once in 3 or 4 years.
- Field hygiene in the form of removal of old infested tubers and weed control.

3.7. LESION NEMATODE, *Pratylenchus penetrans*

3.7.1. Symptoms

High populations of lesion nematodes cause areas of poor growth; plants are less vigorous, turn yellow and cease to grow. Damage is often caused by direct feeding and, usually, only cortical tissues are affected. Large nematode populations cause extensive lesion formation and cortex destruction of unsuberized feeder roots (Mai *et al.*, 1981). Tubers are often attacked, and small lesions are formed on the surface.

3.7.2. Survival

Infected tubers are sources of nematode inoculum and aid in survival of the nematodes.

3.7.3. Management

3.7.3.1. Physical methods: Hot water treatment of infected tubers at 50 °C for 45 to 60 min. aid in reducing nematode spread.

3.7.3.2. Cultural methods

i) Crop rotation: For root lesion nematode control on potatoes, researchers found that forage pearl millet (Canadian Hybrid 101) and marigold (Crakerjack) as rotation crops with potatoes resulted in fewer root lesion nematodes and increased potato yields than rotation with rye (Ball-Coelho *et al.*, 2003).

In areas infested by *P. penetrans*, beet should precede the susceptible potato in the crop rotation system. The cultivation of legumes should be avoided since nematodes breed well in them.

ii) Green manuring: Oil radish as a green manure has dramatically reduced stubby root nematode (*Trichodorus*) and root lesion nematode (*Pratylenchus*) in Idaho potato fields (Anon, 2001).

On-farm research in western Idaho showed that rapeseed green manures decreased soil populations of root lesion nematodes to a greater extent than did Sudan grass green manures. Fall Sudan grass should be plowed down after it is stressed (i.e., the first frost, stopping irrigation). Winter rapeseed and canola should be incorporated in early spring (Cardwell and Ingham, 1996).

3.7.3.3. Host resistance: Potato cv. Russet Burbank is tolerant, while Peconic appears to be less susceptible to the lesion nematode.

3.8. POTATO EARLY DYING COMPLEX, *PRATYLENCHUS PENETRANS*, *VERTICILLIUM DAHLIAE*

3.8.1. Interaction

The lesion nematodes often interact with *V. dahliae* in development of potato early dying disease complex in potato. In the arid and semiarid regions of the world, potato early dying (also known as early dying complex or *Verticillium* wilt) is a common limiting factor in potato production (Rowe *et al.*, 1987). It is endemic in many potato production areas of North America.

Although *V. dahliae* is the most important pathogen in this complex, the root lesion nematode *Pratylenchus penetrans* and the black dot pathogen *Colletotrichum coccodes* are also involved (Otazu *et al.*, 1978; Davis and Huisman, 2001).

3.8.2. Economic importance

When disease develops throughout a field midway through the growing season and then becomes severe during the period of maximum tuber bulking, a significant reduction in tuber size and total marketable yield can result (Rowe and Powelson 2002). In North America, yield reduction in moderately diseased fields can easily be 10 to 15 % and in severely diseased fields it can be as high as 30-50% (Powelson and Rowe, 1993). The economic impact of potato early dying across the potato industry is significant because of the direct losses resulting from low yields (Rowe *et al.*, 1987). Early Dying complex of potato is a major issue in Manitoba.

On potatoes, the *Globodera–Verticillium dahliae* and *Pratylenchus–Verticillium dahliae* disease complexes have become particularly notorious. Early senescence or 'early dying' caused by *V. dahliae* and *V. albo-atrum* is accentuated by populations of *Pratylenchus* spp. (Martin *et al.*, 1981; Wheeler *et al.*, 1992; Bowers *et al.*, 1996; Hafez *et al.*, 1999), *G. rostochiensis* (Evans, 1987), and *G. pallida* (Hide *et al.*, 1984; Storey and Evans, 1987). In terms of yield, Martin *et al.* (1982) calculated that 15, 50 and 150 *P. penetrans* per 100 cm^3 soil in combination with *V. dahliae* would result in 36, 60 and 75% reductions in potato tuber weight, respectively. However, tuber weights were unaffected by the presence of the individual pathogens, except where nematode populations were high (150 *P. penetrans* per 100 cm^3), when a 12% reduction was found. Yield reduction from the *Pratylenchus–V. dahliae* complex has also been reported elsewhere (Botseas and Rowe, 1994), as have other damaging effects such as the disruption of photosynthesis, stomatal conduction, and transpiration (Saeed *et al.*, 1997a; Saeed *et al.*, 1997b). However, the fundamental importance of *P. penetrans* in potato early dying is its ability to activate low populations of *V. dahliae* that would otherwise be inconsequential in disease development (Bowers *et al.*, 1996).

3.8.3. Symptoms

A typical visual symptom of potato early dying is loss of plant vigor during mid- to late summer followed by senescence and death of the crop before normal maturity. Initially the disease causes uneven chlorosis of the lower leaves on a few plants. Areas between the leaf veins turn yellow and later brown. Later wilting of the lower leaflets of the plants may occur, and infection proceeds up the infected stems, which often remain erect. The disease causes vascular discoloration of stems, and their tuber ends and plants may die without reaching maturity. It is common for one or two stems of the plant to die and the other portions of the plant to remain healthy (Rangahau, 2003). The symptoms often occur on one side of the plant or on individual leaves (Powelson and Rowe, 1993). A light brown vascular discoloration in basal stem tissues is usually present in symptomatic plants. In severe cases, plants across an entire field will die over a period of several weeks (Rowe and Powelson, 2002).

A light brown vascular discoloration is often visible at the stem base when sliced. Advanced symptoms usually do not occur until after flowering and may consist of a decline of isolated plants or in severe cases, the entire crop may mature early (Davis 1981; Krikun and Orion 1979; Mace *et al.*, 1981; Rowe 1983). Infected potato tubers with Verticillium show symptoms of light brown discoloration of the vascular ring.

3.8.4. Etiology

A major contributor to this syndrome is the soil-borne fungus *Verticillium*. Two species of the fungus, *Verticillium dahliae* and *V. albo-atrum*, cause the disease. *V. dahliae* grows better at slightly higher temperatures (25-28 °C), whereas *V. albo-atrum* may be the dominant pathogen in cooler production areas (below 25 °C) (Powelson and Rowe, 1993; Rowe *et al.*, 1987). Microsclerotia are formed by *V. dahliae* and melanized hyphae are formed by *V. albo-atrum* (Powelson *et al.*, 1993).

Although the disease is primarily caused by the vascular wilt pathogen *V. dahliae*, coinfection of potato by the root-lesion nematode *Pratylenchus penetrans* an increase both the severity of the disease and tuber loss (LaMondia, 1999; Back *et al.*, 2002). Nematodes are thought to provide entry points for the fungus while feeding on potato roots, or they may reduce the natural host defense mechanism of the potato plant, allowing the fungus to attack roots and infect the host (Rangahau, 2003). Mitchell (1985) found that the effects of *P. penetrans* in combination with *V. dahliae* on symptom expression were additive.

An important interrelationship exists between *Pratylenchus* spp. and *Verticillium* spp. in potato (Morsink and Rich, 1968). In this situation, either pathogen alone may cause disease but damage is greater when nematodes and fungus occur together. Potatoes infected by *V. albo-atrum* also support greater reproduction of *P. penetrans* than fungus-free plants. Most evidence indicates strongly that *Verticillium* wilt promotes increases in reproduction of *Pratylenchus* spp.

A synergistic interaction of *P. penetrans* and *V. dahliae* has been demonstrated to reduce yields and tuber quality of potato, resulting in potato early dying disease. Wounded roots may stimulate the germination of the dormant microsclerotia of the fungus due to increased root exudation or may facilitate access for the fungus to the vascular tissues. Some authors have also suggested that the nematode suppresses the production of phytoalexins. There was a strong reduction of photosynthesis rate following joint infection by both pathogens as compared to infection by either of these pathogens alone.

In early potato dying disease, *P. penetrans* predisposes plants to *V. dahliae* infection. Tests with three population levels of *P. penetrans* and two levels of *V. dahliae*, alone or in all combinations, showed that disease occurred when both pathogens interacted synergistically at population levels that individually had little or no effect. Yield reductions of 25-50% often occurred in the presence of both pathogens. Yield losses were most pronounced in seasons when high temperature stress occurred during tuberization (Rowe *et al.*, 1985).

3.8.5. Management

The severity of PED that develops in each crop is a function of the numbers of *Verticillium* and associated pathogens in the soil at planting, susceptibility of cultivar, and the extent to which the environment favors disease development. Selection of control strategies for potato early dying must be based on an understanding of the biology of the pathogens and the effects of various agricultural practices (Rowe *et al.* 1987). Management practices are aimed at reducing the population of these pathogens in soil or changing the susceptibility of the host. No single practice will control potato early dying. Therefore, integrated management systems are emphasized in which several decisions are implemented throughout multiple-year rotational cropping programs. Potential components of integrated systems are cultural methods like crop rotation, green manures, fertilization, irrigation, and vine removal along with host resistance, pesticides (primarily soil fumigation) and biological controls (Rangahau, 2003). Management practices that modify the rate of stem colonization by *V. dahliae* or increase the host's tolerance to colonization, may also be useful control measures for this disease (Rowe, 1987).

3.8.5.1. Cultural methods: Severity of PED is increased if fertilizer is applied through gravity furrow irrigation, whereas fertigation through a sprinkler system lessens the severity (Davis and Everson, 1986).

i) Crop rotation: Crop rotation has long been considered an important part of the disease management strategies employed for potato early dying. Although 2-to-3-year rotation practices do not result in effective control of the disease, they are useful in the total management scheme. In Idaho, the minimum time to effectively reduce inoculum with grain rotation in moderately infested land is 5-10 years (Davis 1981; Davis and McDole, 1979). In Ohio, potato-wheat rotation has been found to be useful as part of an overall control strategy (Joaquim *et al.*, 1988). Potato yield increases were observed with barley-corn rotation (Davis and McDole, 1979).

The influence of rotation and green manure crops on early dying of potato was investigated in 120 field micro plots infested with both pathogens. A single season of a Polynema marigold rotation crop significantly reduced *P. penetrans* recovery from roots of rye grown after the potato crop and increased marketable tuber yields. Saia oat, Rudbeckia and canola had greater numbers of saprophytic nematodes than potato, and tuber yield was correlated with increased recovery of saprophytes. In the second potato crop after a single year of rotation, tuber yields were higher after Polynema marigold. Canola, while maintaining high lesion nematode numbers, resulted in high tuber yields. Rotation with Rudbeckia and canola resulted in the highest tuber yields.

In fields infested with root-knot nematodes and *V. dahliae,* Scholte (1990) found that a five-year rotation of maize-sugar beet-barley-barley-potato had significantly higher potato yields than continuous potato.

The length of rotation also affects the management of early dying, primarily by affecting the soil populations of *V. dahliae*. Soil populations of *V. dahliae* can be 2- to 4-fold higher

ii) Soil fertility: Davis *et al.* (1996) reported that in Idaho, the disease was controlled after green manure treatments of either Sudan grass or corn and an increase in yield was seen. Spent mushroom compost may be a means of reducing the effects of the disease on superior varieties of potato and increasing the yield when one or both pathogens are present (LaMondia, 1999).

Rangahau (2003) reported that incorporating organic material such as soymeal into soil a few months before planting potatoes to encourage naturally occurring antagonists in the individual paddock will reduce potato early dying disease.

The burning of potato vines may be of value as a first line of defense by delaying the buildup of soil-borne inoculum and improving control of the disease (Mpofu and Hall, 2002). Also, *Verticillium*-free seed may reduce the rate of inoculum increase.

iii) Irrigation: Maintaining plant vigor with careful irrigation and fertility practices is the best management strategy for controlling this disease. The amount and timing of irrigation have been used as a management tool to suppress the disease. Where potatoes are grown under irrigation, the amount of applied water can be reduced early in the season (before tuber initiation) to minimize water use by the crop without sacrificing tuber yield or quality (Rangahau 2003). The severity of potato early dying was greater throughout each season when plants grown in infested soil were irrigated excessively compared to those receiving deficit irrigation (Cappaert *et al.*, 1992).

In 'Russet Burbank' potato, early dying can be suppressed if water is reduced before tuber initiation. In Wisconsin, tuber yield was significantly lower in deficit than in moderate or excessive pre-tuber initiation irrigation regimes (Cappaert *et al.*, 1994).

3.8.5.2. Chemical methods: Methylisothiocyanate has been successfully used for soil fumigation to reduce the populations of soil-borne fungus, *V. dahliae* and alleviate the effects of disease complex involving wilt-inducing fungus (*V. dahliae*) and the lesion nematode (*P. penetrans*) on potato (Rowe *et al.*, 1987).

3.8.5.3. Host resistance: Planting resistant cultivars is the most efficient, effective economical and environmentally sound means of managing potato early dying disease. Planting of resistant cultivars for five seasons can reduce pathogen levels by 60-70% compared with growing only susceptible cultivars (Selvanathan, 2003). Unfortunately, many commercial cultivars do not possess any resistance to potato early dying and even the most resistant cultivars will develop symptoms of the disease under favorable environmental conditions and when the soil inoculum levels are extremely high (Ranghau, 2003). Some commercial potato cultivars do exhibit varying degrees of tolerance to early dying, which can lessen long term pathogen survival. Times change, as do the challenges faced and the strategies required to manage potato early dying both effectively and economically. Successful management will involve five key areas: suppression of soil-borne inoculum, new plant protection chemicals, clean planting stock, enhanced host resistance and improved pathogen monitoring tools (Rowe and Powelson, 2002).

3.9. TOBACCO RATTLE VIRUS/CORKY RINGSPOT OF POTATO (SPRAING)

Corky ringspot disease of potato is caused by tobacco rattle virus (Tobravirus), which is vectored by stubby-root nematodes (*Paratrichodorus* spp. and *Trichodorus* spp.). The condition is referred to as spraing, a Scottish word meaning a bright streak or stripe (de Bokx, 1972).

Tobacco rattle virus is reported in many areas of the world and more than four hundred plant species in 50 plant families are susceptible to infection (Brunt *et al.*, 1996; Dallwitz *et al.*, 1993).

Spraing is the term given to a range of internal and external symptoms in potato tuber flesh consisting of brown, necrotic rings, arcs and flecks. Tobacco rattle virus (TRV) cause these symptoms. In some cases, external concentric rings may be visible on the surface of the tuber. The severity of symptoms in the tuber can range from mild (a few flecks or faint brown lines) to severe (heavy, dark chocolate-colored arcs with a corky texture). A few symptomatic tubers may be found, or the stock could be heavily affected (>50%). Spraing symptoms may not be evident at harvest. Fluctuations in temperature during storage can exacerbate development of the internal tuber symptoms.

3.9.1. Symptoms

Generally, TRV spraing infected tubers produce healthy plants. However, in some cases several healthy and symptomatic, stunted stems will be produced, which is termed as stem mottle. Symptoms may present as pinching and "scorching" of the leaf tip, mottles, distorted/wavy leaf margins and occasionally finely etched lines and rings or chevron type markings reminiscent of PMTV.

Corky ringspot symptoms vary depending on virus strain, potato cultivar, and time of infection. Symptoms often include brown necrotic rings, arcs, and diffuse spots which are considered quality defects and may result in after-harvest devaluation or rejection of either table or processed potatoes.

Corky ringspot symptoms in potato include necrotic rings and pits on the tuber surface and range from diffuse brown spots to concentric rings or arcs of brown, necrotic tissue to dark-brown necrotic tissue which extends through tuber flesh (Mojtahedi *et al.*, 2001). The virus is usually detectable when symptoms are seen but may also be present in asymptomatic tissue (Charlton, 2006).

Several species of *Trichodorus* and *Paratrichodorus* transmit M-type isolates of tobacco rattle virus (Harrison and Robinson, 1986). Stubby-root nematodes are migratory ectoparasites that are mobile during each stage of their life cycle (Stark and Love, 2003) and feed primarily on meristematic cells or root tips which hinders root elongation (Crow, 2005). Damaged root tips may swell, and lateral roots may emerge behind them, resulting in root proliferation.

3.9.2. Biology

Eggs of *Paratrichodorus minor* hatch at 53.3 ± 7.3 degree-days using a basal threshold of 10 °C (DD10C) after egg deposition. Second stage population peaks occurred at 28 DD10C, third stage at 67 DD10C, fourth stage at 109 DD10C and adults at 143 DD10C. Population densities increased in the presence of a suitable host but declined to 33% of the initial level after 300 DD10C in the absence of a host (Schneider and Ferris, 1987). The optimum temperature range for development and reproduction of *P. allius* is 21 to 24 °C (Ayala *et al.*, 1970).

In general, Trichodoridae nematodes prefer coarser textured soils and avoid desiccation stress by migrating vertically to moist soil layers. Vertical movement of Trichodoridae nematodes appears greatest when soil pores are half-full of water and least in waterlogged or dry soil conditions (Decraemer, 1995). Since soil moisture is often maintained below field capacity for optimum yield and quality in potato production, vertical migration may explain the difficulty of consistently finding stubby-root nematodes during routine soil sampling procedures to 30-cm depth in Washington and Oregon (Charlton, 2006).

Stubby-root feeding injury is not important economically and is rarely visible in potato, however, corky ringspot virus can be transmitted at exceptionally low population densities of the nematode (Charlton, 2006).

3.9.3. Spread

Tubers from soil with a history of tobacco rattle virus serve as a reservoir. The virus may spread to daughter tubers when infected tubers are used as seed (Crosslin *et al.*, 1999). Newly formed potato tubers are quite vulnerable to tobacco rattle virus infection and tubers as small as 3-cm in diameter had corky ringspot blemishes in tobacco-rattle infested fields in Florida (Weingartner *et al.*, 1975).

3.9.4. Transmission

Transmission of TRV is via free-living nematodes of the genera *Trichodorus* and *Paratrichodorus*. Because of the biology of the vector TRV tends to be a greater issue on light, sandy soils. Nematodes favor wet conditions and heavy rainfall or over irrigation can exacerbate incidence of the disease in affected areas. The virus has been shown to have a very broad host range and can infect over 400 plant species. These include crops such as beetroot (*Beta vulgaris*) and spinach (*Spinacia oleracea*). In addition to these species the virus can also infect many common field weeds including fathen (*Chenopodium album*), chick weed (*Stellaria media*), knot weed (*Polygonum aviculare*), bind weed (*Polygonum convolvulus*), groundsel (*Senecio vulgaris*) and dandelion (*Taraxacum officinale*).

TRV affected tubers will rarely be systemically infected and will generally grow into healthy plants. Whilst spraing symptoms are still the most visible and economically damaging characteristic of this disease, some commonly grown varieties will become systemically infected. These varieties can function as a reservoir of virus for non-infective nematodes which feed on the infected tubers and acquire the virus. It is important therefore to know whether land contains viruliferous nematodes and the varietal susceptibility to TRV.

3.9.5. Management

3.9.5.1. Cultural methods: The wide host range for both virus and vectors make eradication of corky ringspot in potato fields difficult. Alfalfa (*Medicago sativa*), a rotation crop in Oregon and Washington, and spearmint (*Mentha cardiaca*) do not appear to serve as hosts of tobacco rattle virus (Mojtahedi *et al.*, 2002). Since the virus is lost each time the nematode molts, and since it does not pass through the nematode egg stage, *P. allius* are cleansed of tobacco rattle virus after being reared on these plants for a period of 1 to 3 months (Boydston *et al.*, 2004). Therefore, it may be possible to eliminate tobacco rattle virus from populations of *P. allius* rotation crops that do not harbor the virus. Of course, the crops must be kept free of weed hosts for tobacco rattle for the strategy to be effective (Charlton, 2006).

The main method of control for virus is through field intelligence and effective cultural control. By ensuring that seed tubers are free from spraing diseases, the opportunity for these diseases to spread from field to field can be minimized. The soil bait test offered as a diagnostic service by the Plant Clinic can help growers make informed decisions on which fields to use by identifying fields infected with both the vectors and their respective viral pathogens simultaneously.

3.9.5.2. Chemical control of the nematode vector

Spraing disease of potato checked by methomyl used at 9 kg/ha and dazomet at 170 kg/ha only for one year.

Metham sodium (Vapam HL) delivered through chemigation (water-run) is not effective against corky ringspot of potatoes when used alone (Ingham *et al.*, 2000).

3.10. CONCLUSION

Plant parasitic nematodes constitute one of the major limiting factors for cultivation of potato. Several nematode species are associated with potato and a few of them negatively impact yield. Severe yield losses have been reported in most regions where potatoes are grown. To select suitable nematode management practices, correct diagnosis of the causal agent is of prime importance. Overall, these nematode problems in potatoes are better managed when integrated management practices are used, i.e., regulatory methods (plant quarantine and seed certification), cultural practices (crop rotation, cover crops), genetic control, and ultimately by the use of nematicides. Therefore, for a sustainable cropping of potato, growers, extension services, and researchers must consider these nematodes holistically, the impact they cause and whether these management practices are economically, environmentally, and technically sound.

Integrated nematode management could be tackled with the intelligent planning of nematode-suppressive crop sequences, summer plowing, organic manuring, clean cultivation, adjusting sowing time, water and irrigation management and sensible use of nematicides. The increasing concerns on the ill effects of chemical pesticides on the environment have driven recent research interests on the use of several alternative strategies like botanicals, biopesticides and cropping system research for management of nematodes. Some of the successes have been obtained for managing plant parasitic nematodes with neem-based formulations, fungal formulation of *Purpureocillium lilacinum* and *Trichoderma* spp. and bacterial formulations of *Pasteuria penetrans* and *Pseudomonas fluorescens*. Inconsistent efficacy of the biological control agents in different agro-climatic regions is the main reason for not adopting them widely under field scale. Integration of more than one option can be explored based on their compatibility, economic viability and availability to the farmers. The major challenge ahead for the greater benefit of the farmers is the need for development of holistic approaches for managing field problems including nematodes induced disease complexes with other soil-borne pathogens. There is a need for development of economic, eco-friendly, and sustainable bio-intensive integrated nematode management strategies using all the available components.

Chapter - 4

TOMATO, *Solanum lycopersicum*

Root-knot, reniform and stubby root nematodes are the major limiting factors in the successful production of tomato crop.

4.1. ROOT-KNOT NEMATODES, *Meloidogyne incognita, M. javanica, M. arenaria*

Lal and Ansari (1960) reported *M. arenaria* for the first time on tomato from Bihar, India.

4.1.1. Crop losses

Root-knot nematodes are responsible for 23% annual yield loss in tomato worth Rs. 6035.2 million in India (Walia and Chakraborty, 2018).

M. incognita was responsible for 30.57 to 46.92% loss in fruit yield of tomato (Bhatti and Jain, 1977; Reddy, 1985; Darekar and Mahse, 1988), while *M. javanica* caused 77.5% loss in yield (Anon, 1993).

4.1.2. Symptoms

Infected seedlings exhibit reduced growth, yellowing, wilting and moderate to large number of galls on the root system. Similarly transplanted seedlings in infested field show reduced growth. The infested plants yield less, exhibit yellowing similar to nutrient deficiency and galls can be seen on the root system upon uprooting. The infested plants succumb to wilt and fungal and bacterial diseases quickly.

4.1.3. Interaction with other pathogens

M. javanica increased the extent of damage by pre- and post-emergence phases of damping off caused initially by *R. solani* and *Pythium debaryanum* in tomato (Nath *et al.*, 1984). Pani and Das (1972) have reported the association of root-knot nematode with bacterial wilt of tomato. The growth of tomato plants was significantly reduced in simultaneous inoculations with *M. incognita* and tobacco mosaic virus or when nematode preceded the virus inoculation (Goswami and Chenulu, 1974).

4.1.4. Management

4.1.4.1. Physical methods: For the management of root-knot nematodes and weeds in tomato nursery, soil solarization through tarping with 100-gauge (25µm) LLDPE clear plastic film in summer (in between April 15th and June 15th) for 15 days is recommended. It increased the number of transplantable seedlings by 61.5% giving net profit of Rs.7,661 from 1000 sq. m. (BCR 5.65) and decreased root-knot disease and weeds by 66% and 93%, respectively (Table 4.1).

Table 4.1. Effect of soil solarization of nursery beds on root galling and number of transplantable seedlings of tomato in 1000 sq. m.

Treatment	No. of seedlings	Weight of 100 seedlings	Root-knot index	Net profit	Benefit: cost ratio
With plastic	2,60,400	36 g	1.37	Rs. 7,661	5.65
Without plastic	36,400	26 g	3.97	---	---
% Increase (+)/ decrease (-) over control	(+) 61.5	38	(-) 66	---	---

Rabbing root-knot infested tomato nursery with *bajra* husk @ 7 kg/m^2 a week prior to seeding was found effective for the management of root-knot disease.

Polythene mulching in the main field for 3 weeks prior to transplanting recorded a reduction of nematode population by 52.3% in susceptible variety and 50% in the resistant variety. Polythene mulching gave 32.9% increase in fruit yield over no tillage, followed by plowing and exposure (16.4%).

Red plastic mulch suppresses root-knot nematode damage in tomatoes by diverting resources away from the roots (and nematodes) and into foliage and fruits (Adams, 1997).

In solarized plots, root infestation was intermediate because of the late nematode attack, but no rotting of the root was observed (Cartia *et al.*, 1988). The yield increases averaged 60 percent in the solarized plots over control and were similar to those obtained with methyl bromide treatments.

The reproduction factor in the solarized treatment was 0 on tomato plants for *Meloidogyne* spp., *Tylenchorhynchus* spp., and *Trichodorus* spp. The corresponding RF value in non-solarized soil was 0.89 on the same crop and the same nematodes (Table 4.2).

Table 4.2. Effect of solarization on the control of nematodes on tomato

Treatment	Average number of nematodes per 200 g soil[1]		Reproduction factor (RF) RF = Pf/ Pi
	(Pi)	(Pf)	
Tomato - solarized	80	0	0
Tomato - non - solarized	156	140	0.89

Pi = Initial nematode population; Pf = Final nematode population.

In Japan and other Asian countries, several farmers growing successive crops, such as tomato and melon susceptible to root-knot nematodes use solarization in plastic tunnels for 30 days in summer as an alternative to methyl bromide fumigation (Sano, 2002).

4.1.4.2. Cultural methods: Several cultural methods like selection of nematode-free nursery sites, destruction of infested roots after crop harvest, crop rotation with non-hosts (maize, wheat, sorghum) or antagonistic crops (marigold, mustard, sesame), flooding, fallowing, deep summer plowing and harrowing either alone or in combination proved to be reasonably effective and economical to check multiplication of root-knot nematodes on tomato.

i) Summer plowing: Plowing not only leads to disturbance and instability in nematode communities but also causes mortality by exposing them to solar heat and desiccation. Three summer plowings each at 10 days interval during June (average atmospheric temperature ranging between 40-46 °C) at Hisar (Haryana) led to 96.5 % reduction in *M. javanica* population, while fallowing itself during the same period registered 44.5% reduction (Jain and Bhatti, 1987). Additional use of plastic sheets for covering soil either in nursery beds or in fields further enhanced nematode reduction. Such an approach also helps in reducing the intensity of weeds, fungi, and bacteria in the soil.

Jain and Bhatti (1987) reported that planting of nematode-free seedlings in an infested field which received three summer plowings increased the yield of tomato by 55%.

ii) Crop Rotation: Decrease in *M. incognita* population on tomato occurs when the field was left fallow or following crop rotation with marigold, spinach, and bottle gourd (Khan *et al.*, 1975).

To control *M. incognita* and *M. javanica*, a tomato crop can be followed by peanuts without risk of damage to the peanuts. While the peanuts are growing, the nematodes cannot reproduce. Instead, many of the larvae in the soil die or become non-infective because of starvation and the attacks of predators, fungi, and diseases. If the population is reduced sufficiently, tomatoes can be grown again after peanuts without serious injury (Taylor and Sasser, 1978).

Tomato rotated with cotton-wheat-cotton resulted in low gall index due to *M. javanica* (root-knot index 1-2) and gave maximum yield (38% increase over continuous tomato) (Hashmi and Hashmi, 1990).

Sesamum-tomato sequence produced the least reproduction rate (0.47), least root-knot index (1.33) and highest yield (8.6 t/ha) of tomato which was at par with niger-tomato sequence that had reproduction rate of 0.56, root-knot index of 1.66 and highest yield of 7.55 t/ha of tomato (Sahoo *et al.*, 2004a).

In cropping sequence studies for the management of *M. javanica* on tomato, nematode population decreased to low levels when carrot, capsicum and onion were grown (Kanwar and Bhatti, 1992). The population of *M. incognita* was reduced considerably under groundnut-mustard-tomato cropping sequence (Sharma *et al.*, 1980).

Tomato nursery beds previously planted with trap crop (marigold) gave maximum reduction in root-knot nematode population in soil which also increased the germination of tomato seeds and production of healthier (nematode-free) seedlings (Rangaswamy *et al.*, 1999).

Decrease in *M. incognita* population on tomato occurs following crop rotation with marigold, spinach and bottle gourd (Khan *et al.*, 1975). Crop rotation with non-hosts (maize, wheat, sorghum) or antagonistic crops (marigold, mustard, sesame) is also effective.

Inclusion of rice, maize, groundnut and *Stylosanthes* in rotation sequences-controlled root-knot nematodes and increased yields of tomato. In crop rotation trials, it was found that a crop of groundnut or strawberry grown prior to tomato in soil heavily infested with root-knot nematodes greatly increased yields of tomato. Rotation of tomato with leek (*Allium porrus*), onion, groundnut and sun hemp reduced the population of root-knot nematodes.

Rotation of tomato with bajra was found beneficial for the management of root-knot nematode problem.

Sowing of non-host crop (cauliflower) in *kharif* and disease resistant variety Hisar Lalit of tomato in *rabi* in fields infested with root-knot nematode was found effective.

Cultivation of cowpea as a cover crop, later removing and burning proved beneficial for the management of *M. incognita* in tomato.

iii) Intercropping: Intercropping with marigold or mustard with tomato reduced the damage of root-knot and reniform nematodes. The root-knot development on tomato was low when interplanted with *Tagetes erecta*. The population of *Tylenchorhynchus, Helicotylenchus, Hoplolaimus, Rotylenchulus* and *Pratylenchus* was also markedly reduced (Alam *et al.*,1977; Khan *et al.*, 1971a). α-Terthienyl is the active principle in *Tagetes* spp. which is toxic to these nematodes.

Marigold intercropped with tomato at 1: 4 and 1: 6 ratios and mustard at 1: 2 ratio was found to be effective in reducing root galls, egg masses and nematode population at harvest and gave benefit: cost ratio of 8.36, 7.88 and 3.31, respectively (Rangaswamy *et al.*, 1999).

Interculture of onion with tomato reduced gall index due to *M. javanica* significantly and improved tomato fruit yield. Intercropping with 2 rows of onion with one row of tomato was the most effective treatment in reducing the nematode infection and population and enhancing tomato yield (Ram and Gupta, 2001).

In pots having *Zinnia elegans* with tomato, the population of *M. incognita* and *R. reniformis* declined. The root-knot index on tomato was reduced to 1.08 compared to 3.75 in control. The nematode reproduction factor of *M. incognita* and *R. reniformis* was reduced to 0.88 and 0.80, respectively compared to 3.86 and 2.43, respectively in control (Tiyagi *et al.*, 1986).

Cultivation of tomato with castor results in considerable reduction in number of root-knot galls on tomato (Hackney and Dickerson, 1975).

Mixed cropping of tomato along with onion/garlic reduced the population of root-knot nematode and the fruit yield was also improved by 22% in two years.

iv) Green manuring: Green manuring with naffatia/besharmi or neem leaves (BCR 2.87) or water hyacinth (BCR 2.527) or calotropis (BCR 2.26) or congress grass (*Parthenium hysterophorus*) (BCR 2.09) each at 3 kg/sq. m. 15 days prior to seeding is recommended for economic and effective management of root-knot disease in tomato nursery. Incorporation of green materials of naffatia and neem leaves each at 2 kg/1.44 sq. m. (BCR 15.7) or congress grass and neem leaves each at 2 kg/1.44 sq. m. (BCR 15.6) or calotropis, naffatia, congress grass and neem leaves each at 1 kg/1.44 sq. m. (BCR 14.7) 15 to 20 days prior to tomato seeding gave effective management of root-knot nematodes and production of more healthy seedlings in tomato nursery during kharif season.

v) Biofumigation: Biofumigation with broccoli, *Brassica oleracea* var. *italica* at 14 t ha^{-1} (all plant parts were added to the soil one week before planting and incorporated to a depth of 20 cm with a rotary tiller) could be used efficiently against *M. incognita* in organic tomato production. Biofumigation with broccoli gave lower root-knot index (4.33 as compared to 8.08 in control) and higher yield (2.315 kg $plant^{-1}$ as compared to 1.118 kg $plant^{-1}$ in control) (Kaþkavalci *et al.*, 2009).

To verify the effectiveness of biofumigation strategy in managing nematode *M. incognita*, field trials on tomato were undertaken during the two years in an unheated plastic- house in the southern part of Sicily (Italy). In the first trial, two accessions of mustard species of the *Brassicaceae* family were selected for their glucosinolate content (*Brassica juncea* Sel. ISCI 99 and *Eruca sativa* cv Nemat). These were rotated to ascertain their effectiveness against the root-knot nematode in comparison to the addition of pellets derived from the *Brassicaceae* species. In the second trial the tomato crop was treated with oil extract of *Brassicaceae* species, distributed by drip irrigation when the maximum development of nematode population was registered.

All biofumigation strategies succeeded in reducing the juvenile nematode population in the soil and gall index in the roots at the end of the tomato's growth cycle. All the methods produced a tomato yield significantly higher than with untreated soil. Brassica *juncea* Sel. ISCI 99 acted mainly as a biofumigant while *Eruca sativa* cv Nemat revealed an interesting trap crop effect. Also, the use of pellets derived from *Brassicaceae* species, incorporated into the soil before the transplanting has shown a lower gall index in the roots. The oil extract of *Brassicaceae* species reduced the juvenile nematode population in the soil during the whole tomato cultivation cycle (Colombo *et al.*, 2008).

In a 3-year field study on the effect of biofumigation with 5 different winter-grown crops on *M. incognita* infestation of a summer grown tomato crop in Southern California, Lopez-Perez *et al.* (2008) reported that broccoli reduced tomato root galling and increased yields compared to the other non-*Brassica* treatments, but generally did not lower *M. incognita* soil population levels. They suggested that biofumigation with broccoli tissue prevented an immediate attack of the tomato transplants, allowing plants to become established and rendering them more tolerant to the nematodes.

The *Brassica* crop arugula (*Eruca sativa* var. Nemat) also received interest as a potentially useful biofumigant crop to manage root-knot nematodes, as it appears to act as a trap crop for *M. incognita* (Curto *et al.*, 2006). Positive effects on yields of nematode- susceptible crops grown after arugula cultivation and biofumigation were reported for tomato.

The effects of soil incorporation of dry leaves of *Brassica macrocarpa* four months before transplanting and repeated after three weeks from transplanting was compared with soil incorporation of dry leaves of *B. macrocarpa* only three weeks after transplanting for the management of root-knot nematodes (*Meloidogyne* spp.) and on tomato plant growth and development. The results of both trials showed a significant reduction of root-knot because of soil infestation of nematodes and of *B. macrocarpa* incorporation into the soil. The use of *B. macrocarpa* is permitted for use to control the nematode and to increase fruit yield, above-all in terms of fruit number and weight (Branca and Furtado, 2004).

Three doses of canola, *Brassica napus* seed meal at 4, 6 and 8 tons/feddans used as organic amendment incorporating into the soil were evaluated against root-knot nematode *M. incognita* inhabiting field planting with tomato. Super strain B at all three doses significantly affected root-knot population in soil, gall formation and yield increase (39% over control for the highest dose) (Ameen and Mohamed, 2006).

Four accessions of mustard species (*Brassica juncea* Sel. ISCI 99, *Eruca sativa* cv. Nemat, *Crambe abyssinica* cv. Mario, and *Raphanus sativus* cv. Boss), selected for their glucosinolate content, were rotated with host crops of the southern root-knot nematode, *M. incognita*, to ascertain their effectiveness in managing the nematode, in comparison with a rotation using a susceptible crop *Vicia villosa* cv. Haymaker Plus. All the species used significantly suppressed numbers of the nematode juveniles in the soil and in the roots in comparison with a susceptible intercrop. *B. juncea* and *C. abyssinica* acted mainly as biofumigants while *R. sativus* and *E. sativa* revealed a clear catch crop effect. The results showed a significant tomato yield increase after mustard crops, except for the rotation with *B. juncea*. The beneficial effect of the rotation with mustard species was greater in biocidal crops sown in late spring rather than in late summer (Curto *et al.*, 2006).

In the greenhouse trial, three natural plant products based on *Quillaja saponaria* extract, *Tagetes erecta* oil and *Brassica* defatted meal were applied and compared with a chemical control (Cadusafos) and an untreated one. The results showed a statistically significant effectiveness among treatments, in nematode (*M. incognita*) reproduction factor ($0.9 < R < 11$), root-knot index infestation (2.4-1.9 in the untreated control and in *Q. saponaria*, 1.3-0.5 in *T. erecta*, 1 in *Brassica* defatted meal and 0.4-0.2 in cadusafos) and fruit weight (from 3.8-4.2 kg per plant in the control, *Brassica* defatted meals and cadusafos, to 5.7-4.9 kg per plant in *Q. saponaria* and in *T. erecta*), while the factorial analysis registered a statistical relevance of nematicidal green manure in decreasing the initial nematode population in the soil till safe levels for the following tomato crop (Curto *et al.*, 2006).

Technical feasibility and effectiveness of *Brassica* green manure biofumigation for the control of root-knot nematodes were investigated in a greenhouse experiment on a tomato - melon succession. Green manures of *Brassica juncea* and *Eruca sativa* were compared with a *Brassica* seed meal commercial pellet, alone or combined with *E. sativa* manure, two commercial formulations of quillay or tagetes, chemical nematicide cadusaphos and an untreated soil. In the tomato crop, green manure of *E. sativa* resulted in the lowest *M. incognita* population densities and gall formation, as its results were not significantly different from chemical treatment but were lower than *B. juncea* green manure, the pellet and quillay formulation. This last treatment was statistically like the untreated control. *E. sativa* green manure, either alone or combined with the pellet, provided a tomato yield two-three times higher than control, whereas *B. juncea* and cadusaphos almost doubled the yield of untreated soil.

In the following melon crop, the lowest nematode densities were recorded in soil treated with cadusaphos or the commercial pellet. Green manure of *E. sativa*, alone or combined with the pellet, resulted in a higher suppressivity than *B. juncea*, whereas a limited nematicidal activity emerged for the quillay formulation. The highest melon yield was provided by the seed meal pellet, followed by the combination of *E. sativa* green manure and pellet and by both single green manures (De Mastro *et al.*, 2008b).

The technical applicability and effectiveness for the control of the root-knot nematode *M. incognita* of soil biofumigation treatments with *Brassica juncea* and *Eruca sativa* green manures, *B. juncea* hay, and a *Brassica* seed meal commercial pellet were investigated over two years in field experiments on a carrot - tomato rotation. Biofumigation treatments were compared with the standard nematicide fenamiphos, a quillay-derived non-chemical formulation and an untreated control. At the end of the first cycle, soil population density of *M. incognita* were significantly lower in the plots treated with green manure or hay of *B. juncea* or with green manure of *E. sativa*, either alone or combined with seed meal pellet, compared with untreated control. Nematicidal effects of biofumigation treatments did not differ from that of fenamiphos. Moreover, commercial pellet and *B. juncea* green manure and hay also resulted in carrot yield almost double the control, whereas lower increases were provided by *E. sativa* green manure, either alone or combined with the pellet.

In the following tomato crop the lowest *M. incognita* densities were recorded in soil treated with *B. juncea* hay or with *E. sativa* green manure plus pellet. The hay and green manure of *B. juncea* and integrated treatment with *E. sativa* green manure and pellet resulted in the highest tomato yields, always significantly higher than fenamiphos.

vi) Influence of fertilizers: Increased levels of potash have significantly reduced the number of galls by *M. javanica* in tomato (Gupta and Mukhopadhyaya, 1971).

The growth of tomato increased (35%) and root galling (root- knot index 1 as against 4 in control) due to *M. incognita*, nematode population in soil (80% reduction) and egg mass production reduced (egg mass index 1 as against 4 in control) in soil amended with 20% fly ash (Haq *et al.*, 1985). The increase in plant growth may be ascribed to increased nutrient availability (N, K, Ca, Mg, Na, B, and SO_4) and the reduction in nematode population might be due to toxic compounds (polycyclic aromatic hydrocarbons, dibenzofuran and dibenzo-p-dioxin mixtures) present in fly ash. Row application of fly ash at 0.6 kg/sq. m. enhanced the yield of tomato by 90.4% (Khan and Ghadipur, 2004).

vii) Organic amendments: Application of poultry manure @ 2 t/ha to 3 t/ha 15 days prior to seeding proved beneficial for the management of root-knot disease and getting more numbers of transplantable tomato seedlings. Use of Neem cake @ 750 g per m^2 in nursery beds of tomato proved beneficial.

Soil application of neem and subabul leaves to nursery beds at 0.5 t/ha gave better seedling growth and reduced root galling in tomato (Jain *et al.*, 1988).

Singh and Sitaramaiah (1966) applied finely divided oil cakes to root-knot infested soil and noted a reduction in disease incidence in tomatoes. The oil cakes (neem, groundnut, mustard and castor) were effective in reducing parasitic nematode population (*Hoplolaimus*, *Tylenchorhynchus*, *Meloidogyne* and *Helicotylenchus* species) and in increasing yields of tomato. Singh and Sitaramaiah (1967) reported that application of 5-10% (w/w) of leaves of neem, *Melia azadirach*, *Cassia fistula*, *Sesbania aculeata* and *Crotalaria juncea* to the infested soil reduced the incidence of root-knot nematode on tomato.

Application of neem cake at 15 g/spot or 100 g/m furrow, three weeks prior to transplanting of tomato has been reported to give 36-450% enhanced yield with corresponding reduction in gall index to 2.3 as against 4.3 in control (Anon, 1992).

Best results in respect of root-knot reduction due to *M. javanica* and increase in yield of tomato were obtained by amending the soil with sawdust at 2.5 tons per ha 3 weeks before planting and then applying N through urea at 120 kg per ha (Singh and Sitaramaiah, 1971). Rice hull ash at 2.5 tons per ha increased the yield of tomato by 133 to 317 per cent and reduced root-knot incidence by 46 to 100 percent (Sen and Dasgupta, 1981). Application of poultry manure at 2.0, 2.5 and 3.0 tons/ha 15 days prior to tomato seeding is recommended for effective management of root-knot disease and production of more transplantable seedlings in tomato nursery giving BCR of 4.54, 5.63 and 6.68, respectively.

Olive pomace (FOP) and chicken manure (CM) significantly increased tomato yield and reduced *M. incognita* population in the soil, compared to untreated control, at all the tested dosages (Table 4.3). Combinations of 100 tons/ha olive pomace with 2 and 4 tons/ha chicken manure resulted in a further yield increase and were more suppressive than fenamiphos on *M. incognita*.

Table 4.3. Effect of olive pomace (FOP) and chicken manure (CM) soil amendments on *Meloidogyne incognita* on tomato.

Treatment	Crop yield (tons/ha)	Final nematode population (eggs and J2/ml soil)
Untreated control	44.8 a	49.1 a
Fenamiphos 300 kg/ha	65.0 def	20.7 cdef
FOP 25 t/ha	57.1 bcd	27.6 bc
FOP 50 t/ha	53.0 ab	26.8 cd
FOP 100 t/ha	56.5 bcd	25.5 cde
CM 1 t/ha	53.5 ab	37.9 b
CM 2 t/ha	54.3 bc	26.1 cd
CM 4 t/ha	53.8 bc	20.4 cdef
FOP 25 t/ha + CM 1 t/ha	56.5 bcd	25.4 cde

FOP 25 MT/ha + CM 2 t/ha	56.1 bcd	23.2 cde
FOP 25 t/ha + CM 4 t/ha	60.5 bcd	23.0 cde
FOP 50 t/ha + CM 1 t/ha	60.8 bcd	20.4 cdef
FOP 50 t/ha + CM 2 t/ha	63.3 cde	21.0 cdef
FOP 50 t/ha + CM 4 t/ha	61.0 bcd	20.6 cdef
FOP 100 t/ha + CM 1 t/ha	64.1 def	16.0 def
FOP 100 tT/ha + CM 2 t/ha	73.9 f	14.5 ef
FOP 100 t/ha + CM 4 t/ha	71.8 ef	11.3 f

* Means followed by the same letters in the same column are not significantly different according to Duncan's Multiple Range Test (P = 0.01).

4.1.4.3. Chemical methods

i) Nursery bed treatment: Nursery bed treatment with carbofuran at 2g a.i. /m^2 were effective in increasing seedling growth and reducing root-knot nematode population of tomato (Jain and Bhatti, 1983; Ramakrishnan and Balasubramanian, 1981). In the main field, the above treatment was also effective and increased the fruit yields (55.9% increase in yield of tomato).

Application of Metham sodium 32 EC at 60 ml/m^2 or Carbofuran @ 0.2 g a.i. /m^2 to tomato nursery controlled the root-knot nematode, *M. incognita* and improved the performance of transplants by 80 and 60%, respectively.

In susceptible variety, the nursery treatment with Carbofuran 3.3 g a.i. /m^2 gave 39.2% increase in yield whereas it was only 10.5% in resistant variety.

ii) Seed treatment: Seed treatment of tomato with Abamectin 40% FS @ 0.4 g a.i. /kg seed cause minimum root-knot index (3.75) and 36.9% reduction in soil population of *M. incognita* which reflect in 21.4% increase in yield of tomato.

iii) Bare root-dip treatment: Carbofuran and oxamyl at 1,000 ppm for 15 to 30 min. (Parvatha Reddy and Singh, 1979); fenamiphos at 250 to 750 ppm for 30 min. (Thakur and Patel, 1985); dimethoate at 500 ppm for 6 hr, phosphamidon and dichlofenthion both at 1000 ppm for 8 hr (Jain and Bhatti, 1978) and thionazin at 500 ppm for 15 min (Reddy and Seshadri, 1975) gave effective control of root-knot nematodes on tomato when used as bare-root dips.

Bare root dip treatment of root-knot nematode infested seedlings of tomato using Carbosulfan and Triazophos @ 0.1% and 0.5% for six hours increased the yield and reduced the root-knot index.

iv) Main field treatment: Carbofuran, ethoprophos and fenamiphos each at 1 to 2 kg a.i. per ha were found effective in reducing the root-knot nematode population and in increasing fruit yields of tomato (Ahuja, 1983).

Dichlofenthion was found effective against *M. incognita* in tomatoes (Alam *et al.*, 1973, Saxena *et al.*, 1974).

Application of Carbofuran 3G to nursery at 10 g/m^2 at sowing and 1 kg a.i. /ha in the main field one week after transplanting proved beneficial.

v) Nanoformulations

Application of AgNPs improved tomato plant growth and reduced root-knot nematode (*M. incognita*) infection in comparison to silver nitrate and control treatments. The highest increment of fresh weight as well as the lowest numbers of galls and egg-masses was obtained when tomato plants were treated with AgNP produced by ginger extract at 1 mM (Table 4.4) (El-Deen and El-Deeb, 2018; Ahmed and Bahig, 2018).

Table 4.4. Effect of soil treatment with AgNPs on galls and egg-masses of tomato (index of 0 to 9) caused by *M. incognita* under greenhouse conditions.

Treatments	Conc. (mM)	Fresh plant wt. (g)	No. of galls	RGI	No. of egg-masses	Egg-mass index
Biological AgNP	0.25	48.2 c	4.7 cd	2 ef	0 f	1 d
	0.50	52.0 b	2.0 d	2.0 ef	0 f	1 d
	1.00	55.4 a	1.3 d	1.3 f	0 f	1 d
Chemical AgNP	0.25	41.1 d	6.0 cd	2.7 de	1 ef	1.7 cd
	0.50	37.9 de	10.3 c	3.3 d	3.7 de	2 c
	1.00	23.4 g	20.7 b	4.3 c	5.3 d	2.3 c
$AgNO_3$	0.25	37.5 e	25.0 b	5 c	12 c	4 b
	0.50	33.6 f	28.0 b	5 c	14.7 bc	4 b
	1.00	20.9 g	48.0 a	6 b	6 b	4 b
Control (Check)	---	16.3 h	55.0 a	7 a	24	5 a
CD ($P \leq 0.05$)	---	3.2	8.01	0.85	2.9	0.72

Note. Each value represents the mean of four replicates. Within each concentration, values with the same letter do not differ significantly according to CD ($P \leq 0.05$).

In a pot experiment, all treatments of AgNP and 0.02% TiO_2NP completely controlled *M. incognita*. Treatments of 0.02, 0.01 and 0.005% of AgNP as well as 0.02% of TiO_2NP were toxic to tomato plants and significantly reduced tomato root and stem length and fresh weights in comparison to control.

The egg mass of *M. incognita* on exposure to green synthesized AgNPs were found to be malformed. The juveniles of the root knot nematode appeared to be deformed with oozing out of its body fluid (Surega, 2015).

The green synthesized AgNPs using the most preferred plant source of *Tridax procumbens* with the concentration 0.0005 per cent was found more effective than *P. fluorescens* and carbofuran 3G for the management of *M. incognita* in tomato and increased the yield attributes of the crop. The green synthesized AgNPs had highest effect in checking *M. incognita* population in soil (88.16%), root (92.79), number of females with egg masses (79.80), eggs/egg mass (65.14%), and nematode disease severity in terms of gall index. The significant negative effect of AgNPs on *M. incognita* reflected on tomato plant growth with increased shoot height (26.99%) and weight (13.92%); root length (26.57%) and weight (23.45%) and fruit yield (20.54%) (Surega, 2015).

4.1.4.4. Biological methods

i) Antagonistic bacteria: Nursery treatment with FYM enriched *Pseudomonas fluorescens* formulation @ 20 g/m^2 was found effective.

The plants treated with *P. fluorescens* gave maximum root and shoot length and significantly (P = 0.05) suppressed females per root system (40.52%), J2/g of root (39.80%), galls per root system (41.50%) and egg masses per root system (43.23%) resulting in improved growth over control plants.

Application of *P. fluorescens* at 10 g/m^2 in nursery beds gave good control of root-knot nematodes. The level of infestation of root-knot nematode *M. incognita* on tomato was reduced with fewer galls and egg masses in the soil following root dipping with *P. fluorescens* strain Pf1 (Santhi and Sivakumar, 1995). Similarly, application of the bacterium, *P. chitinolytica* reduced the root-knot nematode infection in tomato crop (Spiegel *et al.*, 1991).

The inoculants Equity (contains 47 strains of bacilli in a liquid formulation), BioYield (contains *Bacillus subtilis* strain GB03 and *B. amyloliquefaciens* strain GB99 in a chitosan carrier), and FZB42 (a strain of *B. amyloliquefaciens*) induced significant reductions in nematode galls per plant on tomato (Cadenaa *et al.*, 2008). AgBlend, containing microbial metabolites, reduced the number of galls. Treatment with each of the inoculants also increased root weight (Table 4.5).

Table 4.5. Effect of PGPR on root weight and root galling due to *Meloidogyne incognita*

Treatment	Root-knot index	Root weight (g)/plant
Control	4.2	15.08
AgBlend	3.3	27.18
BioYield	2.1	18.18
Equity	3.3	22.64
FZB42	2.7	16.77

Mixtures of two PGPR strains plus a formulation carrier chitosan, particularly LS213 (containing *Bacillus subtilis* strain GB03 and *B. amyloliquefaciens* strain GB99 in a chitosan carrier) exhibited a more consistent and significant level of growth promotion on tomato seedlings. The transplant mix delivery system for biological preparations has demonstrated potential to enhance transplant vigor, provide protection against root-knot nematodes and other foliar diseases in the field and enhance yield. These results indicate a synergy in plant growth promotion and induced resistance by the combination of chitosan and the mixtures of two bacterial strains (Reddy *et al.*, 2001).

Two bacterial antagonists, *Bacillus licheniformis* and *Pseudomonas mindocina* and 2 fungal antagonists, *Acrophialophora fusispora* and *Aspergillus flavus* were evaluated for their efficacy as bio-control agents individually and in various combinations against *M. incognita* infecting tomato. Individually, *B. licheniformis* gave best control although when all 4 agents were used, nematode multiplication was reduced by 82.06% (Siddiqui and Husain, 1991).

Two strains of *B. thuringiensis* (Dipel and SAN 415) were evaluated for their bioefficacy against *M. javanica* in tomato. Both the strains suppressed the nematode population and egg masses. They also reduced the percentage hatchability of eggs. SAN 415 was more effective than Dipel and the effect was approximately like the standard nematicide Nemacur (Osman *et al.*, 1988).

Application of *Bacillus thuringiensis* causes 95% mortality of *M. javanica* juveniles due to β-exotoxin production and suppression of gall formation, egg mass production and nematode population in soil.

Field application of a nematicidal *B. thuringiensis* strain to tomato in Puerto Rico reduced galling in roots due to *M. incognita* and increased the yield significantly (Zuckermann *et al.*, 1993). Among different strains tested, strain 868 of *B. thuringiensis* was found effective in reducing root galling and total nematode population (Chahal and Chahal, 1991).

Application of *Pasteuria penetrans* @ 10 g culture (1 x 10^9 spores/m^2) was found to be effective in reducing the nematode population (*M. incognita*) in the main field by 72.3% over control and improved plant growth, fruit yield (7.800 t/ha compared to 3.325 t/ha in control) and nutritional status of tomato fruits (Ravichandra and Reddy, 2008).

Root material containing *Pasteuria penetrans* spores at 212-600 mg powder/kg soil, when broadcast in soil, showed reduction in galling on tomato roots and final nematode population. Tomato yield increased by 20% over untreated check, when *P. penetrans* was introduced to the nursery soil at 1x10^5 spores/g soil. The persistence of *P. penetrans* in the field soil was evident since about 50% of the final juvenile population was encumbered with bacterial spores (Walia and Dalal, 1994). *P. penetrans* is highly resistant to heat, desiccation and has the ability to survive for more than 2 years in soil, qualifies as the most potential biocontrol agent against root-knot nematodes infesting vegetable crops.

Application of organic amendments (cattle manure) to soil inoculated with root-knot nematodes encumbered with spores of *P. penetrans,* improved plant growth and multiplication of both nematode pest and the bacterium (Gomes *et al.*, 2002).

In plastic house experiments in Crete, *P. penetrans* applied at 20,000 or 25,000 spores/g of soil was able to parasitize 65%-75% of juveniles and females of *M. javanica* infecting tomato (Tzortzakakis *et al.*, 1999).

Somasekhar and Gill (1990) studied *P. penetrans* (3.5 g spore/kg soil in pots) against *M. incognita* J_2 on tomato plants and recorded 92% spore attachment on larvae with 65 and 51% reduction in nematode penetration in host root and gall formation, respectively. Reduction of 94% *M. incognita* population by *P. penetrans* at 1 g/kg soil on tomato in pots was obtained by Jayaraj and Mani (1988). Somasekhar and Gill (1991) reported a 69% reduction in gall formation due to *M. incognita* in *P. penetrans* alone treatment.

P. penetrans at 1 x 10^9 spores/m^2 was found to be effective in reducing the nematode population (*M. incognita*) in the main field by 72.3% over control and improved plant growth, fruit yield (7.800 t/ha compared to 3.325 t/ha in control) and nutritional status of tomato fruits (Ravichandra and Reddy, 2008) (Table 4.6).

Table 4.6. Effect of bioagent, botanical and nematicide for the management of *Meloidogyne incognita* infecting tomato

Treatment	No. of galls/ plant	No. of egg masses/plant	Root-knot index	Yield tons/ha
Neemark – 2% at 1.25 l/m^2	43.18	20.40	4.0	5.725
Carbofuran – 3 kg a.i./ha	52.80	19.78	4.0	5.125
P. penetrans -1x10^9 spores/m^2	27.92	12.22	2.8	7.800
Control	122.54	72.04	5.0	3.325
CD (P = 0.05)	1.0715	1.29	0.64	---

P. penetrans at 1 x 10^9 spores/m^2 was found to be effective in reducing nematode population in the main field by 72.3 over inoculated check. The above treatment also improved plant growth characteristics, nutritional status of fruits and plants, and recorded highest yield of 6.8 t/ha compared to inoculated check (3.325 t/ha) (Hemavathi *et al.*, 2009).

The effects of six isolates of PGPR *Pseudomonas putida*, *P. fluorescens*, *Serratia marcescens*, *Bacillus amyloliquefaciens*, *B. subtilis* and *B. cereus*, were studied on tomato plant growth and root-knot nematode reproduction after 45 days from nematode infection. The highest shoot dry weight/g (43.00 g) was observed in plants treated with *S. marcescens* followed by *P. putida* (34.33 g), *B. amyloliquefaciens* (31.66 g), *P. fluorescens* (30.0 g), *B. subtilis* (29.0 g), *B. cereus* (27.0 g) and nematode alone (untreated) 20 g/ plant. The maximum plant height was observed when plant was treated with *S. marcescens*, followed by *P. fluorescens*, *P. putida*, *B. amyloliquefaciens* and *P. putida* 52.66, 50.66, 48 and 48 cm, respectively (Almaghrabi *et al.*, 2013).

The highest fruit yield was observed with *S. marcescens* (319.6 g/plant). The highest number of fruit/plants was observed when plants were treated with *S. marcescens* (10.66), followed by *B. amyloliquefaciens* (8.66), *P. putida* (8), *P. fluorescens* (8) and *B. cereus* (7.66) (Almaghrabi *et al.*, 2013).

On the other hand, the lowest numbers of J_2/10 g of soil (78), galls/root (24.33), egg masses/ root (12.66) and egg/egg masses (280.66) were observed in the plants treated with *S. marcescens* (Almaghrabi *et al.*, 2013).

Avermectins B_1 and B_{2a} (secondary metabolite, *Streptomyces avermitilis*) applied to soil through drip irrigation systems at rates ranging from 0.093 to 0.34 kg a.i./ha applied as a single dose or 0.24 kg a.i./ha applied as three doses each at 0.08 kg on tomatoes against *M. incognita* were as effective as oxamyl and aldicarb at 3.36 kg a.i./ha (Garabedian and Van Gundy, 1983). The aqueous solution of avermectins (250 ml of 0.001%/m^2) significantly reduced root galls of tomato seedlings raised in root-knot infested nursery beds which yielded robust and healthy seedlings with no root infection (Parvatha Reddy and Nagesh, 2002). In a root-knot infested micro-plot study, avermectins at 0.015 kg/ha effectively controlled *M. incognita* on tomato and increased yield by 11 and 8% compared to the untreated and carbofuran treated plots, respectively. (Parvatha Reddy and Nagesh, 2002). Seedling bare root-dip in avermectin 100% gave highest reduction in root-knot index (1.3), soil nematode population (132.8), number of females/g root (12.8), egg masses/g root (5.4) and eggs/egg mass (125.2) in tomato (Jayakumar *et al.*, 2005).

ii) Antagonistic fungi: Application of 'Royal 350' (*Arthrobotrys irregularis* cultured on oat seed medium) at 140 g/m^2 a month before transplantation of tomato resulted in good protection aainst root-knot nematodes.

Among the egg parasites, efficacy of *Purpureocillium lilacinum* (commercially formulated as 'Biocon' in Philippines) has been found to be comparatively higher in suppressing the population of *Meloidogyne* spp. and *R. reniformis* on tomato. Bare root-dip treatment of tomato seedlings in a spore suspension of *P. lilacinum* at 4 x 10^5 spores/ml reduces *M. incognita* infestation in roots by 50% and increases the growth of seedlings by 20%. Application of organic amendments increases the effectiveness of *P. lilacinum* against *M. incognita*.

The control of *M. incognita* on tomato by *P. lilacinum* was tested on a large-scale basis in the province of Herrera, Panama. The fungus-treated plots had higher yield than did the control plots and those treated with carbofuran (Table 4.7) (Jatala, 1985).

Table 4.7. Biological control of *Meloidogyne incognita* by *Purpureocillium lilacinum*

Treatment	Yield (kg/plant)
P. lilacinum	2.13
Carbofuran at 2.5 kg a.i./ha	1.85
Control	1.52
P = 0.05	0.38

Zaki (1994) reported that application of chickpea infested *P. lilacinum* at 4 g/kg soil was optimum for significant reduction of gall index and *M. javanica* population in tomato plants. The egg infection and destruction were 58% and 66%, respectively.

Among four antagonistic fungi (*P. lilacinum, T. harzianum, T. viride* and *Gliocladium virens*) evaluated for the management of *M. incognita* infecting tomato, maximum egg mass and egg parasitization was observed by *P. lilacinum* followed by *T. viride, G. virens* and *T. harzianum*. *T. viride* and *G. virens* comparatively recorded higher root colonization closely followed by *P. lilacinum* (Nagesh *et al.*, 2001) (Table 4.8).

Table 4.8. Effect of antagonistic fungi on egg masses and egg parasitization and root colonization on tomato infected with *Meloidogyne incognita*

Treatment	Fungal parasitization (%)		Root colonization (%)	
	Egg masses	Eggs	Roots in toto	Galled roots
T. harzianum	35.0	28.5	56.0	68.0
G. virens	46.5	34.5	74.0	79.0
T. viride	45.0	37.5	72.0	79.0
P. lilacinum	65.0	49.5	68.0	77.0
Control	0.0	0.0	0.0	0.0
CD (P = 0.05)	4.54	3.77	5.11	6.35

Soil application of talc-based formulation of *P. lilacinum* (8×10^6 spores/g) at 4 g/3 kg soil was found to be highly effective against *M. incognita*. The above treatment gave significant reduction in root galling, egg mass production and final nematode population in soil and roots; coupled with increase in plant growth parameters, fruit yield, egg parasitization and spore density in soil (Shanmuga Priya and Kumar, 2006a, b).

Nursery treatment with FYM enriched *Trichoderma viride* formulation @ 2.5 g/m^2 was found effective.

Sankaranarayanan *et al.* (2000) reported that granular application of *Pochonia chlamydosporia* (formulated on wheat bran, maida and vegetable oil) significantly reduced the number of egg masses of *M. incognita* by 70% and parasitized 70 to 89% egg masses and 63 to 69% eggs on tomato plants. A significant increase in growth of tomato plants was observed with *P. chlamydosporia* treated plants.

Application of granulated product of *P. chlamydosporia* at 10 g/litre soil increased the population densities of *P. chlamydosporia* to 10 cfu/g of soil after 7-14 weeks. The first-generation egg masses produced by *M. javanica* contained between 37 and 82% parasitized eggs (Stirling *et al.*, 1998).

Aspergillus niger was found to be highly pathogenic against *M. incognita* infecting tomato and brinjal (Reddy, 1989).

Hallmann and Sikora (1994) showed biocontrol potential of non-pathogenic *Fusarium oxysporum* against root-knot nematode, *M. incognita*. The reproduction of *M. incognita* in *F. oxysporum* colonized tomato plants was reduced by 50% over *F. oxysporum* free plants.

iii) Arbuscular mycorrhizal fungi: Arbuscular mycorrhizal fungus (AMF), *Glomus fasciculatum* was effective against root-knot nematodes and *G. mosseae* against reniform nematodes on tomato. Organic amendments (oil cakes, calotropis leaves) in combination with AMF enhance the colonization of AMF on tomato roots which further increased plant growth and reduced gall index. Sundaram and Arangarasan (1996) observed that inoculation of *G. fasciculatum* at 10 g/plant was effective in increasing the uptake of nutrients and yield and reducing root galling (10) compared to control, followed by *G. mosseae, Gigaspora margarita* and *Acaulospora laevis*. Application of *G. fasciculatum* (Shimoga banana isolate) at 50 g/sq.m. (200 spores/g of soil) gave minimum number of galls per root system (39.2) at harvest and recorded maximum yield (26.47 t/ha) with an increased benefit: cost ratio (3.34) and decreased nematode population (Kiran Kumar *et al.*, 2004).

Application of *Glomus fasciculatum* @ 200 spores/kg soil in nursery bed of tomato provided more than 50% reduction in root-knot population and 36% increase in yield of tomato (Nagesh *et al.*, 1999) (Table 4.9).

Table 4.9. Effect of *Glomus fasciculatum* spore density on root galling, egg mass production and root colonization on tomato infected with *Meloidogyne incognita.*

Spore density of *G. fasciculatum*/ g soil	Root-knot index (1-5 scale)	% reduction in egg masses/ plant	% root colonization by *G. fasciculatum*	No. of chlamydospores of *G. fasciculatum*/ 10 cm^3 soil
1	3.8	12.8	6.5	19.2
2	3.2	26.2	36.6	42.8
4	2.5	61.0	54.1	77.4
8	2.5	70.0	62.0	98.8
16	2.0	71.0	68.0	124.6
Control	4.0	---	---	---
CD (P = 0.05)	**0.29**	4.75	**6.10**	5.32

Soil application of *G. fasciculatum* (Shimoga banana isolate) at 2 spores/g soil gave significant increase in plant growth parameters, fruit yield, root colonization and spore density in soil of the bioagent coupled with reduction in root galling, egg mass production and final nematode population in soil (Kantharaju *et al.*, 2005).

iv) Predaceous nematodes: Fauzia *et al.* (1998) demonstrated the ability of *Mononchoides longicaudatus* to reduce root galling by *M. incognita* in pot tests, resulting in improved vegetative growth and increased root mass of tomato.

4.1.4.5. Host resistance: Tomato cvs. All Round, Anahu, Anahu R, Atkinson, Beefeater, Beefmaster, Better Boy, Big Seven, Cavalier, Eurocross, Nematox, SL-120, NTR-1, SL-12, Patriot, VFN-8, VFN

Bush, Piersol, Radiant, Nemared, Ronita, Anahu, Bresch, Healani, Kewalo, Campbell-25, Punuui, Arka Vardan, Pelican, Hawaii-7746, Hawaii-7747, VFN 8 and Hisar Lalit have been reported to be resistant to root-knot nematodes. Use of resistant varieties of tomato e.g., Hisar Lalit, Punjab NR-7, SL-120, Mangala, Utkal Pallavi (BT-1), Utkal Dipti (BT-2), Utkal Urbasi (BT-12), and BT-17 gave effective control.

In a field with resistant tomato (Hisar Lalit) grown for a year, the yield was 27% higher than susceptible variety (HS-101) (Kanwar and Bhatti, 1990).

Root galling caused by *M. incognita* was lower (2.95 as compared to 8.08 in control) and the fruit yield was higher (1.785 kg plant^{-1} as compared to 1.118 kg plant^{-1} in control) in tomato plants grafted on the rootstock Beaufort (Kaþkavalci *et al.*, 2009).

Indian Institute of Vegetable Research, Varanasi, has developed grafting technology to mitigate the problem of root-knot nematode infesting tomato. *Solanum torvum* (wild brinjal germplasm) which is identified as resistant against root-knot nematode was used as rootstock to graft with scions of promising tomato varieties cv. Kashi Aman, Kashi Vishesh and Hisar Lalit. Grafted plants were compatible between rootstock and scion and showed significant resistance against root-knot nematode by reducing soil population, reproduction and gall index (Gowda *et al.*, 2017).

The *Mi* gene, which provides effective control against RKN in tomatoes, has been introgressed into cultivated tomatoes and rootstock cultivars (Louws *et al.*, 2010). Grafting of susceptible tomato cultivars on RKN-resistant rootstocks was effective in controlling RKN in fields naturally infested with RKN (Rivard *et al.*, 2010). However, because of temperature sensitivity of the *Mi* gene, such resistance may not be uniformly stable (Cortada *et al.*, 2009).

Diseases controlled by grafting in different solanaceous vegetable crops due to fungi, bacteria, viruses and nematodes are listed in Table 4.10.

Table 4.10. Commonly used rootstocks for tomato and their disease resistance/susceptibility and vigor.

Rootstock	**TMV**	**Fusarium Wilt**		**Verticillium Wilt Race 1**	**Bacterial Wilt**	**Root-knot Nematode**	**Vigor***
		Race 1	**Race 2**				
Anchor[1]	R	R	R	R	R	R	5
Arnold	R	R	R	N/A	N/A	R	N/A
Beaufort[2]	R	R	R	R	S	R	3
Body[3]	R	S	R	R	S	R	5
Estamino	R	R	R	R	S	R	N/A
Maxifort[2]	R	R	R	R	S	R	5
Multifort	S	R	R	S	S	R	N/A
Survivor[1]	R	R	R	R	M	R	5
Aegis[1]	R	R	R	R	M	R	4
Robusta[3]	R	S	R	R	M	R	3
RT-04-105T	S	R	R	R	R	R	N/A
RT-04-106T	R	R	R	S	R	S	N/A

R – Resistant, M – Moderate, S – Susceptible, N/A – Not characterized.

* Vigor is measured on a scale of 1 to 5, where 1 represents poor and 5 represents excellent

Seed supplier [1] Takii Seeds, [2] deRuiter Seeds, [3] Bruinsma Seeds

The hybrid rootstocks i.e., interspecific tomato hybrid rootstock 'Multifort' (*S. lycopersicum* x *S. habrochaites*) and tomato hybrid rootstock 'Survivor' (*S. lycopersicum*) significantly reduced root galling compared with the non-grafted and self-grafted scions by ~ 80.8%. The root galling reduction by 'Survivor' (97.1%) was significantly greater than that by 'Multifort' (57.6%) compared with non-grafted scions (Table 4.11). In general, tomato plants grafted onto 'Multifort' tended to be more vigorous than all other treatments. 'Brandywine' grafted to 'Multifort' resulted in significantly higher ($P<0.05$) yields than the non- and self-grafted 'Brandy-wine' treatments. Grafting with appropriate rootstocks may play an effective role in RKN management during the transition to organic production when high populations of nematodes are present (Barrett *et al.*, 2012).

Table 4.11. Effect of grafting treatments on root-knot nematode galling ratings on tomato cultivars.

Treatment	**Root galling (1-10 scale) in cultivar**	
	Brandywine	**Flamme**
Non-grafted	9.30 a	8.06
Self-grafted	7.30 b	6.12
Grafted on 'Multifort' rootstock	3.88 c	3.48
Grafted on 'Survivor' rootstock	0.54 d	0.00

Most eggplant lines utilized will graft successfully with tomato lines. The Asian Vegetable Research and Development Centre (AVRDC) recommends eggplant accessions EG195 and EG203 that are resistant to flooding, bacterial wilt, root-knot nematode (*Meloidogyne incognita*), *Fusarium* wilt (*Fusarium oxysporum* f. sp. *lycopersici*), and southern blight (*Sclerotium rolfsii*) (Black *et al.*, 2003).

In a grafting study by Owusu *et al.* (2016) against root-knot nematodes, five tomato cultivars were selected with "Big Beef," "Celebrity," and "Jetsetter" being resistant to *Verticillium* wilt, *Fusarium* wilt, nematodes, and tobacco mosaic virus (VFNT), which served as the nematode-resistant rootstocks, and "Tropimech" (VF) and "Big Power" (locally grown nematode-susceptible cultivars) served as scions. Grafted plants had the least nematode populations in the plant house. In field experiments, nematode population levels were lower in "Big Power" that had been grafted on Celebrity, Jetsetter, and Big Beef rootstocks, compared to self-grafted or non-grafted "Big Power". Fruit yields were also higher in the grafted plants utilizing resistant rootstocks than non-grafted plants.

Tomato grafting onto a resistant rootstock of wild brinjal (*S. sisymbriifolium*) under farmers' field conditions at Hemza of Kaski district against root-knot nematodes was undertaken. The root system of the grafted plants was free from gall formations; however, non-grafted plants had an average of 7.5 gall index (GI). Fruit yields significantly ($P > 0.05$) increased by 34.64% in the grafted plants compared with the non-grafted plants (Table 4.12) (Baidya *et al.*, 2017).

Wild *Solanum* rootstocks *S. sisymbriifolium*, *Physalis peruviana*, and *S. torvum* had the least galls per 10 g root (6, 5, 5) and females per g root (2, 2, 2), respectively, and showed the highest level of expression of phenolics and defense-related enzymes viz., peroxidases, polyphenol oxidases, phenylalanine ammonia lyase, and acid phosphatase from leaf samples, compared to the susceptible tomato scion (US-618) (Dhivya *et al.*, 2014).

Table 4.12. Tomato on 'Wild Brinjal' rootstock

Treatment	**Yield (Kg/plant)**	**RKI**
Non-grafted	8.980	8
Grafted on *Solanum sisymbriifolium*	12.064	0
Yield increase over non- grafted	3.104	---
% yield increase	34.64	---

In grafted tomatoes, the rootstocks 'Aloha' and 'Multifort' both reduced root galling (Table 4.13)

Table 4.13. Effect of 'Aloha' and 'Multifort' rootstocks on root galling of tomato

Rootstock	***M. incognita* (J_2/g root)**
Non-grafted 'Florida 47'	2.16
'TX301'	1.52
'Multifort'	0.10
'Aloha'	0.10

In a previous study, two garden eggplant rootstocks, *S. torvum* and *S. aethiopicum* were poor hosts of *M. javanica* and *M. incognita* (Ioannou, 2001).

4.1.4.6. Integrated methods

i) Physical and cultural: Soil application of poultry manure @ 2 t/ha followed by soil solarization (with 100 gauze LLDPE clear film for 15 days in hot summer) is recommended for the control of root-knot nematodes and higher production of tomato transplants in nursery.

Treated nursery + polythene mulching in main field + Neem cake at 200 kg/ha reduced the population of *M. incognita* by 70.9% and increased tomato yield by 39.9%.

ii) Physical and chemicals: For the management of root-knot nematode in tomato, nursery should be covered with polythene sheet for 15 days during May/June after giving light irrigation. At the time of sowing, apply Carbofuran 3G @ 10 g/m^2 in the nursery prior to sowing.

Soil solarization along with application of Neem cake @ 200 kg/ha in the main field increased fruit yield of tomato by 180.6% by reducing infestation by *M. incognita*.

Addition of ammonium phosphate fertilizer to loam soil significantly reduced galling of tomato roots by *M. incognita*. The galling index generally decreased with increasing fertilizer dosage. Soil solarization along with ammonium phosphate fertilizer further reduced nematode galling in all cases (Table 6.17).

Factorial ANOVA showed both fertilizers and solarization to be highly significant in increasing all growth parameters in the loam soil; however, no fertilizer-solarization interactions were found (Table 4.14).

Table 4.14. Effect of ammonium phosphate fertilization and soil solarization on galling of tomato roots by *Meloidogyne incognita*

Treatment and dosage	Root galling[1]	Number of fruits[2]	Fruit fresh wt. (g)	Vegetative fresh wt. (g)
Ammonium phosphate-100 mg/L	2.0	5.9	90.8	604
Ammonium phosphate-100 mg/L + Soil solarization	0.8	14.0	210.4	1102
Ammonium phosphate-200 mg/L	1.0	8.7	151.8	538
L	0.3	14.5	206.4	1080
Ammonium phosphate-300 mg/L	0.9	12.3	175.1	690
Ammonium phosphate-300 mg/L + Soil solarization	0.2	21.4	382.6	1582
Ammonium phosphate-400 mg/L	0.7	18.5	321.5	1162
Ammonium phosphate-400 mg/L + Soil solarization	0.0	26.1	502.7	1668
Control	3.5	11.7	132.1	638
Soil solarization only	1.8	15.0	215.7	1078
LSD	0.7	8.0	171.9	437
Factorial ANOVA Fertilizer	0.01[2]	0.01[3]	0.01	0.01
Soil solarization	0.01	0.01	0.01	0.01
Fertilizer x Soil solarization	NS[3]	NS	NS	NS

1 Based upon 0-4 rating scale.
2 Significance level.
3 NS = no significant difference.

The mean maximum temperature at 15 cm depth in the solarized plot was 42°C (9°C higher than non-covered control soil), and the mean minimum was 34°C (7°C higher than the control). The maximum temperatures at 15 cm depth in the fine sandy loam plot were 44°C and 36°C in solarized and control plots, respectively.

There were no differences in the number of surviving plants in either soil. Factorial ANOVA showed both fertilizers and solarization to be highly significant in increasing all growth parameters in the loam soil; however, no fertilizer-solarization interactions were found. No significant differences in tomato growth due to fertilizers were found in the sandy loam plot. Solarization, however, was highly significant in increasing numbers of fruit and fruit fresh weight (Tables 4.15).

Table 4.15. Fresh market tomato (cv. Early Pak 7) yield in soil treated with ammonia-based fertilizers and/or solarization.

Soil Treatment[1]	Loam soil			Fine sandy loam soil		
	Number of fruits[2]	Fruit fresh wt. (g)	Vegetative fresh wt. (g)	Number of fruits	Fruit fresh wt. (g)	Vegetative fresh wt. (g)
Aqua ammonia	5.9	90.8	604	22.7	398.3	1386
Aqua ammonia + Solarization	14.0	210.4	1102	30.5	493.6	1384
Urea	8.7	151.8	538	26.0	490.6	1130
Urea+ Solarization	14.5	206.4	1 080	30.7	605.0	1246
Ammonium Sulphate	12.3	175.1	690	22.6	404.8	1160
Ammonium Sulphate + Solarization	21.4	382.6	1582	32.3	552.0	1660
Ammonium Phosphate	18.5	321.5	1 162	27.9	359.7	1754
Ammonium Phosphate + Solarization	26.1	502.7	1 668	28.3	533.0	1766
Non-treated Control	11.7	132.1	638	22.4	360.3	1180
Solarization	15.0	215.7	1 078	33.5	574.4	1448
LSD	8.0	171.9	437	11.9	247.6	670
Factorial analysis of variance Fertilizers	0.01[3]	0.01	0.01	NS	NS	NS
Solarization	0.01	0.01	0.01	0.01	0.01	NS
Fertilizers[3] x Solarization[4]	NS	NS	NS	NS	NS	NS

[1] All fertilizers were applied at 305 kg N/ha.

[2] All values are given on a per-plant basis.

[3] P = 0.05 or 0.01; NS = no significant differences.

[4] Interaction of treatments.

iii) Physical and host resistance: In tomato soil solarization along with the use of resistant variety Hisar Lalit, significantly reduced *M. incognita* infestation.

iv) Physical, cultural and chemical/host resistance: Solarization coupled with treated nursery (Carbofuran @ 0.3 g a.i./m^2) or resistant variety (NR-7) along with Neem cake @ 200 kg/ha effectively controlled the root-knot nematode and increased the yield. This package has been demonstrated at farmer's field at Benikheda village of Panagar, where the average increase in yield was 43%.

v) Physical, cultural and chemicals: The effect of combinations of a susceptible (sunflower), moderately susceptible (watermelon) or resistant (maize) hosts, and the successive tomato crop on the *Meloidogyne* population density and its control was studied. Data showed that the population density of *Meloidogyne* spp. was greatly influenced by the preceding summer crops and cropping system. The greatest percentage of nematode control was obtained when a resistant crop preceded tomato. When tomato was preceded by a susceptible host, the nematode population density was at a maximum level. However, significant nematode control was observed when soil solarization was introduced to the system, ranging from 40-51 percent. The result was the same, whether soil solarization was applied before or after the susceptible host, or both. Percent nematode control increased when the soil was chemically treated, and particularly when solarization was included in the system, it ranged between 75 and 88 percent.

Nematode control was between 62 percent and 83 percent when a moderately susceptible host preceded tomato. Suppression of nematode populations was more pronounced in this case than in the two above-mentioned cases, i.e., when a susceptible or a resistant host preceded tomato. Nevertheless, nematode population reductions after chemical treatment of soil were not as great, because nematode control in this rotation was generally high (87 to 96 percent). It was reported that solarization consistently surpassed fumigation in reducing nematode population density only within the top 7.3 cm of soil (Heald and Robinson, 1987).

Significant nematode control (75 to 88%) was observed when soil solarization was introduced to the cropping system (a resistant crop preceding tomato), ranging from 40-51 percent. Percent nematode control increased when the soil was chemically treated, and particularly when solarization was included in the system (Table 4.16) (Heald and Robinson, 1987).

Table 4.16. The effect of short-term soil solarization and cropping system on the integrated control of *Meloidogyne* spp. under laizemeter conditions

Treatment*	Egg masses/5g roots (% of control)	Females/5g roots (% of control)	Nematode control (%)
S-T	100	100	0
M-T	41	35	62
R-T	0	0	100
Z-S-T	64	55	40
Z-M-T	29	26	73
Z-R-T	0	0	100
S-Z-T	60	55	43
M-Z-T	23	21	78
R-Z-T	0	0	100
Z-S-Z-T	50	48	51
Z-M-Z-T	18	17	83
Z-R-Z-T	0	0	100
S-Tr	20	30	75
M-Tr	9	17	87

R-Tr	0	0	100
Z-S-Tr	11	21	84
Z-M-Tr	4	6	95
Z-R-Tr	0	0	100
S-Z-Tr	15	25	80
M-Z-Tr	1	10	95
R-Z-Tr	0	0	100
Z-S-Z-Tr	10	15	88
Z-M-Z-Tr	3	6	96
Z-R-Z-Tr	0	0	100

*S = Susceptible host; M = Moderately susceptible host; R = Resistant host; Z = Solarization; T = Tomato; and Tr = Tomato + Chemical treatment.

vi) Bioagents and botanicals: In nursery, integration of *Pasteuria penetrans* (at 28 x 10^4 spores/m^2), *P. lilacinum* (at 10 g/m^2 with 19 x 10^9 spores/g) and Neem cake (at 0.5 kg/m^2) gave maximum increase in plant growth and the number of seedlings/bed. Integration of Neem cake with *P. lilacinum* resulted in maximum parasitization of eggs, while integration of Neem cake with *P. penetrans* gave highest parasitization of *M. incognita* females. In field, planting of tomato seedlings (raised in nursery beds amended with Neem cake + *P. penetrans*) in pits incorporated with *P. lilacinum* (at 0.5 g/plant) gave least root galling and nematode multiplication rate and increased fruit weight and yield of tomato (Table 4.17) (Parvatha Reddy *et al.*, 1997).

Table 4.17. Effect of neem cake, *Pasteuria penetrans* and *Purpureocillium lilacinum* on root galling, nematode multiplication rate and yield of tomato infected with *Meloidogyne incognita.*

Treatment		**Root-knot index**	**Nematode multiplication rate**	**Yield (kg)/6 m^2**
Nursery (m^2)	**Main field (per plant)**			
Neem cake -1 kg	*P. lilacinus* – 0.5 g	3.4	3.6	9.168
Neem cake -1 kg	*P. penetrans* (28 x 10^4 spores)	3.2	3.4	9.312
P. lilacinum – 20 g	*P. penetrans* (28 x 10^4 spores)	3.0	3.1	9.504
P. penetrans (28 x 10^7 spores)	*P. lilacinus* – 0.5 g	2.9	3.0	9.624
Neem cake – 0.5 kg + *P. lilacinum* – 10 g	*P. penetrans* (28 x 10^4 spores)	2.5	3.0	9.672
Neem cake – 0.5 kg + *P. penetrans* (28 x 10^4 spores)	*P. lilacinus* – 0.5 g	2.0	2.7	9.984
Neem cake – 0.5 kg + *P. lilacinum* – 10 g + *P. penetrans* (28 x 10^4 spores)	---	2.6	3.1	9.600
Control	---	4.6	3.8	8.352
CD (P = 0.05)		0.14	0.21	0.100

Soil drenching with 5% Wellgro (neem formulation) solution (at 200 ml/seed pan) along with *P. lilacinum* (at 2 x 10^4 spores/ml) gave maximum increase in seedling weight and highest reduction in root galling. Roots of tomato seedlings raised in the above treatment dipped in 5% Wellgro solution mixed with *P. lilacinum* spores (at 2 x 10^4 spores/ml) for 20 min. when transplanted in pots gave maximum increase in plant growth, root colonization and parasitization of egg masses of *M. incognita* by the bioagent and highest reduction in root galling and final nematode population both in soil and roots (Rao and Parvatha Reddy, 1993a).

Bare root-dip treatment of tomato seedlings in 10% castor leaf extract mixed with *P. lilacinum* spores (at 1.5 x 10^8 spores /ml) for 20 min. significantly increased the plant growth and reduced root galling and final nematode population. The above treatment also gave significant increase in parasitization of eggs and egg masses and propagule density of *P. lilacinum* in roots (Rao *et al.*, 1999). Similarly, root-dip treatment of tomato seedlings in 5 and 10% neem leaf extract mixed with *P. lilacinum* spores (at 6 x 10^4 spores /ml) for 30 min. gave significant reduction in root galling and final nematode population. Significant increases in root colonization, parasitization of eggs and egg masses and propagule density of the bioagent in soil were also noticed in the above treatment (Table 4.18) (Parvatha Reddy *et al.*, 1998).

Table 4.18. Effect of tomato root dip treatment in botanicals mixed with spores of *Purpureocillium lilacinum* on root galling, fecundity, root colonization and egg parasitization.

Treatment	Root-knot index	Propagule density on root (cfu/g)	Propagule density in soil (cfu/g)	% egg parasitization	Eggs/egg mass
Purpureocillium lilacinum	3.2	3242	3043	46	415
Neem leaf suspension 5%	3.0	---	---	---	376
Neem leaf suspension 5% + *P. lilacinum*	2.5	4650	2756	52	334
Neem leaf suspension 10%	2.8	---	---	---	362
Neem leaf suspension 10% + *P. lilacinum*	2.2	4935	3276	57	297
Pongamia leaf suspension 5%	3.5	---	---	---	406
Pongamia leaf suspension 5% + *P. lilacinum*	3.0	3456	2458	48	395
Pongamia leaf suspension-10%	3.3	---	---	---	402
Pongamia leaf suspension-10% + *P. lilacinum*	2.9	3567	2366	50	387
Control	4.3	---	---	---	439
CD (P = 0.05)	0.34	549.56	313.25	6.48	32.57

Application of *Pupureocillium lilacinum* (2 x 10^6 cfu/g) @ 50 g/m^2 in nursery bed (0.5 ton/ha) +

P. lilacinum @ 5 kg along with 2.5 ton of FYM /ha in the main field prior to planting tomato proved effective for suppressing soil nematode population and enhancing yield by 17.2-19.3%.

The integration of castor leaves with *P. lilacinum* increased the efficiency of nematode control (*M. javanica*) in tomatoes. The residual effect on the second crop of tomato revealed the reduction in root-knot index (29.6 to 356.8%) and J_2 population in soil (26.4 to 57.9%) (Zaki, 1998). *P. lilacinum* in combination with castor leaves reduced the nematode population up to 89% and increased plant growth and yield in tomatoes (Zaki and Bhatti, 1991).

Incorporation of neem cake (20 g/pot) along with *P. chlamydosporia* (10 ml/pot containing 4 x 10^5 spores/ml) gave maximum increase in plant growth and significant reduction in root galling and nematode population both in soil and roots. The above treatment also gave highest parasitization of eggs and egg masses of *M. incognita* and maximum propagule density of the bioagent in soil and roots (Rao *et al.*, 1998b) (Table 4.19).

Table 4.19. Effect of integration of *Pochonia chlamydosporia* with oil cakes on root galling and parasitization of *Meloidogyne incognita* by the bioagent on tomato.

Treatment	Dose/ plant	Root-knot index	% egg masses parasitized	% eggs parasitized	Root colonization (cfu/g)	Spore density in soil (cfu/g)
P. chlamydosporia	10 ml	2.5	56	60	5346	8732
Neem cake	10 g	2.8	---	---	---	---
Neem cake	20 g	2.5	---	---	---	---
Pongamia cake	10 g	3.0	---	---	---	---
Pongamia cake	20 g	2.8	---	---	---	---
Neem cake + *P. chlamydosporia*	10 g + 10 ml	1.7	69	75	7349	12346
Neem cake + *P. chlamydosporia*	20 g + 10 ml	1.5	73	78	8145	13043
Pongamia cake + *P. chlamydosporia*	10 g + 10 ml	2.3	61	63	5821	9353
Pongamia cake + *P. chlamydosporia*	20 g + 10 ml	2.0	62	65	6146	9801
Control	---	4.1	---	---	---	---
CD (P = 0.05)	---	0.13	6.4	7.1	496	1649

Application of *Pochonia chlamydosporia* (100 ml/seed pan containing 4 x 10^5 spores/ml) with neem leaves (150 g/seed pan) increased seedling weight and colonization of roots with the bioagent. Tomato seedlings raised in the above treatment transplanted in pots resulted in maximum increase in plant growth, least root galling, nematode population both in soil and roots and highest parasitization of eggs, egg masses and propagule density of *P. chlamydosporia* in roots (Parvatha Reddy *et al.*, 1999).

Incorporation of neem cake (20 g/pot) along with *P. chlamydosporia* (10 ml/pot containing 4 x 10^5 spores/ml) gave maximum increase in plant growth and significant reduction in root galling and nematode population both in soil and roots. The above treatment also gave highest parasitization of eggs and egg masses of *M. incognita* and maximum propagule density of the bioagent in soil and roots (Rao *et al.*, 1998b).

The efficacy of neem cake along with *P. fluorescens*, against root-knot and reniform nematodes (*Meloidogyne incognita* and *Rotylenchulus reniformis*) were evaluated on tomato. Application of *P. fluorescens* with neem cake helped in the increased root colonization of *P. fluorescens* in tomato. Root colonization of *P. fluorescens* was more when the PGPR was applied along with neem cake in comparison to its application along with vermi-compost, in tomato (Table 4.20). Both neem cake and vermicompost when enriched with *P. fluorescens* are able to support the growth and multiplication of this PGPR. However, when it is applied along with neem cake it has an increased bio-efficacy. Neem cake application is changing the root surface phenomenon and making it more favorable for higher colonization of this PGPR. It was observed that higher the root colonization of *P. fluorescens*, higher was its bio-efficacy (Rao, 2011).

Table 4.20. Effect of application of *P. fluorescens* enriched neem cake or vermi-compost on root population densities of *M. incognita* and *R. reniformis*, fruit yield and root colonization by the bioagent in tomato (mean of 5 replicates)

Treatments*	No. of *M. incognita*/ 10 g root	No. of *R. reniformis*/ 10 g root	Increase in fruit yield (%)	Root colonization of *P. fluorescens* (cfu/g×10^6)
T1	24.03	21.23	8	3.0
T2	15.58	14.62	10	6.3
T3	17.80	17.05	15	4.2
T4	12.46	21.52	26	8.3
T5	16.47	16.10	21	5.3
T6	23.35	21.23	6	0.0
T7	29.37	26.10	4	0.0
T8	20.47	18.79	12	3.6
T9	21.36	19.84	10	4.5
T10	44.50	34.80	-	0.0
CD – 5%	3.46	2.84	-	0.76

* T1-seedlings of tomato treated with *P. fluorescens* (CFU 2×10^8/ml); T2 - untreated seedlings sown in the plots mixed with neem cake enriched with *P. fluorescens* 20 g/m^2; T3 - untreated seedlings sown in the plots mixed with vermicompost enriched with *P. fluorescens* 100 g/m^2; T4- seedlings treated with *P. fluorescens* sown in plots mixed with neem cake enriched with *P. fluorescens*; T5- seedlings treated with *P. fluorescens* sown in plots mixed with vermicompost enriched with *P. fluorescens*; T6- Untreated seedlings transplanted in plots mixed with neem cake alone 20g/m^2; T7- Untreated seedlings transplanted in plots mixed with vermicompost alone 20g/m^2; T8- seedlings treated with *P. fluorescens* transplanted in plots mixed with neem cake alone 20g/m^2; T9- seedlings treated with *P. fluorescens* transplanted in plots mixed with vermicompost alone 20g/m^2; T10 - untreated seedlings transplanted in the plots without any treatment, served as control.

Integration of neem cake (40 g/plant) with *T. harzianum* (at 5 g/plant with 4 x 10^8 spores/g) was effective in increasing the plant growth and reducing root galling and final population of *M. incognita* infecting tomato. The above treatment also gave maximum reduction in number of eggs/egg mass and increased root colonization, spore density in soil and parasitization of adult females with *T. harzianum* (Table 4.21) (Rao *et al.*, 1997c). Similar results were also reported by Kumar and Khanna (2006).

Table 4.21. Effect of *Trichoderma harzianum* and neem cake on root galling, fecundity, root colonization and parasitization of adult females of *Meloidogyne incognita* on tomato

Treatment	Dose/ plant (g)	Gall index	No. of eggs/ egg mass	Root colonization of *T. harzianum* (cfu/g)	Spore density of *T. harzianum* in soil (cfu/g)	Parasitization (%) of adult females by *T. harzianum*
T. harzianum	5	7.6	376	12,578	9,473	25
Neem cake	20	7.0	344	---	---	---
Neem cake	40	6.5	325	---	---	---
T. harzianum + Neem cake	5+20	5.1	312	17,247	12,742	37
T. harzianum + Neem cake	5+40	4.7	304	23,742	17,438	52
Control	---	8.8	437	---	---	---
CD (P=0.05)	---	1.26	22.57	1,743.46	1,437.89	3.26

Combined application of *P. penetrans* + *T. viride* + neem/castor cake at 1/3 doses was significantly superior compared to their individual applications in terms of increased plant growth and reduced root galling, egg mass production and final populations of *M. incognita*. Similarly, total parasitization by bioagents was higher in plants treated with these bioagents in terms of number of juveniles encumbered and adult females infected with *P. penetrans,* and egg masses parasitized by *T. viride*, compared to their individual applications (Rangaswamy *et al.*, 1999).

Soil application of *Trichoderma viride* @ 20 g/m^2, mixed with Neem cake/FYM/ Vermicompost @ 100 g/m^2 in the beds was found promising for the management of root-knot nematodes (*Meloidogyne* spp.) in tomato.

Application of *Aspergillus niger* and *P. lilacinum* along with mustard cake gave maximum reduction in nematode population both in root and soil with enhanced plant vigor. *A. niger* being toxic agent, kills the second stage juveniles present in the rhizosphere; while *P. lilacinum* being the egg parasite, invade the eggs of *M. incognita* which escaped from the toxicity of *A. niger*. As a result, there is an overall reduction in root and soil population. Further, addition of mustard cake also helps to maintain the general plant health in addition to possessing nematicidal properties (Goswami *et al.*, 1998) (Table 4.22).

Table 4.22. Effect of *Aspergillus niger, Paecilomyces lilacinus* and mustard cake on biomass and multiplication of *Meloidogyne incognita* infecting tomato

Treatment	Biomass	No. of galls/ plant	No. of egg masses / plant	No. of eggs/ egg mass	Nematode popn./500 g soil
Aspergillus niger	9.8	34	18	272	470
Purpureocillium lilacinum	10.2	42	23	262	580
A. niger+ P. lilacinum	12.2	26	16	264	270
A. niger + *P. lilacinum* + Mustard cake	12.6	22	10	255	290
Control	9.0	98	72	320	2880
CD at 5%	2.33	7.24	3.24	27.18	162.80

Nursery bed treatment with *P. fluorescens* (with 1 x 10^9 spores /g) and *Pochonia chlamydosporia* (10 ml/pot containing 4 x 10^6 spores/ml) each at 20 g/m^2 and field application of 5 tons of FYM enriched with the above bioagents each at 5 kg, significantly reduced root-knot nematodes in tomato by 76% over untreated check. The yield increase was up to 21.7% with benefit: cost ratio of 4.9.

Application of PGPR (*Azotobacter chroococcum, Bacillus subtilis* and *Pseudomonas putida*) alone and in combination with cattle manure resulted in a significant increase in the growth of nematode-inoculated plants. The highest increase (79%) in the growth of nematode-inoculated plants was observed when *P. putida* was used with cattle manure. *P. putida* resulted in a high reduction in galling and nematode multiplication, followed by *B. subtilis* and *A. chroococcum*. *P. putida* with cattle manure was useful to achieve greater biocontrol of *M. incognita* on tomato (Siddiqui and Futai, 2009).

vii) Arbuscular mycorrhizal fungi and botanicals: In nursery, integration of neem cake (at 500 g/m^2) with *Glomus mosseae* (at 250 g/m^2 containing 16 chlamydospores/g) significantly reduced *M. incognita* population in soil, root galling, egg mass production, fecundity and produced vigorous tomato seedlings with increased root colonization with *G. mosseae*. The seedlings raised in the above treatment when planted in the main field significantly reduced root galling and increased fruit yield, root colonization with *G. mosseae* and chlamydospore population in soil (Table 4.23) (Parvatha Reddy *et al.*, 1998; Rao *et al.*, 1995).

Table 4.23. Effect of integration of *Glomus mosseae* with oil cakes on root galling, nematode reproduction, root colonization, spore density and yield of tomato infected with *Meloidogyne incognita*.

Treatment	Dose/ plant	Root gall index	Yield (kg)/ plant	% AMF colonization of roots	No. of chlamydospores /100 cm^3 soil
Control	---	4.4	1.44	---	---
Castor cake	50 g	3.9	2.00	---	---
Karanj cake	50 g	4.1	2.06	---	---
Neem cake	50 g	3.6	2.23	---	---
G. *mosseae*	500 cm^3 soil	3.4	2.19	29.5	146

Castor cake + *G. mosseae*	50 g + 500 cm^3 soil/m^3	3.4	2.53	38.7	208
Karanj cake + *G. mosseae*	50 g + 500 cm^3 soil/m^3	3.0	2.69	52.5	238
Neem cake + *G. mosseae*	50 g + 500 cm^3 soil/m^3	2.8	2.98	64.6	260
Carbofuran	0.1 g	3.0	2.33	---	---
CD (P = 0.05)	---	0.27	0.43		

Inoculation of *G. mosseae* or *G. fasciculatum* in the nursery beds amended with neem cake/castor cake/neem leaf/calotropis leaf reduced the infestation of root-knot and reniform nematodes to the maximum extent in tomato. Amendment of botanicals in the nursery beds increased the multiplication of endomycorrhizae and colonization of tomato root which in turn could protect the crop from these nematodes in the main field resulting in increased yields (Rao *et al.*, 1996; 1997).

Application of calotropis leaves at 400 g/ m^2 along with *G. fasciculatum* (250 g/ m^2 containing 25-30 chlamydospores/g) in nursery beds gave significant reduction in root galling and fecundity of *M. incognita*. The above treatment also gave maximum increase in plant growth and root colonization with *G. fasciculatum* (Rao *et al.*, 1996). Organic amendments (oil cakes, calotropis leaves) in combination with AMF enhance the colonization of AMF on tomato roots which further increased plant growth and reduced gall index.

viii) Bio-agents and arbuscular mycorrhizal fungi: Interestingly, integration of bio-agent (*P. lilacinum*) and endomycorrhizae *(G. mosseae/G. fasciculatum)* have culminated in the successful management of *M. incognita* infecting tomato. This phenomenon facilitated standardization of a strategy wherein inoculation of mycorrhizae and bio-agent in the nursery beds protected the seedlings of tomato from the attack of *M. incognita*. Further, these mycorrhizal seedlings (colonized either with *G. mosseae* or *G. fasciculatum*) can be given a root-dip treatment with spore suspension of *P. lilacinum* for 5-10 minutes for the effective management of these nematodes in the main field after transplanting (Rao *et al.*, 1993).

Integration of *G. deserticola* with *P. chlamydosporia* gave effective management of *M. incognita* in tomato, increased seedling weight, root colonization with *G. deserticola* and *P. chlamydosporia* and parasitization of eggs of *M. incognita* with *P. chlamydosporia* and reduced root galling, egg mass production and fecundity of root-knot nematodes (Table 4.24) (Rao *et al.*, 1997).

Table 4.24. Effect of integration of *Glomus deserticola* with *Pochonia chlamydosporia* on root galling, egg mass production, root colonization and egg parasitization on tomato infected with *Meloidogyne incognita.*

Treatment	No. of galls/10 seedlings	No. of egg masses/10 seedlings	% Root colonization by *G. deserticola*	Root colonization by *P. chlamydosporia* (cfu/g)	% Eggs parasitized by *P. chlamydosporia*
G. deserticola	67	38	42	---	---
P. chlamydosporia	82	63	---	10946	52
G. deserticola + *P. chlamydosporia*	61	34	39	11459	68

Control	84	67	---	---	---
CD (P = 0.05)	9.36	7.22	5.45	843.76	7.34

PGPR (*Bacillus* sp.and *B. polymyxa*) increased colonization of roots by AM fungi (*Glomus versiforme* and *G. mosseae*), and AM fungi increased numbers of PGPR in the rhizosphere. Dual inoculations of AM fungi plus PGPR provided greater control of *M. incognita* and greater promotion of plant growth than single inoculations, and the best combination was *G. mosseae* plus *Bacillus* sp. The results indicate that specific AM fungi and PGPR can stimulate each other and can interact to suppress root-knot disease development (Liu *et al.*, 2012).

ix) Bio-agents, arbuscular mycorrhizal fungi and botanicals: Inoculation of *G. mosseae* in neem leaf/neem cake amended nursery beds followed by the root-dip treatment of mycorrhizal seedlings of tomato in spore suspension of *P. lilacinum* gave effective management of *M. incognita* under field conditions (Rao *et al.*, 1995).

In nursery, integration of neem cake (at 500 g/m^2) with *G. fasciculatum* (200 g/m^2 containing 15 spores/g) gave maximum increase in plant growth and highest root colonization with *G. fasciculatum* and least root galling. Integration of neem cake (at 500 g/m^2) with *T. harzianum* (at 100 g/ m^2 with 4 x 10^8 spores/g) gave maximum parasitization of eggs with *T. harzianum*. In field, planting of tomato seedlings raised in nursery beds treated with neem cake + *G. fasciculatum* in pits incorporated with *T. harzianum* (at 0.5 g/ plant with 4 x 10^8 spores/g) was effective in increasing tomato fruit yield, root colonization with *G. fasciculatum* and egg parasitization with *T. harzianum* and significant reduction in root galling and multiplication rate of *M. incognita* (Table 4.25) (Parvatha Reddy *et al.*, 1998).

Table 4.25. Effect of integration of neem cake with *Trichoderma harzianum/ Glomus fasciculatum* on root galling, root colonization by *G. fasciculatum*, egg mass parasitization by *T. harzianum* and yield of tomato infected with *Meloidogyne incognita*.

Treatment	Dose/m^2	Root-knot index	% Root colonization by *G. fasciculatum*	Egg mass parasitization by *T. harzianum*	Yield (kg) /plant
Neem cake	1 kg	3.1	---	---	1.57
T. harzianum	100 g	3.2	---	24.6	1.43
G. fasciculatum	6000 spores	2.6	29.7	---	1.72
Neem cake + *T. harzianum*	500 g + 50 g	2.4	---	36.8	2.22
Neem cake + *G. fasciculatum*	500 g + 3000 spores	2.0	41.6	---	2.64
Control	---	4.2	---	---	1.10
CD (P = 0.05)	---	0.21	---	---	0.38

Seed treatment with talc-based formulation of *P. fluorescens* was able to increase the growth of the seedlings of tomato. Seed treatment with talc-based formulation of *P. fluorescens* and talc based or neem formulation of *P. fluorescens* with coco-peat treatment increased control of *M. incognita*. This resulted in the production of an increased number of nematode free seedlings of tomatoes. There was a significant increase in the production of nematode free tomato seedlings in *G. mosseae* with coco-peat treatment + seed treatment with talc-based formulation of *P. fluorescens* + neem-based formulation of *P. fluorescens* with coco-peat treatment. There appears to be synergistic interaction among *P. fluorescens*, *G. mosseae* and neem cake and complementing each other resulting in additive effect on the shoot and root growth. Neem cake helped in the increased root colonization of *P. fluorescens* and *G. mosseae* in tomato. When these seedlings were transplanted in the main field, there was significant reduction in the infestation of *M. incognita*, *M. javanica* (root-knot nematodes), *Rotylenchulus reniformis* (reniform nematode) on tomato (Table 4.26) (Gavaskar *et al.*, 2011).

Table 4.26. Effect of application of *P. fluorescens*, *G. mosseae* and coco peat on growth of tomato seedlings, infestation of *M. incognita* and root-colonization with bioagents (mean of 5 replicates)

Treatments*	Seedling height (cm)	% nematode free seedlings	Root colonization of *P. fluorescens* (cfu/g×10^6)	Root colonization of *G. mosseae* (%)
T1	19.04	87	1.56	-
T2	20.52	90	3.42	-
T3	21.60	91	4.82	-
T4	21.28	85	0.00	27
T5	23.85	92	2.25	-
T6	21.35	90	4.94	-
T7	22.86	92	6.39	-
T8	24.27	94	5.68	34
T9	25.16	97	8.26	42
T10	22.37	93	4.66	37
T11	24.46	94	23.76	40
T12	17.26	86	15.20	-
CD – 5%	2.66	-	1.72	-

*T1 - seed treatment with talc based formulation of *P. fluorescens*, T2 - talc based formulation of *P. fluorescens* with coco-peat treatment, T3 - neem based formulation of *P. fluorescens* with coco-peat treatment, T4 - *G. mosseae* with coco-peat treatment, T5 - seed treatment with talc based formulation of *P. fluorescens* + *G. mosseae* with coco-peat treatment, T6 - seed treatment with talc based formulation of *P. fluorescens* + talc based formulation of *P. fluorescens* with coco-peat treatment, T7 - seed treatment with talc based formulation of *P. fluorescens* + neem based formulation of *P. fluorescens* with coco-peat treatment, T8 - seed treatment with talc based formulation of *P. fluorescens* + talc based formulation of *P. fluorescens* with coco-peat treatment + *G. mosseae* with coco-peat treatment, T9 - seed treatment with talc based formulation of *P. fluorescens* + neem based formulation of *P. fluorescens* with coco-peat treatment + *G. mosseae* with coco-peat treatment, T10 - talc based formulation of *P. fluorescens* with coco-peat treatment + *G. mosseae* with coco-peat treatment, T11 - neem based formulation of *P. fluorescens* with coco-peat treatment + *G. mosseae* with coco-peat treatment, T12 - control without seed and coco-peat treatment.

x) Two bioagents: Tomato seedlings raised in nursery beds treated with *P. chlamydosopria* and *P. penetrans* when transplanted in the main field significantly lowered root-knot index, number of eggs per egg mass, nematode population in roots and soil; increased root colonization and egg parasitization with *P. chlamydosporia*, infection of *M. incognita* females with *P. penetrans*; and increased tomato fruit yield (Rao *et al.*, 1998).

Seed treatment with *Pseudomonas fluorescens* at 10 g/kg seed 24 hours before sowing and seed treatment with *Azospirillum* at 100 g/kg just before sowing in nursery and main field treatment with *P. fluorescens*, *Azospirillum* and *Phosphobacterium* each at 2 kg/ha was found effective in reducing nematode population in soil and root by 48.9% and 57.8%, respectively over control. Shanthi (2010) found that there was 58.9% increase in fruit yield and increased benefit: cost ratio to 4.2 by the above treatment.

Integration of two bio-agents, *P. lilacinum* and *P. chlamydosopria* resulted in combined and complimentary effects for the successful management of *M. incognita* infecting tomato (Rao and Parvatha Reddy, 1992c).

In nursery, integration of *P. chlamydosopria* (200 g/m^2 containing 16 chlamydospores/g) with *P. penetrans* at 1 g mixed with 100 g of sand/m^2 (containing 10^5 spores/10 mg) was effective in increasing the seedling weight and root colonization with *P. chlamydosopria* and adult females of *M. incognita* adequately infected with *P. penetrans*. The tomato seedlings raised in the above treatment when transplanted in the field gave maximum reduction in root galling, fecundity, nematode population both in soil and roots and increased root colonization with *P. chlamydosporia*, infection of *M. incognita* females with *P. penetrans* and tomato fruit yield (Table 4.27) (Rao *et al.*, 1998a).

Table 4.27. Effect of integration of *Pochonia chlamydosopria* with *Pateuria penetrans* on root galling, root colonization and egg parasitization by *P. chlamydosopria* and *Meloidogyne incognita* female parasitization by *P. penetrans* on tomato

Treatment	Root-knot index	Yield /6 m^2 (kg)	Root colonization by *P. chlamydosopria*	% eggs parasitized by *P. chlamydosopria*	% females parasitized by *P. penetrans*
P. chlamydosopria	5.8	7.4	32110	43	---
P. penetrans	6.6	6.3	---	---	27
P. chlamydosopria + *P. penetrans*	4.7	8.2	29260	57	25
Control	8.0	5.1	---	---	---
CD (P = 0.05)	0.78	0.88	2348	4.72	NS

Maheshwari and Mani (1988) also observed that the population of *M. incognita* and *M. javanica* were suppressed effectively and yields of tomato were greater when *P. penetrans* and *P. lilacinum* were applied together. Dubey and Smart (1987) noted greater nematode control (*M. incognita*) when *P. lilacinus* and *P. penetrans* were applied together.

The combination treatment with *T. harzianum* + *T. viride* each at 50 g (4 x 10^8 cfu/g) considerably increased the plant growth, yield and reduced the root galls and soil nematode population (Hassan and Sobita Devi, 2004).

Combined soil application of *P. lilacinum* and *Aspergillus niger* at the time of transplanting tomato is very effective in reducing root-knot nematodes (Goswami and Sharma, 1999). Similar findings were observed by Goswami and Mittal (2000) when *A. fumigatus* was applied to the soil along with *T. viride* on tomato infested with *M. incognita*. Here, in addition to the reduction in nematode population, improvement in plant growth was also observed.

P. penetrans has been used in combined applications with *Pochonia chlamydosporia* to control *M. incognita* on tomato (de Leij *et al.*, 1992). The bioagents gave maximum reduction in root galling on tomato seedlings.

Gautam *et al.* (1995) reported that combined application of *Bacillus subtilis* and *P. lilacinum* suppressed root-knot nematode population beyond application of agents individually. Biocontrol agents suppressed the numbers of root galls, females, eggs and second stage juveniles.

Two bacteria *Bacillus licheniformis* and *Pseudomonas mindocina* and 2 fungi Acrophialophora *fusispora* and *Aspergillus flavus* were evaluated for their efficacy as bio-control agents individually and in various combinations against *M. incognita* infecting tomato. Individually, *B. licheniformis* gave best control although when all 4 agents were used, nematode multiplication was reduced by 82.06% (Siddiqui and Husain, 1991).

Consortial formulation of biocontrol agents *viz.*, *Pseudomonas fluorescens* Pf 128 and *Bacillus subtilis* Bbv 57 recorded the highest defense enzymatic activity (peroxidase, polyphenol oxidase and phenylalanine ammonia lyase) and lowest nematode population in tomato roots compared to other strains either alone or in combination (Sankarimeena *et al.*, 2012).

The greatest increase in growth of nematode-inoculated tomato plants and reduction in nematode galling due to *M. incognita* was observed when *Paenibacillus polymyxa* was used with *P. lilacinum* or *P. chlamydosporia.*

xi) Bioagents and chemicals: *Pseudomonas fluorescens* applied in the nursery bed @ 20 g/m^2 along with Carbofuran @ 1.5 g a.i. /m^2 has shown promising results against root-knot nematode in bringing down the population and increasing the yield of tomato.

Maheshwari *et al.* (1987) reported that application of *P. penetrans* in combination with carbofuran, miral, sebufos and phorate significantly improved plant growth of tomato by greatly reducing galling due to *M. javanica.*

Somasekhar (1991) reported that application of both *P. penetrans* and carbofuran together at half their original doses (2.5 g of *P. penetrans* preparation per kg soil and the chemical at 3 kg a.i./ ha, respectively) had a synergistic effect on plant growth characters and reduction in root galling.

Integration of *P. chlamydosopria* with carbofuran recorded maximum plant growth parameters and minimum gall index and nematode population. The above treatment also recorded maximum number of fruits/plant and yield/plant. Maximum parasitization of *M. incognita* with *P. chlamydosopria* was also noticed when the bioagent and the chemical were integrated (Gopinatha *et al.*, 2002).

Oxamyl increased the efficacy of *P. penetrans* in trials against *M. javanica* infection of tomato crop and the effects on nematode control were additive (Tzortzakakis and Gowen, 1994).

Taba *et al.* (2006) combined the nematicide, fosthiazate with the nematode-trapping fungus, *Monacrosporium ellipsosporum* in a granular application, which effectively controlled *M. incognita* on tomato plants and established the fungus in soil.

xii) Botanicals and chemicals: Application of Carbofuran 3G @ 10 g or Phorate 10G @ 3 g/m^2 at seeding in the nursery beds and spot application of Carbofuran 3G @ 33 kg or Phorate 10G @ 10 kg + Neem cake @ 400 kg/ha at the base of plant 10 days after transplanting was found effective for the control of root-knot nematode.

Integration of chopped castor leaves (40 to 60 g/kg soil) 1 or 2 weeks before transplanting with application of carbofuran at 2 kg a.i./ha at transplanting, significantly reduced number of galls due to *M. javanica* and enhanced the growth of tomato (Dutt and Bhatti, 1986).

Integration of soil application of castor leaves with inorganic fertilizer (75 kg N/ha) enhanced plant growth of tomato and reduced *M. javanica* infestation (95% and 78% reduction in root galls and egg mass production, respectively) (Zaki and Bhatti, 1989).

In tomato, application of carbofuran at 1 kg a.i. per ha in combination with neem cake and urea each at 10 kg N per ha at transplanting, produced maximum yield with lowest gall index (2.5) and nematode population, 90 days after planting (Routaray and Sahoo, 1985).

Application of carbofuran at 1 kg a.i. /ha in the nursery beds followed by neem cake at 400 kg/ha in the main field increased yield and reduced the gall index (Singh and Gill, 1998) (Table 4.28).

Table 4.28. Integrated management of root-knot nematodes in tomato with carbofuran and neem cake.

Treatment/Dose	% Reduction in root galling	% Increase in yield
Carbofuran (1 kg a.i./ha) + Neem cake at 400 kg/ha	77.00	61.50
Carbofuran (1 kg a.i./ha) + Urea (23.8 kg/ha) + Neem cake at 200 kg/ha	67.90	49.20
Fenamiphos (1 kg a.i./ha) + Neem cake at 400 kg/ha	74.90	39.70
Fenamiphos (1 kg a.i./ha) + Urea (23.8 kg/ha) + Neem cake at 200 kg/ha	67.90	31.43

Integration of chopped castor leaves (40 to 60 g/kg soil) 1 or 2 weeks before transplanting with application of carbofuran at 2 kg a.i./ha at transplanting, significantly reduced number of galls due to *M. javanica* and enhanced the growth of tomato (Ravi Dutt and Bhatti, 1986).

xiii) Cultural and chemicals: Adoption of INM technology involving summer plowing and fallowing, growing non-host crops like wheat, maize, sorghum, bajra; intercrops like marigold, sun hemp; green manuring with sun hemp; application of FYM and Neem cake; and use of nematicides like Phorate 10G and Carbofuran 3G proved beneficial.

Combinations of deep plowing (up to 20 cm) and nursery bed treatment with carbofuran at 0.4 g per m^2 and main field treatment with carbofuran at 1 kg a.i./ha proved effective in the control of root-knot nematodes in tomato which also registered maximum yield (102.6% higher yield over control) (Jain and Bhatti, 1985).

xiv) Physical and botanicals: Integration of soil solarization of nursery beds (with LLDPE transparent film of 25 μ for 15 days) with soil incorporation of Calotropis leaves at 4 kg/1.44 m^2 resulted in increased plant height (29.9 cm compared to 20 cm in control), fresh shoot weight (176.49g compared to 100g in control), transplantable seedlings (546/ m^2 compared to 297 in control) with minimum root-knot index (1.83 compared to 3.85 in control) (Patel *et al.*, 2006). Soil solarization with clear LLDPE film (25μ) for 15 days in hot summer in combination with poultry manure at 2 t/ha proved effective in the management of nematodes and higher production of transplants in tomato nursery with cost benefit ratio of 1: 3.10.

xv) Biofumigation and grafting: For 2007 autumn and 2008 spring, root galling caused by *M. incognita* was lower (1.90 in autumn and 2.95 in spring) in the grafted tomato plants on the rootstock Beaufort than the other treatments. In terms of yield, among the tested treatments, grafting (0.922 kg/plant in autumn and 1.785 kg/plant in spring) and biofumigation with *R. communis* in autumn (0.924 kg/plant) and biofumigation with broccoli in the spring (2.315 kg/plant) gave higher yields.

The overall results showed that biofumigation with broccoli and grafting could be used efficiently against *M. incognita* in organic agriculture production.

The treatment of grafting with Beaufort was placed statistically ($P<0.05$) in a different group compared to the untreated plots, reducing dramatically root galling. However, the highest root galling indices were found in the soil solarization treatment (4.63±0.45) in autumn and in the control plots (8.08±0.34) in the spring season (Table 4.29).

Table 4.29. Effect of different treatments on root galling of tomato plants infected with *Meloidogyne incognita.*

Treatment	Galling index (1-10 scale)	
	2007 autumn	2008 spring
Control	3.50 ± 0.32 b*	8.08 ± 0.34 c
Grafted plants	1.90 ± 0.19 a	2.95 ± 0.36 a
Biofumigation with Broccoli	3.55 ± 0.43 b	4.33 ± 0.36 b
Tagetes erecta	2.28 ± 0.29 a	6.70 ± 0.43 c
Soil solarization	4.63 ± 0.45 c	---
Biofumigation with castor	2.83 ± 0.33 ab	---
BioAct	---	7.48 ± 0.39 c
Nestin	---	7.25 ± 0.27 c

* Standard error of mean is given after ±

In terms of the reproduction ratio, there was a difference between the seasons. In autumn, the lowest reproduction ratio of root-knot nematodes in the soil was assessed from the treatments of biofumigation with *R. communis* (0.42) and broccoli (0.67), followed by control (1.37) and grafting (1.94). However, with the root galling index evaluation, the highest reproduction ratio was obtained in the soil solarization plots (3.40)

In the spring, the reproduction ratio of root-knot nematodes in the soil was the lowest with the treatment of BioAct (0.36) and followed by grafting (1.32) and Netisin (1.36) whereas the highest ratio was in the control plots (6.62) (Table 4.30).

Table 4.30. The effect of different treatments on reproduction ratio of *Meloidogyne incognita* in 100 cm^3 soil

Treatment	2007 autumn			2008 spring		
	Pi	Pf	Rr	Pi	Pf	Rr
Control	465 b	635 b	1.37	65 a	430 a	6.62
Grafted plants	80 a	155 a	1.94	155 a	255 a	1.32
Biofumigation with Broccoli	455 b	305 ab	0.67	215 a	455 a	2.12
Tagetes erecta	80 a	260 ab	3.25	90 a	205 a	2.28
Soil solarization	100 a	340 ab	3.40	---	---	---
Biofumigation with castor	455 b	190 b	0.42	---	---	---

BioAct	---	---	---	290 a	105 a	0.36
Nestin	---	---	---	175 a	175 a	1.34

Reproduction ratio (Rr) = Final population (Pf)/ Initial population (Pi)

There was a significant difference in terms of yield between the seasons. In autumn, the tomato yield recorded from all treatments was statistically placed at the same significance level ($P<0.05$). The lowest tomato yield was obtained from the *T. erecta* plots (0.771±0.15 kg/plant), however, the highest yield was in the control treatment (0.989±0.81 kg/plant), followed by the grafting (0.922±0.76 kg/plant) and biofumigation with *R. communis* plots (0.924±0.90 kg/plant) ($P<0.05$). While the yield was the lowest in the control (1.118±0.21 kg/plant) in the spring, placing statistically ($P<0.05$) in a different group compared to all other treatments. The highest yield was recorded only from the biofumigation with broccoli (2.315±0.11 kg/plant) followed by the grafting (1.785±0.14 kg/plant) at a statistically different significance level ($P<0.05$) (Table 4.31).

Table 4.31. The effect of different treatments on yield assessment of tomato plants

Treatment	**Tomato fruit yield (kg/plant)**	
	2007 autumn	**2008 spring**
Control	0.989 ± 0.81 a	1.118 ± 0.21 c
Grafted plants	0.922 ± 0.76 a	1.785 ± 0.14 ab
Biofumigation with Broccoli	0.785 ± 0.15 a	2.315 ± 0.11 a
Tagetes erecta	0.771 ± 0.15 a	1.487 ± 0.27 bc
Soil solarization	0.780 ± 0.13 a	---
Biofumigation with castor	0.924 ± 0.90 a	---
BioAct	---	1.437 ± 0.21 bc
Nestin	---	1.592 ± 0.14 bc

* Standard error of mean is given after ±

In conclusion, grafting of tomato on Beaufort rootstock or biofumigation with broccoli-controlled *M. incognita* and increased tomato yield to some extent. Therefore, these treatments could be used for root-knot nematode control in organic greenhouse tomato production.

xvi) Biofumigation and solarization: Fewer root-knot nematode eggs were recovered on the tomatoes grown in solarized soil than in the non-solarized soil. Fewer root-knot eggs (*M. javanica*) were found in the soil amended with Dwarf Essex (*Brassica napus*) than with Aspen (*B. napus*). The tomato plants grown in the rapeseed amended soil were larger than those grown in the unamended soil.

Root-knot nematodes were generally suppressed by the incorporation of any organic material whether *Brassica* or not, and suppression was often unrelated to GSL content. *Brassica* tissues rarely outperformed other forms of green manure unless experimental conditions favored the impacts of volatile compounds such as in sealed containers. In an experiment in which the solid, liquid and vapor phases were separated and applied individually or in combination; revealed that all these components had activity against nematodes. A significant outcome of the nematode work was the revelation that among brassicas only some radish lines (*Raphanus sativus*) were poor hosts of the nematodes and during their growth could reduce populations to a similar extent as traditional sorghum green manures (Wang *et al.*, 2008).

Tsror *et al.* (2006) reported much improved control of *Meloidogyne* spp. on tomato by combining solarization with biofumigation compared to solarization alone.

xvii) Biofumigation, botanicals and trap cropping: In commercial greenhouse trials in Spain, an integrated management system was developed, including biofumigation with sheep manure and mushroom residue and the cultivation of short-cycle vegetables acting as trap crops. Using this strategy, initial very high levels of *M. incognita* were reduced to near zero in the main susceptible tomato crop (Bello, 1998).

4.2. INTERACTION OF ROOT-KNOT NEMATODE WITH FUSARIUM WILT

The root-knot nematode, *Meloidogyne incognita* and Fusarium wilt, *Fusarium oxysporum* f. sp. *lycopersici/F. solani* are frequently associated with tomato resulting in considerable damage to the crop.

4.2.1. Interaction

Jenkins and Coursen (1957) induced wilting in *Fusarium* wilt-resistant tomato variety 'Chesapeake' only when root-knot nematodes were present along with fungal inoculum. Furthermore, when *M. hapla* was combined with the fungus, only 60% of the plants wilted, whereas *M. incognita acrita* promoted wilt in 100% of the plants.

Bhagawati and Goswami (2000) reported the interaction of *M. incognita* and *Fusarium oxysporum* f. sp. *lycopersici* on tomato. It was found that when either or both pathogens were inoculated simultaneously or nematode was inoculated 10 days prior to inoculation of the fungus, the symptoms were visible by 20 days as against no such symptoms in plants where fungus was inoculated 10 days prior to nematodes. Nematode inoculation prior to fungus and simultaneous inoculation of both the pathogens resulted in significantly higher intensity of wilt after 40-60 days of inoculation. The number of galls and egg masses and nematode population in soil was significantly reduced when simultaneously inoculated or prior inoculation of nematodes.

Young plants inoculated with both the pathogens collapsed earlier before crop maturity. The nematode, by impairing water and nutrient availability, disrupts root function and plant growth processes. These effects combined with the vascular blocking due to the wilt fungi can be particularly severe, and if widespread, result in total crop failure. Giant cells in root galls produced by the root-knot nematode provide the substrate suitable for rapid colonization of the fungal pathogen. Root-knot nematodes can thus significantly enhance disease development and yield loss, elevating primary, or secondary pathogens to major pest status.

The effect of the nematode in combination with the fungus enhanced the suppression of plant growth than that of the fungus alone. Inoculation of the nematode and fungus exhibited a synergistic effect on growth retardation of plants. Maximum reduction in plant height (33.08 cm) was observed when nematode and fungus were inoculated simultaneously.

Inoculation of *M. incognita* juveniles prior to *F. oxysporum* f. sp. *lycopersici* inoculation advanced the wilt of tomato by one week, while nematode inoculation after fungus did not increase wilting (Table 4.32) (Nagesh *et al.*, 2006).

Table 4.32. Progress of Fusarium wilt in the presence of *M. incognita*

Treatment	Wilt index at weekly intervals starting 10 DAP					
	1	2	3	4	5	6
M. incognita (*Mi*)	0	0	0	0	0	0
F. oxysporum f. sp. *lycopersici* (*Fo*)	0	0	+	++	++	-
Mi + *Fo* (*Mi* inoculated 7 days prior to *Fo*)	0	+	+	++	++	-
Fo + *Mi* (*Fo* inoculated 7 days prior to *Mi*)	0	0	+	++	++	-

+ = Wilting of young leaves, ++ = Wilting of whole plant, - = Dead plant

4.2.2. Management

4.2.2.1. Chemical methods: Application of Metham sodium through drip irrigation was found effective against root-knot and *Fusarium* sp. complex of tomato (Roberts *et al.*, 1988).

4.2.2.2. Biological methods

i) Antagonistic bacteria: Root infection caused by *M. javanica* and the soil-borne root-infecting fungi *F. oxysporum* and *F. solani* was effectively suppressed following application of *Pseudomonas aeruginosa.* Biocontrol and growth promoting potential of the bacterium was enhanced when soil was kept at 50% or 75% moisture holding capacity. Siddiqui and Ehteshamul-Haque (2001) found that increase in plant growth was optimal at an inoculum level 2.5×10^8 cfu/ml of the bioagent, *P. aeruginosa.*

ii) Antagonistic fungi: Soil application of *T. harzianum* at 50 kg/ha (2×10^8 cfu/g) was effective in improving the plant growth parameters and fruit weight followed by *P. fluorescens*, Carbofuran, *A. niger*, *T. virens*, and Neem seed powder. Carbofuran was found highly effective against *M. incognita*, Topsin-M against *F. solani* and *T. harzianum* was effective against both the pathogens (Table 4.33) (Haseeb *et al.*, 2006).

Table 4.33. Effect of bioagents, botanicals and chemicals for the management of disease complex in tomato

Treatment	Fruit weight (g)	Reproduction factor	Root-knot index	% wilt incidence
Uninoculated control	230	---	---	---
Inoculated control	103	8.8	2.6	50
Neem seed powder	180	2.8	1.0	15
Neem leaves	155	4.5	1.5	23
Murraya leaves	150	5.5	1.7	23
P. fluorescens	198	4.4	4.5	9
T. harzianum	203	3.3	1.2	6
T. virens	185	5.1	1.6	12
P. lilacinus	172	1.4	0.7	18
A. niger	195	4.3	1.5	11
Farm yard manure	130	8.4	2.5	43

Mint manure	142	6.9	2.0	30
Carbofuran	197	1.1	0.6	10
Topsin-M	173	8.0	2.3	5
CD (P = 0.05)	8.639	0.22	0.12	1.03

Paenibacillus polymyxa GBR-462, GBR-508 and *P. lentimorbus* GBR-158 strains showed the strongest antifungal and nematicidal activities. These three strains used in pot experiment reduced the symptom development of the disease complex caused by *M. incognita* and *F. oxysporum* f. sp. *lycopersici* on tomato (wilting and plant death), and increased plant growth. The control effects were estimated to be 90–98%, and reduced root gall formation by 64–88% compared to the untreated control (Table 4.34) (Son *et al.*, 2009).

Table 4.34. Integrated management of root-knot nematode with Fusarium wilt disease complex

Treatments	**Wilt severity Index***	**CE (%)**	**Gall severity Index****	**CE (%)**
FO+MI	4.1a	-	5.0a	-
FO+MI+GBR-508	0.1b	97.6	0.6c	88.0
FO+MI+GBR-462	0.3b	92.7	1.0b	78.0
FO+MI+GBR-158	0.4b	90.2	1.8b	64.0
Untreated	0.0	-	0.0	-

In a column, means followed by a common letter are not significantly different by Duncan's multiple range test at P = 0.05.

CE = Control effect.

* Wilt severity was graded using a 0–5 scale, where 0 = no visible symptoms; 1 = epinasty and chlorosis/wilting of primary leaves; 2 = chlorosis/wilting of second and third leaves, primary leaves may be lost; 3 = wilting above third leaves, second and third leaves may be lost; 4 = wilting of whole plant, and 5 = plant died.

**Severity of root galling was assessed on a 0–5 rating scale, in which 0 = 0–10% of galled roots; 1 = 11–20%; 2 = 21–50%; 3 = 51–80%; 4 = 81–90%; and 5 = 91–100%.

4.2.2.3. Host resistance: Abawi and Barker (1984) found that the monogenic resistance in tomato cvs. (Florida, MH-1, and Manapal) against *F. oxysporum* f. sp. *lycopersici* race 1 remained unaffected in the presence of *M. incognita*.

4.2.2.4. Integrated methods

i) Bioagents and botanicals: *T. harzianum* controlled wilt in the presence of *M. incognita* for six weeks after transplanting. Tomato roots that received *P. lilacinum* along with *T. harzianum* and Neem cake were free from root-knot nematodes (*M. incognita*) and did not wilt due to *F. o.* f. sp. *lycopersici* till harvest. The roots were also free from *Fusarium* infection. The above treatment also reduced the % of wilt (10% compared to 90% in control), root-knot index (1.8 compared to 4.4 in control) and increased root colonization with bioagents (74% compared to 0% in control), parasitization of egg masses (56% compared to 0% in control) and eggs/egg mass (44% compared to 0% in control) (Table 4.35) (Nagesh *et al.*, 2006).

Table 4.35. Effect of integration neem cake and bioagents on the management of *Meloidogyne incognita* infecting tomato

Treatment	Healthy plants (%)	Root-knot index	Colonization by bioagents		
			Root Colonization (%)	% Colonization of egg masses/eggs	
				egg masses	eggs
Control	10	4.4	---	---	---
P. lilacinum	30	3.6	48	48	46
T. harzianum	40	4.0	50	44	32
Neem cake	10	4.2	---	---	---
P. lilacinum + Neem cake	60	2.8	62	62	58
T. harzianum + Neem cake	70	3.8	68	36	32
P. lilacinum + *T. harzianum* + Neem cake	90	1.8	74	56	44
CD (P=0.05)	8.87	1.78	3.66	4.11	3.22

ii) Bioagents and cultural: Deep plowing and exposing soil to hot sun in summer, removal and burning of crop debris, soil application of *T. viride* and *P. lilacinum* help in controlling the disease complex.

iii) Bioagents, cultural and host resistance: Deep plowing and exposing soil to hot sun in summer, removal and burning of crop debris, soil application of *T. viride* and *P. lilacinum*, use of wilt resistant varieties like Utkal Pallavi, Utkal Deepti, Utkal Kumari, Utkal Urbasi, *etc.* help in controlling the disease complex.

iv) Arbuscular mycorrhizal fungi and botanicals: Combined application of the endomycorrhiza, *Glomus etunicatum* along with mustard cake was found more effective in managing the disease complex caused by *M. incognita* and *F. oxysporum* f. sp. *lycopersici* on tomato as compared to their individual application (Bhagawati *et al.*, 2000).

v) Two bioagents: Efficacy of *Pseudomonas aeruginosa* alone or in combination with *Purpureocillium lilacinum* was evaluated for the control of root-knot nematode and root-infecting fungi under field conditions. The efficacy of *P. aeruginosa* alone or in combination with *P. lilacinum* significantly suppressed soil-borne root-infecting fungi such as *Macrophomina phaseolina, Fusarium oxysporum, Fusarium solani, Rhizoctonia solani* and the root-knot nematode, *Meloidogyne javanica*. Siddiqui *et al.* (2000) reported that the parasitization of eggs and females of *M. javanica* was significantly better with *P. lilacinum* than *P. aeruginosa*.

4.3. INTERACTION OF ROOT-KNOT NEMATODE WITH ROOT ROT

4.3.1. Interaction

Abu-Elamayem *et al.* (1978) observed that damping-off of tomatoes was more severe in soil infested with both *M. javanica* and *R. solani* than with the fungus alone. *M. javanica* increased the extent of damage by pre- and post-emergence phases of damping off caused initially by *R. solani* in tomato (Nath *et al.*, 1984).

Plants in untreated field soil or in sterilized soil inoculated with both organisms, developed a root rot in about 42 days. If the nematode preceded the fungus by three weeks, the root rot was more severe and appeared within 14-21 days. The fungus penetrated either directly or through ruptures in the root created by the mature female nematode. *R. solani* colonized nematode giant cells and root xylem cells. Vascular discoloration occurred both in roots and stem, however, no fungus was isolated from stems.

M. incognita predisposed roots to *R. solani* which resulted in severe root rot and subsequent plant death. Tomato plants inoculated with either *R. solani* or *M. incognita* alone were free from root decay for the entire period of six-week study. Three weeks after nematode and fungus inoculation, black sclerotia of *R. solani* were visible on nematode- induced galls, while non-galled portions on the same root systems were free of sclerotia. Prior to root rot development, *R. solani* demonstrated marked preference for root gulls on nematode infected roots. It is hypothesized that the leakage of nutrients from the root was responsible for attracting the fungus to the galls and for initiating sclerotial formation.

Five weeks after inoculation, distinct brown lesions were observed only on the galls of plants inoculated with both *M. incognita* and *R. solani*. The lower leaves of plants were chlorotic and suffered premature leaf drop.

M. incognita and *R. solani* are also frequently associated with tomatoes causing considerably greater damage to this crop (Haseeb, 2003). Highest reduction in fruit yield, plant height, fresh and dry weight was observed in plants inoculated with nematode-fungus simultaneously followed by nematode 7 days prior to fungus (Table 4.36) (Kumar and Haseeb, 2009). The fact that prior inoculation of nematode caused more damage indicates that the roots are predisposed by *M. incognita* for subsequent damage by *R. solani*.

Table 4.36. Interaction of *Meloidogyne incognita* and *Rhizoctonia solani* on plant growth and fruit yield of tomato cv. K-25.

Treatment	Fruit yield (kg/plant)	Plant height (cm)	Fresh plant weight (g)	Dry plant weight (g)
Untreated uninoculated	2.530	67.5	200.5	38.8
M. incognita (4000 J_2/4 kg soil)	1.280	50.8	155.0	29.2
R. solani (10g mycelial mat/4 kg soil)	1.440	53.7	168.0	31.7
M. incognita + *R. solani* simultaneously	0.487	26.7	94.8	17.6
M. incognita 7 days prior to *R. solani*	0.658	31.8	112.0	21.0
R. solani 7 days prior to *M. incognita*	0.821	40.0	124.7	23.4
CD (P = 0.05)	---	2.41	7.68	1.46

The highest reproduction rate (14.9) and root-knot index (3.5) was observed in plants inoculated with nematode alone followed by nematode inoculation 7 days prior to the fungus (Table 4.37). The highest root infection by the fungus was observed in simultaneous inoculation followed by nematode inoculation 7 days prior to the fungus (Kumar and Haseeb, 2009).

Table 4.37. Effect of *Meloidogyne incognita* and *Rhizoctonia solani* on nematode multiplication, root galling and fungus infection on tomato cv. K-25.

Treatment	Final nema popn. (soil + roots)	Reproduction factor	Root-knot index	% root infection by R. solani
Untreated uninoculated	---	---	---	---
M. incognita (4000 J_2/4 kg soil)	59,500	14.9	3.50	---
R. solani (10g mycelial mat/4 kg soil)	---	---	---	30.0
M. incognita + *R. solani* simultaneously	38,700	9.7	2.00	63.5
M. incognita 7 days prior to *R. solani*	45,550	11.4	2.50	55.0
R. solani 7 days prior to *M. incognita*	35,850	9.0	1.35	48.5
CD (P = 0.05)	2808.55	0.64	0.13	2.65

4.3.2. Management

4.3.2.1. Chemical methods: Soil fumigation usually decreased disease severity due to *R. solani* and *M. incognita* disease complex in tomatoes (Golden and Van Gundy, 1975).

Abu-Elamayem *et al.* (1978) reported that the combined use of CGA 12223 and benomyl provided significant reduction in the disease intensity.

4.3.2.2. Biological methods

i) Antagonistic bacteria: Root infection caused by *M. javanica* and the soil-borne root-infecting fungus *R. solani* was effectively suppressed following application of *Pseudomonas aeruginosa*. Biocontrol and growth promoting potential of the bacterium was enhanced when soil was kept at 50% or 75% moisture holding capacity. An inoculum level 2.5 × 10^8 cfu/ml of *P. aeruginosa* was optimal for the enhancement of plant growth (Siddiqui and Ehteshamul-Haque, 2001).

ii) Antagonistic fungi: *Trichothecium roseum* was found most effective in mitigating the adverse effect of *R. solani* and *M. incognita* followed by *Trichoderma viride*. The effect of *T. roseum* was more pronounced when the plants were inoculated both at seedling emergence and transplantation stages (Arya and Saxena, 2009).

4.3.2.3. Integrated methods

i) Two bioagents: Efficacy of *Pseudomonas aeruginosa* alone or in combination with *Purpureocellium lilacinum* was evaluated for the control of root-knot nematode and root-infecting fungi under field conditions. The biocontrol fungus and the bacterium significantly suppressed soil-borne root-infecting fungi such as *Macrophomina phaseolina, F. oxysporum, F. solani, R. solani,* and the root-knot nematode, *M. javanica*. *P. lilacinum* parasitized eggs and females of *M. javanica* and this parasitism was not significantly influenced in the presence of *P. aeruginosa* (Siddiqui *et al.*, 2000).

P. aeruginosa-Bacillus subtilis treatment was the most effective in the suppression of root-rot disease complex with enhancement of plant growth (Siddiqui and Ehteshamul- Haque, 2001).

4.4. INTERACTION OF ROOT-KNOT NEMATODE WITH BACTERIAL WILT

4.4.1. Interaction

Pani and Das (1972) have reported the association of root-knot nematode with bacterial wilt of tomato. Haider *et al.* (1987) reported the significant reduction in the root-knot index and larval development of *M. incognita* in soil where *Ralstonia solanacearum* was present. *R. solanacearum* and *M. incognita* alone as well as in different combinations reduced plant growth and yield significantly with the nematode followed by the bacterium combination showing the maximum reduction in growth.

Disease severity was greatly increased when wounding was done just before inoculation with *R. solanacearum* and decreased when wounding was carried out six days before inoculation (Abd-El-Ghafar and Abd-El-Kader, 1997).

Napiere (1980) and Napiere and Quinio (1980) found that wilt disease development occurred earlier and with a higher mortality rate in both wilt-resistant and susceptible tomato cultivars grown in *R. solanacearum* and *M. incognita* – infested soil.

4.4.2. Management

4.4.2.1. Integrated methods

i) Bioagents and botanicals: Treatment of nursery bed with *Pseudomonas fluorescens* (10^9 cfu/g) and *Trichoderma harzianum* (10^6 cfu/g) each at the rate of 20 g/m^2 and subsequent application of 5 tons of farmyard manure enriched with 5 kg each of *P. fluorescens* (10^9 cfu/g) and *P. lilacinum* (10^6 cfu/g) per hectare, significantly reduced *R. reniformis* and *M. incognita* in tomato roots by 74% and 70%, respectively; reduced the incidence of bacterial wilt; and increased the yield by 24.2%. Rao *et al.* (2009) reported that the benefit: cost ratio worked out to be 4.4 when both the bioagents used for the management of the disease complex.

4.5. RENIFORM NEMATODE, *ROTYLENCHULUS RENIFORMIS*

4.5.1. Crop losses

R. reniformis was responsible for 42.25 to 49.02% loss in fruit yield of tomato (Subramanyam *et al.*, 1990).

4.5.2. Symptoms

General symptoms include reduced root system, leaf chlorosis, overall stunting of host plants, reduced yields, and plant longevity. Female nematodes and their eggs are often visible when plant roots are viewed under a dissecting microscope.

4.5.3. Management

4.5.3.1. Cultural methods: Soil amendment with fresh glyricidia leaves, sawdust and chicken manure controlled the root-knot and reniform nematode population on tomato (Castillo, 1985). The yield increases and the nematode suppressive effect of fresh chicken manure surpassed even fenamiphos (Table 4.38).

Table 4.38. Effect of soil amendment with organic materials on *M. incognita* and *R. reniformis* population and tomato fruit yield.

Treatment	No. of nematodes/300 ml soil & 1 g roots			Yield (t/ha)
	No. of weeks after treatment			
	2	6	11	
Fenamiphos at 10 kg a.i./ha	96	366	465	31.1
Rice straw at 10 t/ha	215	500	1921	29.1
Glyricidia leaves at 10 t/ha	201	533	1885	27.3
Saw dust at 10 t/ha	342	859	958	25.3
Chicken manure at 10 t/ha	17	169	1078	32.6
Control	552	816	2515	22.2

4.5.3.2. Chemical methods: Application of metham sodium (250 l/ha) or methomyl/ thionazin (4-16 kg/ha) was effective in reducing the reniform nematode population and in increasing tomato yields (Sivakumar *et al.*, 1979).

Reddy and Seshadri (1972) reported that thionazin at 4 kg a.i. per ha proved effective against *R. reniformis* infecting tomato.

4.5.3.3. Biological methods

i) Antagonistic bacteria: Multiplication rate of *R. reniformis* was reduced significantly (44.5%) when soil drenching with *B. subtilis* (isolate Bs_t) was done a week before nematode inoculation. But it was not so in the case of treatments that received simultaneous inoculations of the bacterium and the nematode. These results indicated that *B. subtilis* could induce systemic resistance in tomato against *R. reniformis* (Niknam and Dhawan, 2001a).

ii) Antagonistic fungi: The green muscardine fungus, *Metarhizium anisopliae* at 9 g/kg soil gave significant increase in plant growth parameters and root colonization with the bioagent and reduction in number of females, egg mass production, eggs per egg mass and final nematode population in soil (Tribhuvaneshwar *et al.*, 2008).

iii) Arbuscular mycorrhizal fungi: The arbuscular mycorrhizal fungus (AMF) *Glomus fasciculatum* increased the resistance in tomato plants against *R. reniformis* infestation. Inoculation of tomato transplants with the fungus significantly reduced the juvenile penetration and development on AMF treated plants than control (Sitaramaiah and Sikora, 1982). The development of gelatinous matrix was delayed and few eggs/egg mass were produced. Data showed that *G. fasciculatum* adversely affected *R. reniformis* during several phases of its life cycle.

4.5.3.4. Host resistance: Balasubramanian and Ramakrishnan (1983) observed that Kalyanpur Selection-I, Kalyanpur Selection -III, and CA-121 exhibited resistant reaction.

4.5.3.5. Integrated methods

i) Bioagents and chemicals: Integration of a bio-agent, *P. lilacinum* with carbofuran at 1 kg a.i./ha was found effective for the management of reniform nematode, *R. reniformis* infecting tomato (Table 4.39) (Parvatha Reddy and Khan, 1988).

Table 4.39. Effect of *Purpureocillium lilacinum* and carbofuran on *Rotylenchulus reniformis* infecting tomato

Treatment	Dose	Mature females in root/plant	Total nema population/ plant	Reproduction factor	% males
P. lilacinum	0.5 g/plant	29	3893	3.8	45.2
P. lilacinum	1.0 g/plant	9	3257	3.3	53.1
P. lilacinum	2.0 g/plant	14	2232	2.2	49.2
P. lilacinum + Carbofuran	2.0 g/plant + 2 kg a.i./ha	9	1537	1.5	56.1
Carbofuran	2 kg a.i./ha	8	1160	1.2	65.3
Control	---	57	6669	6.7	42.8
CD (P=0.05)	---	5.8	2011	---	15.9

ii) Bioagents and Botanicals: Nursery bed treatment with *P. fluorescens* (with 1 x 10^9 spores/g) and *P. chlamydosporia* (with 1 x 10^6 spores/g) each at 20 g/ m^2 and field application of 5 tons of enriched FYM with the above bioagents each at 5 kg significantly reduced reniform nematodes in tomato by 72% over control. The yield increase was up to 21.7% with benefit: cost ratio of 4.9.

iii) Botanicals and chemicals: Application of neem cake in the nursery at 100 g/m^2 followed by carbofuran at 1 kg a.i./ha in the main field significantly reduced the soil population of *R. reniformis* and enhanced the fruit yield of tomato by 67% (Anitha and Subramanian, 1998).

4.6. STUBBY ROOT NEMATODE, *PARATRICHODORUS MINOR*

4.6.1. Distribution

From India, it has been reported from Delhi and Uttar Pradesh (Prasad and Dasgupta, 1964; Siddiqi, 1962).

4.6.2. Symptoms

Stubby-root nematodes feed on the roots of tomato plants and cause the roots to appear stunted and stubby. They feed primarily on root tips, causing elongation of roots to stop, and making it more difficult for tomatoes to receive adequate water and nutrients. Symptoms of stubby-root nematodes are more severe in sandy soils and generally appear as odd-shaped patches.

4.6.3. Host range

P. minor has been reported from the rhizosphere of apple, citrus, sugarcane, potato, vegetables, cereals, rose and fruits from India.

4.6.4. Life cycle

Males are very rare, and reproduction is evidently by parthenogenesis. The life cycle on tomato is completed in 21-22 days at 22 °C and in 16-17 days at 30 °C with a tenfold or more increase in 60 days (Rhoade and Jenkins, 1957).

4.6.5. Management

4.6.4.1. Physical methods: Soil solarization using clear plastic mulch or photo selective (black) polyethylene film reduce populations of *P. minor* on tomato as effectively as fumigation with a mixture of methyl bromide and chloropicrin in warm climates.

4.6.4.2. Cultural methods: *Asparagus officinalis* would not support populations of *P. minor* for more than 40 to 50 days. Tomato, normally a good host of this nematode, supported only a low population when asparagus was growing in the same pots. Glycoside (asparagusic acid) is the active principle involved which is toxic to *P. minor* and several other nematode species.

4.7. CONCLUSION

Plant parasitic nematodes constitute one of the major limiting factors for cultivation of tomatoes. Several nematode species are associated with tomatoes and few of them negatively impact yield. Severe yield losses have been reported in most regions where tomatoes are grown. To select suitable nematode management practices, correct diagnosis of the causal agent is of prime importance. Overall, these nematode problems in tomato are better managed when integrated management practices are used, i.e. regulatory methods (plant quarantine and seed certification), cultural practices (crop rotation, cover crops), genetic control, and ultimately by the use of nematicides. Therefore, for a sustainable cropping of tomato growers, extension services, and researchers must consider these nematodes holistically, the impact they cause, and whether these management practices are economically, environmentally and technically sound.

Integrated nematode management could be tackled with the intelligent planning of nematode-suppressive crop sequences, summer plowing, organic manuring, clean cultivation, adjusting sowing time, water and irrigation management and sensible use of nematicides. The increasing concerns on the ill effects of chemical pesticides on the environment have driven recent research interests on the use of several alternative strategies like botanicals, biopesticides and cropping system research for management of nematodes. Some of the successes have been obtained for managing plant parasitic nematodes with neem-based formulations, fungal formulation of *Purpureocillium lilacinum* and *Trichoderma* spp. and bacterial formulations of *Pasteuria penetrans* and *Pseudomonas fluorescens*. Inconsistent efficacy of the biological control agents in different agro-climatic regions is the main reason for not adopting them widely under field scale. Integration of more than one option can be explored based on their compatibility, economic viability and availability to the farmers. The major challenge ahead for the greater benefit of the farmers is the need for development of holistic approaches for managing nematodes induced disease complexes with other soil-borne pathogens. There is a need for development of economic, eco-friendly, and sustainable bio-intensive integrated nematode management strategies using all the available components.

Chapter - 5

BRINJAL, *Solanum melongena*

Root-knot and reniform nematodes are the major nematode problems in successful cultivation of brinjal crop.

5.1. Root-knot nematodes, *Meloidogyne incognita, M. javanica*

5.1.1. Crop losses

The estimated brinjal fruit yield losses ascribed to *M. incognita* was found to be 27.30 to 48.55% (Bhatti and Jain, 1977; Darekar and Mahse, 1988; Parvatha Reddy and Singh, 1981).

Root-knot nematodes are responsible for annual yield loss of 21% in brinjal amounting to Rs. 3499.12 million in India (Walia and Chakraborty, 2018).

5.1.2. Symptoms

- » Under severe infection, stunted growth and unthrifty development of plants occurred.
- » The leaves are yellowish green to yellow, tend to droop and sudden wilting if wilt organisms are present.
- » The leaf margin was scorched inwards from the margin.
- » The most characteristic symptom of root-knot infection is the formation of galls of varying size on main and feeder roots.
- » In advanced stages, the galled tissues decay and are invaded by other pathogenic and saprophytic organisms.
- » The infested roots eventually rot and affected plants die.

5.1.3. Interaction with other pathogens: *M. incognita* and *Ozonium taxanum* var. *parasiticum* together acted synergistically by reducing the germination of brinjal to 28 per cent (Nath *et al.*, 1976).

The combined effect of *M. incognita* and mycoplasma-like organisms (little leaf) on growth of brinjal was more than their individual effects (Dhawan and Sethi, 1977).

5.1.4. Management

5.1.4.1. Physical methods: Solarization of nursery beds with 25 μm transparent plastic film for 15 days during hot summer is effective for raising healthy seedlings.

Burning of paddy husk or saw dust on infested nursery proved to be reasonably effective.

5.1.4.2. Cultural methods

i) Summer plowing: Plowing not only leads to disturbance and instability in nematode communities but also causes mortality by exposing them to solar heat and desiccation. Generally, 2 to 3 summer plowings each at 10 days interval during April-May (40-46 °C) have been recorded to reduce 96.5 % of *M. javanica* population while fallowing alone during the same period registered 44.5% reduction. Additional use of plastic sheets for covering soil either in nursery beds or in fields further enhanced nematode reduction. Such an approach also helps in reducing the intensity of weeds, fungi, and bacteria in the soil.

The maximum reduction in *M. javanica* population and increase in fruit yield was recorded in brinjal by soil plowing + covering with polythene sheet, followed by soil plowing + exposure to sun (Anon, 1989a).

ii) Crop rotation: Crop rotation with non-solanaceous crops such as paddy, pigeon pea, gingelly, marigold, and groundnut help in reducing the buildup of nematodes in soil. Sheela *et al.* (2002) recorded 47% reduction in the root-knot nematode population and 22% increase in fruit yield by rotating brinjal with sweet potato (cv. Sree Bhadra). Crop rotation with sorghum, wheat and chilli reduced the root-knot nematode population. The population of *M. incognita* larvae decreased under brinjal-wheat, brinjal-chilli-wheat, and red gourd-brinjal- mustard. The root-knot larvae were eliminated in marigold-marigold-fallow and chilli-cauliflower-cauliflower. Garlic following brinjal also brought about considerable reduction in larvae of root-knot nematode. The population levels of *M. incognita* on highly infested land dropped to zero after 18 months of continuous cultivation of *Panicum maximum* (Netscher, 1983).

Decrease in *M. incognita* population on brinjal occurs by following crop rotation with marigold, spinach and bottle gourd (Khan *et al.*, 1975).

iii) Intercropping: Intercropping brinjal with marigold, onion, garlic or resistant tomato (SL-12) improved plant growth and reduced number of galls and final soil population of nematode (Jain *et al.*, 1990). Marigold (cv. Calcutta Yellow)-brinjal cropping sequence has been recommended to the farming community for management of *M. javanica* affecting brinjal (Singh, 1991).

iv) Trap cropping: Growing of trap crop like knol-khol with brinjal but removed and destroyed before new generation of nematode could emerge helps in reducing the root-knot nematode population (Ayyar, 1926).

v) Influence of fertilizers: Row application of fly ash at 0.6 kg/ m^2 enhanced the fruit yield by 27.7% and inhibited the reproduction of *M. incognita* (Khan and Ghadipur, 2004).

vi) Organic amendments: Nursery bed treatment with Neem cake at 200 g/ m^2 gave maximum reduction in nematode population and resulted in yield of 2.075 t/ha (Sheela and Nisha, 2004).

Enhancement of fruit yield and reduction in the nematode infestation was effectively achieved by row incorporation of saw dust at 210 g/3 m row and spot application of mustard cake at 25 g/spot. Similarly, Hazarika (1990) found that spot application of Castor/ Mustard/Neem cake at 15 g/spot and furrow application of the above cakes at 100 g/3 m furrow reduced the infestation of

M. incognita and increased the yield of brinjal. The yield was maximum in spot application of Neem cake along with maximum reduction of *M. incognita* population in soil.

Soil amendment with chopped shoot parts of jack, *Ficus carica, F. elastica, F. glomerata, Ipomea fistulosa, Nerium indicum, Tabernaemontana coronaria* gave good control of *M. incognita* on brinjal (Siddiqi *et al.*, 1987).

Soil amendment with pure latex and chopped shoot parts of *Calotropis procera* significantly reduced the root-knot development caused by *M. incognita* and the population of *Hoplolaimus indicus, Tylenchorhynchus brassicae,* and *T. filiformis* and improved the plant growth of brinjal (Alam, 1986; Maqbool *et al.*, 1982).

vii) Time of planting: In early transplanted (April) brinjal, the rate of multiplication of *M. incognita* was low, in comparison to late transplanted crop (May-June); the latter significantly reduced yield in comparison to early crops (April). Galling was also comparatively higher in the late sown crops as compared to crops sown in April. The study revealed that the crop sown at the right time is more tolerant to the *Meloidogyne* population as compared to the late crop.

5.1.4.2. Chemical methods: Treatment of brinjal nursery beds with carbofuran at 3g a.i. /m^2 led to 48% increase in yield and at 6g a.i. /m^2 gave 60.8% higher yield.

Carbofuran, ethoprophos and fenamiphos each at 1 to 2 kg a.i. per ha were found effective in reducing the root-knot nematode population and in increasing fruit yields of brinjal (119 to 145 per cent) (Singh *et al.*, 1978).

Alam *et al.* (1973) reported that oxamyl and dichlofenthion were found effective for the management of root-knot nematodes infecting brinjal.

Application of Metham sodium 32 EC at 80 ml/m^2 or Carbofuran at 0.2 g/m^2 controlled *M. incognita* in brinjal nursery and increased the fruit yield.

Treatment of nursery bed of brinjal with Dazomet/Phorate/Carbofuran @ 0.6 g a.i. /m^2 reduced the root-knot nematode.

Bare root dip treatment of brinjal seedlings with Zolone (Phosalone 35 EC) or Monocrotophos (Monocil 35 EC) or Carbosulfon (Marshall 25 ST) @ 0.05% for 6 hours before transplanting reduced the population of *M. incognita* and increased the yields substantially.

5.1.4.3. Biological methods:

i) Antagonistic bacteria: Application of *Pasteuria penetrans* in nursery soil of brinjal resulted in 31 and 36% suppression in root galling when seedlings were transplanted to sterilized and nematode-infested soils, respectively. The results were also indicative of easy dissemination of *P. penetrans* to the main field through infected seedlings (Walia *et al.*, 1992).

Nursery bed treatment with *Pasteruia penetrans* @ 10 g culture ($1x10^8$ spores/g)/m^2 before sowing was found effective for managing root-knot nematode, *Meloidogyne incognita* infecting brinjal and increasing the yield.

B. thuringiensis reduced the number of galls/plant and total nematode population and increased dry weight of shoot, chlorophyll content and nitrate reductase activity of leaves. Among different strains of *B. thuringiensis* tested, strain 868 proved better than other strains (Table 5.1) (Chahal and Chahal, 1999).

Table 5.1. Effect of different strains of *Bacillus thuringiensis* for the control of *Meloidogyne incognita* in brinjal.

Treatment	Dry shoot wt. (g)	No. of galls/ plant	Total nematode popn.	Chlorophyll content of leaves (mg/g)	Nitrate reductase activity of leaves (Nm NO_2/h/g)
Control	6.15	143	24,478	1.09	1.59
B. t strain 868	8.87	72	12,581	1.21	1.83
B. t strain 869	8.91	79	15,688	1.18	1.75
B. t - 1	7.57	75	13,495	1.11	1.69
B. t – 2	7.82	78	12,792	1.15	1.82
B. t – 3	8.40	92	16,297	1.20	1.72
B. t – 4	7.20	85	14,212	1.17	1.64
B. t - 5	7.00	98	15,782	1.21	1.71

Chahal and Chahal (1986) reported that *Azotobacter chroococcum* caused significant inhibition of hatching of egg masses of *M. incognita* and did not allow the larvae to penetrate into the roots of brinjal to form galls. *A. chroococcum* was also responsible for reduction in soil and root population of root-knot nematodes and root galling. Among different strains of *A. chroococcum* tested, strain 23 proved better than other strains (Chahal and Chahal, 1988; 1999).

Nursery bed treatment with *Pseudomonas fluorescens* @ 20 g/m^2 gave effective control.

Nursery bed treatment with *Bacillus macerans* at 25 g/m^2 + soil drench (2% solution) ten days after sowing gave maximum reduction in nematode population and maximum yield of 3.825 t/ha (Sheela and Nisha, 2004).

Avermectins (secondary metabolites of *Streptomyces avermitilis*) effectively reduced root-knot nematode infection in brinjal under field conditions (Parvatha Reddy and Nagesh, 2002).

ii) Antagonistic fungi: Application of 'Royal 350' (*Arthrobotrys irregularis* cultured on oat seed medium) at 140 g/m^2 a month before transplantation of brinjal resulted in good protection against root-knot nematodes. Colombo *et al.* (1995) found that distribution of a commercial formulation containing mycelium and viable spores of *Arthrobotrys oligospora* and *A. superba* before transplanting at fortnightly intervals until harvest caused reduction of the second stage juveniles of *M. incognita* in the soil during the growing season of eggplant.

Among the egg parasites, efficacy of *Purpureocellium lilacinum* (commercially formulated as 'Biocon' in Philippines) has been found to be comparatively higher in suppressing the population of *Meloidogyne* spp. and *R. reniformis* on brinjal. Soil application of *P. lilacinum* at 20 g/sq. m. to brinjal nursery reduced infection by *M. incognita* by 50% and improved growth of seedlings by 30%.

Sharma and Trivedi (1987) tried different doses of *P. lilacinum* against *M. incognita* on brinjal in pots and reported 75 g dose to be most effective giving 91% nematode control.

Soil application of *P. lilacinum* (2.9 x 10^6 spores/ml) at 4 ml/kg soil gave significantly highest plant growth, yield, root colonization and spore density in soil; and reduced root galling, egg mass production and final nematode population both in soil and roots of brinjal at 4% moisture level and

25 °C temperature (Krishnamoorthi and Kumar, 2007) (Table 5.2).

Table 5.2. Effect of *Purpureocellium lilacinum* at different moisture levels and temperatures for the management of *Meloidogyne incognita* infecting brinjal

Treatment	Gall index	Yield (g/ plant)	Parasitized egg masses/plant	Parasitized eggs/egg mass	cfu/g soil (10^4)
Moisture level – 4%	3	324	38.2	105.2	50.0
Moisture level – 6%	4	253	33.2	90.7	44.1
Moisture level – 8%	5	207	26.8	72.4	41.0
CD (P = 0.05)	---	4.93	3.44	3.31	---
Temperature – 25°C	3	307	37.2	102.0	54.0
Temperature – 30°C	4	242	32.2	84.7	49.0
Temperature – 35°C	5	191	29.2	72.0	41.7
CD (P = 0.05)	---	4.90	2.81	3.17	---

Application of *Pochonia chlamydosporia* at 50 g/m^2 in the nursery beds significantly increased the plant growth parameters (42 and 50 % increase in height and weight of seedlings, respectively over control) and root colonization by the bioagent on transplants. When the above seedlings were transplanted in the main field, there was reduction in root galling index (5.2 compared to 7.8 in control), number of egg masses in 5 g root (20 compared to 49 in control), number of juveniles in 5 g root (62 compared to 162 in control) and number of juveniles in 100 cc of soil (67 compared to 152 in control) and increased plant growth and yield in eggplant at harvest (Naik, 2004).

At harvest, enhancement of brinjal plant growth and yield (7.4 kg/6 m^2 as compared to 5.2 kg/6 m^2 in control) and reduction in number of juveniles in 100 ml soil (67 compared to 152 in control), number of juveniles in 5 g root (62 compared to 162 in control), root galling index (5.2 compared to 7.8 in control), and number of egg masses in 5 g root (20 compared to 49 in control) were observed when brinjal seedlings raised in nursery treated with *P. chlamydosporia* at 50 g/m^2 were transplanted in field (Table 5.3) (Rao *et al.*, 2003).

Table 5.3. Effect of *Pochonia chlamydosporia* for the management of *Meloidogyne incognita* infecting brinjal

Treatment/Dose (g/m^2)	Root-knot index	Yield (kg/6 m^2)	% egg parasitization by bioagent	Root colonization (cfu/g)	Spore density (cfu/g soil)
P. chlamydosporia - 25	6.5	7.4	42.59	38480	32580
P. chlamydosporia - 50	5.2	8.2	59.35	52690	47870
P. chlamydosporia - 100	5.1	8.5	65.37	54680	45690
Control	7.8	5.2	---	---	---
CD (P = O.05)	0.37	0.86	9.23	4676.34	4216.87

Aspergillus niger was found to be highly pathogenic against *M. incognita* infecting tomato and brinjal (Reddy, 1989).

iii) Arbuscular mycorrhizal fungi: *Glomus fasciculatum* when inoculated in the nursery protected brinjal seedlings from root-knot nematode infestation. When transplanted in the main field, yield of the plant too was increased significantly (Asha, 1996). Prior establishment of AMF *G. fasciculatum* and *G. mosseae* tended to mitigate the effect of *M. incognita* on brinjal (Sharma and Trivedi, 1994).

Brinjal plants inoculated with *G. fasciculatum* ten days before *M. incognita* inoculation significantly decreased root galling, egg mass production and final nematode population in soil. The above treatment also increased crude protein, P, K, total sugars, total phenols, and total free amino acids in the roots (Borah and Phukan, 2006).

Nursery bed treatment with *G. fasciculatum* @ 50 g culture (200 spores/g)/m^2 before sowing proved beneficial for managing root-knot nematode, *Meloidogyne incognita* infecting brinjal and increasing the yield.

5.1.4.4. Host resistance: Use of resistant varieties for e.g., Utkal Madhuri (BB-44), Utkal Jyoti (BB-13), Utkal Anushree (BB-45C), Ghatikia white, and KS-224 (for Bihar) gave effective control.

Cultivar identified from Bangalore Centre as resistant to *M. incognita* race-1 includes Matti Gulla (Brinjal).

Brinjal cvs. Vijaya and Banaras Giant were reported to be moderately resistant to *M. incognita.*

The rootstocks *Solanum torvum* and *S. sisymbriifolium* showed resistant reaction against root-knot nematode, *Meloidogyne incognita.* Three cultivated eggplant varieties viz., Sufala, Singnath, and Kazla were grafted on *S. torvum* and *S. sisymbriifolium.* The highest grafting success was 95% in the case of *Solanum torvum* with Sufala and the lowest (85%) in *Solanum sisymbriifolium* with Singnath. The success of grafting was not affected significantly because of scion and or of rootstocks. The grafted plants showed resistant reaction against *M. incognita* while the scion plants showed susceptibility in the root-knot nematode sick beds. The grafted plants also showed resistant reaction against *M. incognita* under the field conditions. The grafted plants also outyielded compared to the scion plants. The grafting combination *Solanum torvum* with Sufala gave the highest yield compared to other grafting combinations and non-grafted plants (Table 5.4) (Rahman *et al.*, 2002).

Table 5.4. Effect of different rootstocks on root-knot nematode resistance in three eggplant cvs.

Treatments	Fruit yield (tons/ha)	No. of galls/g root
S. torvum + Singnath	35.3 bcd	10.40
S. sisymbriifolium + Singnath	33.3 d	6.42
S. torvum + Sufala	39.2 a	7.80
S. sisymbriifolium + Sufala	38.0 ab	5.56
S. torvum + Kazla	36.5 abc	6.20
S. sisymbriifolium + Kazla	36.4 cd	7.92
Singnath	13.4 f	36.20
Sufala	15.6 f	40.62

Kazla	21.4 e	39.42
CD (P = 0.05)	7.36	---

In a column, means having a common letter does not differ significantly at 5% level.

Grafting of eggplant onto root-knot nematode resistant rootstock, usually *Solanum torvum*, is used in high value production systems in some European and Asian countries. This method is an effective way to prevent nematode damage and build-up.

Grafting of eggplant cv. Aristotle onto root-knot nematode resistant rootstocks like Charleston Hot, Carolina Wonder, Charleston Belle, Mississippi Nemaheart and Carolina Cayenne significantly reduced population of root-knot juveniles and root galling (Table 5.5).

Table 5.5. Effects of grafting Brinjal cv. Aristotle on root-knot nematode resistant rootstocks, *Meloidogyne incognita*

Treatment	***M. incognita*/g root**	**RKI**
Aristotle non-grafted	16.73 b	3.28 b
Aristotle self-grafted	43.29 a	5.18 a
Aristotle grafted on Charleston Hot	8.75c	0.18 c
Aristotle grafted on Carolina Wonder	1.97 c	0.02 c
Aristotle grafted on Charleston Belle	0.22 c	0.24 c
Aristotle grafted on Mississippi Nemaheart	1.72 c	0.10 c
Aristotle grafted on Carolina Cayenne	1.11 c	0.12 c

5.1.4.5. Integrated methods

i) AMF and chemicals: Integration of AM fungus *G. fasciculatum* at 25 g/m^2 (600 spores) with carbofuran at 0.5 kg a.i. /ha gave significant increase in plant growth characters and yield with corresponding decrease in root galling and egg mass production, followed by *T. harzianum* at 2 g/kg soil + carbofuran at 0.5 kg a.i. /ha (Table 5.6) (Saikia and Borah, 2008).

Table 5.6. Effect of integration of AMF and carbofuran for the management of *Meloidogyne incognita* infecting brinjal

Treatment	**% increase in yield**	**No. of galls/ plant**	**Egg masses/ plant**	**Nema popn./200 ml soil**	**Nema popn. in roots**
G. fasciculatum- 25 g/m^2	70.12	51.73	29.30	265	2486
P. penetrans - 1 g/kg soil	56.49	77.86	61.93	324	3372
T. harzianum -2 g/kg soil	55.84	138.93	116.86	310	3078
Carbofuran – 2 kg a.i./ha	43.50	40.76	33.83	147	1608
G. fasciculatum- 25 g/m^2 + Carbofuran - 0.5 kg a.i./ha	108.44	20.63	12.30	170	1727

P. penetrans - 1 g/kg soil + Carbofuran - 0.5 kg a.i./ha	68.83	33.63	22.43	214	2320
T. harzianum -2 g/kg soil + Carbofuran - 0.5 kg a.i./ha	80.51	24.06	16.43	176	2187
Control	---	366.10	211.30	527	5558
CD (P = 0.05)	---	4.944	17.33	16.6	18.55

ii) Physical and cultural: Soil solarization of nursery beds (using 100-gauge LDPE clear film for 15 days) and application of neem cake at 200 g/sq. m. proved most effective in reducing root-knot nematode infestation and increasing yield in brinjal. Solarization of nursery beds for 15 days in summer and application of poultry manure at 200 kg/ha gave maximum yield (Jain and Gupta, 1991).

Integrating summer plowing with soil solarization with polythene mulching effectively reduced *M. javanica* population in brinjal (Table 5.7) (Jain and Gupta, 1991).

Table 5.7. Integrated management of root-knot nematodes by summer plowing/ fallowing and soil solarization in brinjal.

Treatment	Reduction in nematode population (%)		Yield in kg/sq. m.	
	Hisar	Vellayani	Hisar	Vellayani
Plowing + Fallowing for 15 days	87.00	66.20	28.98	---
Plowing + Covering with polythene sheet for 15 days	94.90	77.00	18.84	78.00
No plowing + No covering	77.60	---	---	---

Soil solarization of nursery beds (using 100-gauge LDPE clear film for 15 days) and application of neem cake at 200 g/sq.m. proved most effective in reducing root-knot nematode infestation and increasing yield in brinjal. Solarization of nursery beds for 15 days in summer and application of poultry manure at 200 kg/ha gave maximum yield.

iii) Botanical and chemical: Application of Neem cake @ 500 kg/ha 15 days before sowing and Carbofuran @ 1 kg a.i. /ha at sowing is effective in brinjal (Saikia *et al.*, 2007).

iv) Bioagents and botanicals: Integration of *P. chlamydosporia* (at 100 ml/seed pan containing 1.2 x 10^4 spores/ml) in castor cake (at 40 g/seed pan) amended soil was effective in increasing the seedling weight and colonization of roots with the bioagent. Brinjal seedlings raised in the above treatment transplanted in pots gave maximum increase in plant growth, root colonization and parasitization of eggs of *M. incognita* by *P. chlamydosporia*. The above treatment also gave least root galling and final nematode population both in soil and roots (Table 5.8) (Rao and Parvatha Reddy, 1993b).

Table 5.8. Effect of integration of castor cake and *Pochonia chlamydosporia* on root galling, final nematode population, root colonization and egg parasitization in eggplant infected with *Meloidogyne incognita*

Treatment	Root-knot index	Final nema popn. (soil + roots)	*P. chlamydosporia* root colonization (cfu/g)	% parasitization of egg masses by *P. chlamydosporia*
Castor cake	3.2	3474	---	---
P. chlamydosporia	2.7	2863	3578	53
Castor cake + *P. chlamydosporia*	2.0	1746	5102	71
Control	4.3	6497	---	---
CD (P = 0.05)	0.25	349.75	415.35	5.32

Borkakaty (1993) observed that inoculation of *P. lilacinum* at 4 g/kg of soil in combination with mustard oil cake at 0.5 and 1.0 t/ha increased plant growth with corresponding decrease in number of galls, egg masses and eggs/egg mass of *M. incognita* on brinjal.

Application of 10% neem cake extract (at 20 ml/pot) mixed with spores of *P. lilacinum* (at 1 x 10^6 spores/ml) was effective in increasing plant growth and reducing root galling and final nematode population both in soil and roots. The above treatment also gave maximum parasitization of egg masses of *M. incognita* and spore density of *P. lilacinum* in soil (Rao and Parvatha Reddy, 1994).

Significant reduction in root galling and final nematode population of *M. incognita* were observed in brinjal seedlings which were given bare root-dip treatment in 10% neem leaf suspension mixed with spores of *P. lilacinum* (at 4 x 10^5 spores/ml) for 30 min. Significant increases were also observed in root colonization, parasitization of eggs and spore density of *P. lilacinum* in soil (Table 5.9) (Rao *et al.*, 1997b).

Table 5.9. Effect of root dip treatment with *Purpureocellium lilacinum* and neem leaf suspension on root galling, root colonization, spore density in soil and parasitization of eggs of *Meloidogyne incognita* infecting brinjal

Treatment	Root-knot index	Final nema popn. (soil + roots)	*P. lilacinum* root colonization (cfu/g)	*P. lilacinum* spore density in soil (cfu/g)	Parasitization of eggs by *P. lilacinum* (%)
P. lilacinum	2.8	3832	32540	20560	53
NLS - 5%*	3.1	5346	---	---	---
NLS - 10%*	2.7	4056	---	---	---
P. lilacinum + NLS - 5%	2.4	2964	36680	25640	59
P. lilacinum + NLS-10%	2.1	2356	39460	27690	64
Control	4.2	8258	---	---	---
CD (P = 0.05)	0.46	785.32	3050.76	2850.78	4.59

*Neem leaf suspension

Root-dip treatment of brinjal seedlings in neem cake extract based formulation of *P. lilacinum* (at 5 x 10^6 spores/ml) for 20 min. and planted in pots gave significant increase in plant growth, root colonization, propagule density in soil and parasitization of eggs of *M. incognita* by *P. lilacinum* and drastic reduction in root galling, fecundity and final nematode population in soil and roots (Rao *et al.*, 1998a).

Application of *P. lilacinum* (2x10^6 cfu/g) @ 50 g/m^2 in nursery bed + main field application of *P. lilacinum* (2x10^6 cfu/g) @ 5 kg along with 2.5 tons of farmyard manure/ha prior to transplanting was found effective for the management of root-knot nematode in brinjal.

Application of *P. lilacinum* @ 25 kg spore dust (10^9 conidia/g) with carrier/ha at the time of transplanting + Poultry manure @ 10 t/ha one week prior to transplanting or Mustard cake @ 2 t/ha one week prior to transplanting or *P. lilacinum* @ 25 kg spore dust (10^9 conidia/g) with carrier/ha at the time of transplanting + Neem cake @ 2 t/ha one week prior to transplanting proved beneficial.

Application of castor cake extract based formulation of *T. harzianum* (at 500 ml/m^2 containing 9.9 x 10^3 spores/ml) to nursery beds of brinjal was effective in producing vigorous seedlings (with maximum seedling weight) with least root galling. The above treatment also increased root colonization and parasitization of *M. incognita* females by *T. harzianum* (Table 5.10) (Rao *et al.*, 1998c).

Table 5.10. Effect of oil cake-based *Trichoderma harzianum* on root galling, root colonization, egg parasitization and yield of eggplant infected with *Meloidogyne incognita*

Treatment	No. of galls/100 seedlings	*T. harzianum* root colonization (cfu/g)	% Parasitization of females by *T. harzianum*
Neem cake extract (NCE) – 10%	62	---	---
Castor cake extract (CCE) – 10%	58	---	---
Pongamia cake extract (PCE) – 10%	64	---	---
10% NCE based *T. harzianum*	60	13746	43
10% CCE based *T. harzianum*	57	15625	51
10% PCE based *T. harzianum*	61	12242	42
T. harzianum (grown on paddy seeds)	69	9,648	30
Control	93	---	---
CD (P = 0.05)	8.56	94.75	4.26

Integration of neem seed powder/neem cake with *P. lilacinum* gave effective control of root-knot and reniform nematodes infecting brinjal (Table 5.11).

Table 5.11. Integrated management of *Meloidogyne incognita* and *Rotylenchulus reniformis* in brinjal using botanicals and biocontrol agents

Treatment	No. of galls/ plant	Nema popn. in soil (100 ml)	
		M. incognita	*R. reniformis*
Neem seed powder (NSP)	25	125	200
Neem cake (NC)	27	125	250

Purpureocellium lilacinum (*Pl*)	31	150	258
Pochonia chlamydosporia (*Pc*)	32	167	242
NSP + *Pl*	10	100	167
NSP + *Pc*	13	108	208
NC + *Pl*	14	108	192
NC + *Pc*	18	117	208
Control	41	200	375

The treatment of *P. fluorescens* at 10 kg/ha + neem cake at 1 t/ha was found effective in reducing *M. incognita* population (49.28%), number of root galls/egg masses (68.45%) and gall index (58.5%) and increasing the yield (30.66%) of brinjal with 4.64 benefit: cost ratio (Table 5.12) (Zore *et al.*, 2013)

Table 5.12. Effect of *Pseudomonas fluorescens* on root knot nematode and fruit yield of brinjal

Treatment	**Gall index**	**Yield (t/ha)**	**Benefit: cost ratio**
Soil application of *P. fluorescens* at 20 kg/ha at transplanting	4.06	18.444	4.67
Soil application of *P. fluorescens* at 20 kg/ha + neem cake at 1 t/ha	3.33	20.741	4.64
Untreated control	8.85	15.873	---
CD (P = 0.05)	0.26	13.12	---

v) Bioagents, botanicals and chemicals: Soil application of poultry manure at 10 t/ha a week prior to transplanting + *P. lilacinum* (10^9/conidia/g) at 25 kg spore dust with carrier/ha at the time of transplanting; or neem cake at 2 t/ha a week prior to transplanting + *P. lilacinum* (10^9 /conidia/g) at 25 kg spore dust with carrier/ha at the time of transplanting; or carbofuran at 2 kg/ha in 2 equal splits (one at the time of transplanting and the other after 75 days) + *P. lilacinus* (10^9/conidia/g) at 25 kg spore dust with carrier/ha at the time of transplanting considerably reduced root galling and also gave higher brinjal fruit yield over control (Table 5.13) (Vyas *et al.*, 2009) .

Table 5.13. Effect of different organic amendments, bioagent and nematicide for the management of *Meloidogyne incognita* infecting brinjal

Treatment	**Gall index (0-5 scale)**	**Yield (tons/ha)**	**B: C Ratio**
P. lilacinum at 25 kg/ha + poultry manure at 10 t/ha	2.0	42.088	18.5
P. lilacinum at 25 kg/ha + Mustard cake at 2 t/ha	2.0	44.801	7.7
P. lilacinum at 25 kg/ha + Neem cake at 2 t/ha	1.8	41.745	5.8
P. lilacinum at 25 kg/ha + carbofuran at 2 kg a.i. /ha	1.9	41.498	13.6
P. lilacinum at 25 kg/ha	2.5	39.941	77.6
Poultry manure at 10 t/ha	2.3	40.000	18.5
Mustard cake at 2 t/ha	1.9	41.810	6.7

Neem cake at 2 MT/ha	1.9	39.954	5.2
Carbofuran at 2 kg a.i./ha	2.0	40.530	14.9
Control	3.9	28.107	---

Nursery bed treatment with either *Trichoderma viride* enriched with FYM @ 2.5 g/m^2 or *Pseudomonas fluorescens* @ 20 g/m^2 gave effective control.

vi) Bio-agents and arbuscular mycorrhizal fungi: Brinjal seedlings raised in seed pans treated with *G. mosseae* (at 100 g/seed pan containing 28-32 chlamydospores/g) dipped in *P. lilacinum* spore suspension (containing 4 x 10^5 spores/ml) for 5 min. and transplanted in pots gave maximum increase in plant growth, root colonization, propagule density in soil of both *G. mosseae* and *P. lilacinum* and parasitization of eggs of *M. incognita*. The above treatment also gave least root galling, fecundity and final nematode population in soil and roots (Table 5.14) (Rao *et al.*, 1998).

Table 5.14. Effect of integration of *Glomus mosseae* and *Purpureocellium lilacinum* on root galling, final nematode population, root colonization and egg parasitization in eggplant infected with *Meloidogyne incognita*

Treatment	Root-knot index	Final nematode popn.	*P. lilacinum* root colonization (cfu/g)	% Root colonization by *G. mosseae*	Parasitization of eggs by *P. lilacinum* (%)
G. mosseae	3.0	4876	---	62	---
P. lilacinum	2.7	3252	35240	---	52
G. mosseae + *P. lilacinum*	2.3	2564	37270	64	65
Control	4.3	7746	---	---	---
CD (P=0.05)	0.32	541	875	NS	5

vii) Bioagents, AMF and botanicals: Inoculation of *G. fasciculatum* in the castor cake amended nursery beds followed by the root-dip treatment of mycorrhizal seedlings of brinjal in spore suspension of *P. lilacinus* was found effective for the management of *M. incognita* (Rao *et al.*, 1993b).

In nursery, soil application of neem cake at 400 g/m^2 along with *G. mosseae* (at 500 g/ m^2 containing 26-32 chlamydospores/g) and *P. lilacinum* (at 2 liters/ m^2 containing 6 x 10^5 spores/ml) resulted in production of healthy and vigorous brinjal seedlings colonized with *G. mosseae* and *P. lilacinum* and with least root galling. In field, transplanting of brinjal seedlings raised in nursery treated with Neem cake at 400 g/m^2 along with *G. mosseae* (at 500 g/m^2 containing 26-32 chlamydospores/g) and *P. lilacinum* (at 2 liters/m^2 containing 6 x 10^5 spores/ml) gave maximum reduction in root galling, egg mass production, fecundity and final nematode population in soil and roots; and enhanced fruit yield (7.8 kg/6 m^2 as 5.3 kg/6 m^2 in control), root colonization with *G. mosseae* and *P. lilacinum* and egg parasitization with *P. lilacinum* (Table 5.15) (Rao and Parvatha Reddy, 2001).

Table 5.15. Effect of integration of *Glomus mosseae, Purpureocellium lilacinum* and neem cake on root galling, root colonization, egg parasitization and yield of eggplant infected with *Meloidogyne incognita*

Treatment	Root-knot index	Yield (kg)/ 6 m^2	*P. lilacinum* root colonization (cfu/g)	% Root colonization by *G. mosseae*	Parasitization of eggs by *P. lilacinum* (%)
G. mosseae	7.2	6.2	---	47	---
Neem cake	6.6	5.6	---	---	---
P. lilacinum	7.0	5.3	41660	---	31
Neem cake + *G. mosseae*	5.9	7.3	---	58	---
Neem cake + *P. lilacinum*	5.7	6.9	59600	---	36
G. mosseae + *P. lilacinum*	5.4	6.5	44760	56	40
Neem cake + *G. mosseae* + *P. lilacinum*	4.8	7.8	51382	50	51
Control	8.7	5.3	---	---	---
CD (P = 0.05)	0.49	0.89	1377.21	6.49	4.52

viii) Bioagents, AMF and chemicals: Integration of AM fungus *G. fasciculatum* at 25 g/m^2 (600 spores) with carbofuran at 0.5 kg a.i./ha gave significant increase in plant growth characters and yield with corresponding decrease in root galling and egg mass production, followed by *T. harzianum* at 2 g/kg soil + carbofuran at 0.5 kg a.i. /ha (Saikia and Borah, 2008) (Table 5.16).

Table 5.16. Effect of integration of AMF/bioagents and carbofuran for the management of *Meloidogyne incognita* infecting brinjal

Treatment	% increase in yield	No. of galls/ plant	No. of egg masses/ plant	Nema popn. /200 ml soil	Nema popn. in roots
G. fasciculatum- 25 g/m^2	70.12	51.73	29.30	265	2486
P. penetrans - 1 g/kg soil	56.49	77.86	61.93	324	3372
T. harzianum -2 g/kg soil	55.84	138.93	116.86	310	3078
Carbofuran – 2 kg a.i./ha	43.50	40.76	33.83	147	1608
G. fasciculatum- 25 g/m^2 + Carbofuran - 0.5 kg a.i./ha	108.44	20.63	12.30	170	1727
P. penetrans - 1 g/kg soil + Carbofuran - 0.5 kg a.i./ha	68.83	33.63	22.43	214	2320
T. harzianum -2 g/kg soil + Carbofuran - 0.5 kg a.i./ha	80.51	24.06	16.43	176	2187
Control	---	366.10	211.30	527	5558
CD (P = 0.05)	---	4.944	17.33	16.6	18.55

ix) Two bioagents: Combined soil application of *P. lilacinum* and *Aspergillus niger* at the time of transplanting brinjal is very effective in reducing root-knot nematodes.

Integration of highly toxic fungus, *Aspergillus niger* (kills most of the infective second stage juveniles) and an egg parasite, *Cladosporium oxysporum* (invades and kills the eggs in egg sac) both at half the doses significantly reduced *M. incognita* population and exhibited better plant growth than when either of the fungal bioagents in brinjal (Goswami and Singh, 2001).

In eggplant, the nursery seedling stand index was good with seed treatment with *P. fluorescens* (50 g/kg) + soil application (10 g/m^2 seed bed) followed by soil application of neem based *T. harzianum* + *P. fluorescens* and neem based *P. fluorescens*, where the crop stand index was 4.6 and 4.0, respectively.

x) Arbuscular mycorrhizal fungi and botanicals: Brinjal seedlings raised in nursery treated with *G. fasciculatum* (at 250 g/m^2 containing 16 chlamydospores/g) planted in pots amended with castor cake (applied at 10 g/kg soil 15 days before transplanting) resulted in maximum plant growth, root colonization with *G. fasciculatum* and chlamydospore density in soil and least root galling, fecundity and final nematode population in soil and roots (Table 5.17) (Rao *et al.*, 1998).

Table 5.17. Effect of integration of castor cake and *Glomus fasciculatum* on root galling, final nematode population, root colonization and egg parasitization in eggplant infected with *Meloidogyne incognita*

Treatment	Root-knot index	Final nema popn. (soil + roots)	No. of eggs/egg mass	% root colonization by *G. fasciculatum*	*G. fasciculatum* chlamydospore density in 10 g soil
G. fasciculatum	6.2	5349	356	61	53
Castor cake – 5 g	5.7	4756	426	---	---
Castor cake – 10 g	5.1	3431	458	---	---
G. fasciculatum + Castor cake – 5 g	4.4	2934	295	73	72
G. fasciculatum + Castor cake – 10 g	4.1	2346	263	78	80
Control	8.3	9398	584	---	---
CD (P = 0.05)	1.25	814.36	27.52	9.95	10.32

Integration of *G. fasciculatum* and neem cake at 0.5 t/ha was found to be effective in increasing plant growth parameters and yield (30 t/ha compared to 17 t/ha in control) and in reducing the root-knot index (2.6 compared to 4.8 in control) and final nematode population (173.3/250 ml soil compared to 510.8/250 ml soil in control) (Borah and Phukan, 2004).

xi) Botanicals and chemicals: Application of Carbofuran 3G @ 3 kg a.i. /ha in the nursery beds prior to seeding and Neem cake @ 500 kg/ha 10 days before transplanting in the main field was found effective.

Application of Carbofuran 3G @ 10 g or Phorate 10G @ 3 g/m^2 at seeding in the nursery beds and spot application of Carbofuran 3G @ 33 kg or Phorate 10G @ 10 kg + Neem cake @ 400 kg/ha at the base of plant 10 days after transplanting proved beneficial for the control of root-knot nematode.

xii) Bioagents and chemicals: Split application of *T. harzianum* at 50 kg/ha (10^8 cfu/g) (before transplanting and 45 days after transplanting) + carbofuran at 16.5 kg/ha was effective in increasing the brinjal fruit yield (Haseeb *et al.*, 2004).

Application of *Pseudomonas fluorescens* at 10g/m^2 + Carbofuran at 0.15g a.i./m^2 effectively controlled initial root-knot nematode (*M. incognita*) population in the nursery which in turn protected the brinjal crop in the main field also and increased the yield by 17 to 22% (Sheela *et al.*, 2002).

Application of carbofuran at 1.0 kg a.i./ha in the nursery beds along with neem cake at 400 kg/ha increased yield (68.10%) and reduced root galling (68.10%) (Singh and Gill, 1998)

5.2. INTERACTION OF ROOT-KNOT NEMATODE WITH BACTERIAL WILT

5.2.1. Interaction

Eggplant is prone to many soil-borne diseases among which the bacterial wilt (*Ralstonia solanacearum*) in combination with root-knot nematode (*Meloidogyne incognita*) takes heavy toll every year all over the world (Naik *et al.*, 2003). Combined inoculation of the nematode and the bacterium had a greater pathogenic effect than pathogens inoculated individually (Ravichandra *et al.*, 1990; Savitha and Lingaraju, 1996).

The root-knot nematode, *M. incognita*, present along with *R. solanacearum*, greatly increases the incidence and severity of bacterial wilt of brinjal (Table 5.18).

Table 5.18. Effect of *M. incognita* and *R. solanacearum* alone and in combination on bacterial wilt incidence in brinjal (Husain and Bora, 2009).

Treatment	No. of galls/root system	% wilt incidence
M. incognita at 1000 J2/kg soil	108.30	0.00
M. incognita at 2000 J2/kg soil	110.00	0.00
R. solanacearum at 25 ml/kg soil	0.00	25.00
R. solanacearum at 50 ml/kg soil	0.00	25.00
M. incognita at 1000 J2/kg soil + *R. solanacearum* at 50 ml/kg soil	86.10	100.00
M. incognita at 2000 J2/kg soil + *R. solanacearum* at 50 ml/kg soil	43.30	100.00
M. incognita at 1000 J2/kg soil + *R. solanacearum* at 25 ml/kg soil	89.00	75.00
M. incognita at 2000 J2/kg soil + *R. solanacearum* at 25 ml/kg soil	91.20	75.00
Control	0.00	0.00

5.2.2. Management

5.2.2.1. Biological methods: Significant reduction in the final soil nematode population and wilt disease incidence in brinjal was obtained by soil application of biological control agents like *Pseudomonas fluorescens* at 5 g/kg soil and *T. harzianum* at 3 g/kg soil. The highest reduction in bacterial population in soil was observed in the treatment of *P. fluorescens* (Barua and Bora, 2008). Further, significant increase in the antagonist population in soil and reduction in root galling, egg mass production and final population of *M. incognita* and *R. solanacearum* in soil were observed with combined application of wheat bran formulations of *P. fluorescens* and *T. harzianum* (Table 5.19) (Barua and Bora, 2009).

Table 5.19. Effect of different treatments on root galling, egg mass production and population of nematodes and bacteria in brinjal

Treatment/Dose	No. of galls/ plant	Egg masses/ plant	Nema. popn. in soil	Nema. popn. in root	Bact. popn. (10^8 cfu/g soil)
T. harzianum – 1g/kg soil	69	44	355	59	21
T. harzianum – 2g/kg soil	60	36	298	52	27
T. harzianum – 3g/kg soil	52	31	282	46	15
P. fluorescens – 5g/kg soil	64	40	309	54	13
P. fluorescens – 10g/kg soil	56	34	290	49	7
P. fluorescens – 15g/kg soil	44	28	235	40	4
Carbofuran – 1.5 kg a.i./ha	26	16	176	24	21
Streptocycline – 500 ppm	83	38	452	70	11
Neem cake – 15 g/kg soil	39	24	226	96.8	11
Control	92	53	484	416	42
CD (P = 0.05)	0.093	0.129	80.00	11.1	2.75

Root dip of brinjal seedlings in *P. fluorescens* (5 g/liter of water for 30 min) gave maximum suppression of root-knot nematode population in soil followed by *P. chlamydosporia* (soil application at 10 g/plant + root dip at 5 g/liter of water for 30 min), Neem cake, *P. fluorescens* (soil application at 10 g/plant + root dip at 5 g/liter of water for 30 min), *P. chlamydosporia* as soil application (10 g/plant), *P. fluorescens* as soil application (10 g/plant) and *P. chlamydosporia* as root dip (5 g/liter of water for 30 min) (Dhawan *et al.*, 2008).

Nursery bed treatment with *P. chlamydosporia* at 100 g/m^2 gave reduced root galling; and increased seedling height, seedling weight and root colonization by *P. chlamydosporia* (Table 5.20) (Naik *et al.*, 2003). The seedlings colonized by *P. chlamydosporia* were highly vigorous and reached the main field when transplanted.

Table 5.20. Effect of *P. chlamydosporia* on plant growth and management of root-knot nematodes infecting brinjal under nursery conditions

Treatment	Plant height (cm)	Seedling weight (g)	*P. chlamydosporia* root colonization (cfu/g)
P. chlamydosporia at 25 g/m^2	18.8	2.1	13,560
P. chlamydosporia at 50 g/m^2	21.6	2.4	19,780
P. chlamydosporia at 100 g/m^2	23.7	2.7	21,470
Control	15.2	1.6	---
CD (P = 0.05)	3.34	0.48	785.79

The seedlings raised in nursery beds treated with *P. chlamydosporia* at 100 g/m^2 when transplanted in the main field gave reduced root galling, egg mass production and egg parasitization; and increased fruit yield and root colonization by *P. chlamydosporia* (Table 5.21) (Naik *et al.*, 2003).

Table 5.21. Effect of *P. chlamydosporia* on plant growth and management of root-knot nematodes infecting brinjal under field conditions

Treatment*	Root-knot index	No. of egg masses/ 5g root	% egg parasitization by bioagent	Fruit yield (kg/6m^2)	*P. chlamydosporia* root colonization (cfu/g)
T_1	6.5	25	42.59	7.4	13,560
T_2	5.2	20	59.35	8.2	19,780
T_3	5.1	18	65.37	8.5	21,470
T_4	7.8	49	0.00	5.2	---
CD (P = 0.05)	0.37	6.84	9.23	0.86	785.79

*T_{1-} *P. chlamydosporia* at 25 g/m^2, T_{2-} *P. chlamydosporia* at 50 g/m^2, T_{3-} *P. chlamydosporia* at 100 g/m^2, T_{4-} Control.

5.2.2.2. Integrated methods

i) Bioagents and botanicals: The nursery seedling stand was good in seed treatment with *P. fluorescens* at 50 g/kg + nursery bed treatment with *P. fluorescens* at 50 g/m^2 followed by nursery bed treatment with *T. harzianum* + *P. fluorescens*, and Neem based *P. fluorescens* (Naik *et al.*, 2003).

Seed treatment with talc-based formulation of *P. fluorescens* at 50 g/kg gave least mortality of brinjal plants (13.18%) followed by seed treatment with Neem based formulation of *P. fluorescens* at 50 g/kg (16.66%) compared to 45.29% mortality in control plants (Table 5.22) (Naik *et al.*, 2003).

Table 5.22. Effect of bioagents on mortality of brinjal plants

Treatment	Mortality of plants (%)
Seed treatment with talc-based formulation of *P. fluorescens* at 50 g/kg	13.18 (21.30)
Seed treatment with neem-based formulation of *P. fluorescens* at 50 g/kg	16. 66 (24.12)
Soil application of talc-based formulation of *P. fluorescens* + *T. harzianum* at 25 + 25 g/bed	25.45 (30.33)
Soil application of neem-based formulation of *P. fluorescens* + *T. harzianum* at 25 + 25 g/bed	22.13 (28.04)
Control	45.29 (44.60)
CD (P = 0.05)	4.212

ii) Two bioagents: Results related to root-knot and wilt disease complex management in eggplant under field conditions revealed that the combination formulation of *T. harzianum* and *P. chlamydosporia* performed very well. The yield was 2.024 kg/3 m^2 followed by combination of *P. fluorescens* + *P. chlamydosporia* where the yield was 1.782 kg and *T. harzianum* + *P. fluorescens* (1.680 kg). Root gall index (RGI) and wilt disease incidence due to *Ralstonia solanacearum* (WDI) was minimum in combination treatment of *T. harzianum* + *P. chlamydosporia* followed by *P. fluorescens* + *P. chlamydosporia* and *T. harzianum* + *P. fluorescens* where the RGI was 1.8, 1.81 and 1.93, respectively and that of WDI was 22.18%, 25.13% and 27.28%, respectively (Table 5.23) (Naik, 2004).

Table 5.23. Effect of integration of bioagents on root galling, wilt disease incidence and fruit yield of brinjal

Treatment	No. of fruits / plant	Fruit yield (kg/3m²)	Root-knot index	Wilt disease incidence (%)
Neem + wheat bran formulation (8:2) of *T. harzianum*	36.64	1.614	2.23	33.25 (35.24)
Liquid broth formulation of *P. fluorescens*	38.10	1.657	1.12	31.23 (33.96)
Neem + pongamia + wheat bran formulation (4:4:2) of *P. chlamydosporia*	39.47	1.733	2.00	30.68 (33.65)
T. harzianum + *P. fluorescens*	39.00	1.680	1.93	27.28 (31.50)
T. harzianum + *P. chlamydosporia*	42.97	2.024	1.80	22.18 (28.11)
P. fluorescens + *P. chlamydosporia*	40.64	1.782	1.81	25.13 (30.07)
Neem + wheat bran	34.70	1.571	3.10	46.21 (42.82)
Neem + pongamia + wheat bran	37.60	1.471	3.28	48.30 (44.03)
Control	28.30	1.133	4.20	98.00 (81.87)
CD (P = 0.05)	3.514	0.245	0.15	3.21

The nursery seedling stand was good in seed treatment with *P. fluorescens* at 50 g/kg + nursery bed treatment with *P. fluorescens* at 50 g/m² followed by nursery bed treatment with *T. harzianum* + *P. fluorescens*, and neem-based *P. fluorescens* (Naik *et al.*, 2003) (Table 5.24).

Table 5.24. Effect of integration of bioagents on nursery stand of brinjal seedlings.

Treatment	Nursery stand of seedlings*
Seed treatment with talc-based *P. fluorescens*	4.0
Nursery bed treatment with neem-based *P. fluorescens*	3.6
Seed treatment with talc-based *P. fluorescens* + Nursery bed treatment with neem based *P. fluorescens*	4.6
Nursery bed treatment with *T. harzianum* + *P. fluorescens*	4.0
Control	3.0

* 3 – Poor, 4 – Good, 5 – Very good.

Integration of seedling dip for 15 min in talc-based formulation of *P. fluorescens* + *P. chlamydosporia*

(5 + 5 g/liter) with soil application of the above bioagents gave maximum survival of brinjal plants under field conditions (53.33%) compared to 33.50% survival in control (Table 5.25) (Naik *et al.*, 2003).

Table 5.25. Effect of integration of bioagents on survival of brinjal plants under field conditions

Treatment	Survival of plants (%)
Seedling dip treatment for 15 min in talc-based formulation of *P. fluorescens* at 10 g/litre	43.33 (41.15)
Seedling dip treatment for 30 min in talc-based formulation of *P. fluorescens* at 10 g/litre	43.33 (41.15)
Seedling dip treatment for 15 min in talc-based formulation of *P. fluorescens* at 10 g/litre + soil drenching	46.66 (43.05)
Seedling dip treatment for 15 min in talc-based formulation of *P. fluorescens* + *T. harzianum* at 5 + 5 g/litre + soil application	53.33 (46.89)
Control	33.50 (35.37)
CD (P = 0.05)	4.92

The highest reduction of *M. incognita* and *R. solanacearum* population in soil was observed in combined application of *T. harzianum* and *P. fluorescens*. *P. fluorescens* was proved to be more promising followed by *T. harzianum* in suppressing the population of *R. solanacearum*, while *T. harzianum* proved more effective than *P. fluorescens* in reducing root galling. The population of both the bioagents increased in the soil (Table 5.26) (Barua and Bora, 2009).

Table 5.26. Effect of *T. harzianum* and *P. fluorescens* alone and in combination for the management of disease complex

Treatment	No. of galls/root system	Popn. of *R. solanacearum* (x 10^8 cfu/ml)	Popn. of *T. harzianum* (x 10^7 cfu/ml)	Popn. of *P. fluorescens* (x 10^8 cfu/ml)
T. harzianum	36.00	18.20	42.00	0.00
P. fluorescens	45.60	6.60	0.00	40.00
T. harzianum + *P. fluorescens*	29.40	2.40	23.00	23.00
Inoculated control	93.80	22.60	0.00	0.00
Uninoculated control	0.00	0.00	0.00	0.00
CD (P = 0.05)	0.43	1.00	0.75	0.72

iii) Bioagents, botanicals, chemicals and cultural methods: Integration of two summer plowings during May-June, half recommended dose each of Carbofuran (0.75 kg a.i. /ha), Neem cake (7.5 g/spot), Streptocycline (250 ppm at 30 ml/spot) and full dose of *T. harzianum* (150 g/spot) improved plant growth parameters and fruit yield with corresponding decrease in the nematode reproduction rate (*M. incognita*), bacterial population in the soil and bacterial wilt incidence (*R. solanacearum*) in brinjal (Table 5.27) (Hussain and Bora, 2008).

Table 5.27. Effect of Integration of different practices for the management of disease complex in brinjal

Treatment	No. of galls/ root system	No. of egg masses/root system	% wilt incidence	Yield (kg/4m^2)
2 Summer plowings + full dose of carbofuran (1.5 kg a.i./ha)	65	59	27.77	3.21
2 Summer plowings + full dose of neem cake (15 g/spot)	71	69	27.77	3.10
2 Summer plowings + full dose of streptocycline (500 ppm at 30 ml/ spot)	128	112	38.88	2.43
2 Summer plowings + full dose of *T. harzianum* (150 g/spot)	67	63	27.77	3.10
2 Summer plowings + ½ dose carbofuran + full dose *T. harzianum*	35	26	5.53	3.90
2 Summer plowings + ½ dose neem cake + full dose *T. harzianum*	57	47	11.10	3.30
2 Summer plowings + ½ dose streptocycline + full dose *T. harzianum*	59	55	24.99	3.20
2 Summer plowings + ½ dose carbofuran + ½ dose neem cake + ½ dose streptocycline	28	20	5.53	4.20
2 Summer plowings + ½ dose carbofuran + ½ dose neem cake + ½ dose streptocycline + full dose *T. harzianum*	24	17	5.53	4.90
Control	362	207	66.66	1.6
CD (P = 0.05)	6.69	6.75	11.15	0.83

5.3. INTERACTION OF ROOT-KNOT NEMATODE WITH FUSARIUM WILT

5.3.1. Interaction

Brinjal was not susceptible to *Fusarium oxysporum* unless *Meloidogyne incognita* was also present (Smits and Noguera, 1982). Both the pathogens are responsible for wilt disease complex.

5.3.2. Management

5.3.2.1. Integrated methods

i) Bioagents and chemicals: Treatment of seedlings with carbofuran at 1 kg a.i./ha + *T. harzianum* at 50 kg/ha (having 10^8 cfu/g) was found highly effective in increasing the number and fresh weight of brinjal seedlings/bed and maximum reduction in root-knot index due to *M. incognita*. However, minimum percent root infection by *F. solani* was observed in carbofuran at 1 kg a.i./ha + bavistin at

1 kg a.i./ha treated seedlings (Kumar *et al.*, 2009).

Studies were undertaken to determine the effect of bio-control agents *viz.*, *Aspergillus niger*, *Trichoderma harzianum*, *Purpureocellium lilacinum*, and *Pseudomonas fluorescens* alone at 100 kg/ha each having 10^8 cfu/g culture, pesticides *viz.*, carbofuran and bavistin each at 2 kg a.i. /ha and organic amendment *viz.*, neem seed powder at 500 kg/ha alone and in combination (at half dose of each treatment material) against root-knot-wilt disease complex caused by *M. incognita* and *F. solani* on brinjal. Results indicated that all the treatments alone and in combination significantly improved the number and fresh weight of the seedlings and reduced the root-knot index and percent root infection. Carbofuran + *T. harzianum* was found highly effective in increasing the number and fresh weight of seedlings/bed. Maximum reduction in root-knot index was also observed in carbofuran + *T. harzianum* treated seedlings. However, minimum percent root infection by *F. solani* was observed in carbofuran + bavistin treated seedlings (Kumar *et al.*, 2009).

5.4. RENIFORM NEMATODE, *ROTYLENCHULUS RENIFORMIS*

5.4.1. Symptoms

Symptoms are primary expressed as minimized root systems, leaf chlorosis, nutrient defciency, overall stunting, and reduced yields of the infected plants. Reduction in plant longevity negatively afects harvest duration. Female nematodes and their egg masses are often visible when plant roots are viewed under a dissecting microscope.

5.4.2. Life cycle

The lifecycle of reniform nematode is frequently less than three weeks in warm regions (Abd-Elgawad, 2020).

5.4.3. Survival

The nematode can survive for more than two years without its host under stressed conditions, i.e., in dry soil. In such a case, *R. reniformis* can live in a dormant state stimulated by drought and the nematode is almost dehydrated with diminished metabolic activity; a condition called anhydrobiosis that helps *R. reniformis* to live without water for long time (Wang, 2019).

5.4.4. Management

5.4.2.1. Cultural methods: Soil amendment with chopped shoot parts of jack, *Ficus carica, F. elastica, F. glomerata, Ipomea fistulosa, Nerium indicum, Tabernaemontana coronaria* gave good control of *R. reniformis* on brinjal (Siddiqi *et al.*, 1987).

5.4.2.2. Chemical methods: Bare root dip treatment with carbofuran/turbofos at 500 or 1,000 ppm for 15 to 30 min. in brinjal economized the reniform nematode management by reducing the quantity of nematicide required (Prasad and Krishnappa, 1981).

5.4.2.3. Biological methods: Lone *et al.* (2014) found that the simultaneous inoculation of *R. reniformis* at 5000 infective juveniles of *Steinernema* sp./500 g soil drastically decreased the injury.

5.4.2.3. Integrated methods

i) Bioagents and chemicals: Integration of a bio-agent, *P. lilacinum* with carbofuran at 1 kg a.i. /ha was found effective in the management of reniform nematode, *R. reniformis* infecting brinjal (Table 5.28) (Parvatha Reddy and Khan, 1989).

ii) Botanicals, bioagents and AMF: Inoculation of *G. fasciculatum* in the castor cake amended nursery beds followed by the root-dip treatment of mycorrhizal seedlings of brinjal in spore suspension of *P. lilacinum* was found effective for the management of *R. reniformis* (Rao *et al.*, 1993b).

Table 5.28. Effect of *Purpureocellium lilacinum* and carbofuran on *Rotylenchulus reniformis* infecting brinjal

Treatment	Dose	Mature females in root/plant	Total nema popn./plant	Reproduction factor	% males
P. lilacinum	1 g/plant	12.00	1086	1.09	27.00
P. lilacinum	2 g/plant	7.33	834	0.83	48.79
P. lilacinum	4 g/plant	5.33	315	0.31	60.00
P. lilacinum + Carbofuran	2.0 g/plant + 0.5 kg a.i./ha	6.67	704	0.70	55.02
P. lilacinum + Carbofuran	2.0 g/plant + 1.0 kg a.i./ ha	4.67	389	0.39	34.81
Carbofuran	0.5 kg a.i. /ha	8.67	961	0.96	33.33
Carbofuran	1.0 kg a.i./ha	10.00	551	0.55	30.77
Control	---	25.50	2550	---	28.54
CD (P=0.05)	---	7.80	95.00	---	---

5.5. CONCLUSION

Plant parasitic nematodes constitute one of the major limiting factors for cultivation of brinjal. Several nematode species are associated with brinjal and few of them negatively impact yield. Severe yield losses have been reported in most regions where brinjal is grown. To select suitable nematode management practices, correct diagnosis of the causal agent is of prime importance. Overall, these nematode problems in brinjal are better managed when integrated management practices are used, i.e., regulatory methods (plant quarantine and seed certification), cultural practices (crop rotation, cover crops), genetic control, and ultimately by the use of nematicides. Therefore, for a sustainable cropping of brinjal, growers, extension services, and researchers must consider these nematodes holistically, the impact they cause and whether these management practices are economically, environmentally and technically sound.

Integrated nematode management could be tackled with the intelligent planning of nematode-suppressive crop sequences, summer plowing, organic manuring, clean cultivation, adjusting sowing time, water and irrigation management and sensible use of nematicides. The increasing concerns on the ill effects of chemical pesticides on the environment have driven recent research interests on the use of several alternative strategies like botanicals, biopesticides and cropping system research for management of nematodes. Some of the successes have been obtained for managing plant parasitic nematodes with neem-based formulations, fungal formulation of *Purpureocillium lilacinum* and *Trichoderma* spp. and bacterial formulations of *Pasteuria penetrans* and *Pseudomonas fluorescens*. Inconsistent efficacy of the biological control agents in different agro-climatic regions is the main reason for not adopting them widely under field scale. Integration of more than one option can be explored based on their compatibility, economic viability and availability to the farmers. The major challenge ahead for the greater benefit of the farmers is the need for development of holistic approaches for managing field problems including nematodes induced disease complexes with other soil-borne pathogens. There is a need for development of economic, eco-friendly, and sustainable bio-intensive integrated nematode management strategies using all the available components.

Chapter - 6

CHILLI AND BELL PEPPER, *Capsicum annuum*

The root-knot nematode is the major limiting factor in the successful production of chilli and bell pepper crops.

6.1. ROOT-KNOT NEMATODES, *Meloidogyne incognita, M. javanica*

6.1.1. Crop losses

M. incognita was responsible for 24.54 to 28.00% loss in fruit yield of chillies (Singh *et al.*, 2003).

Root-knot nematodes are responsible for annual yield loss of 15% in chilli and 10% in bell pepper amounting to Rs. 744.90 million and Rs. 52.92 million, respectively (Walia and Chakraborty, 2018).

6.1.2. Symptoms

- » Root-knot nematodes induce much smaller galls on the roots of chillies.
- » Stunting of plant growth and rapid wilting of plants especially under dry growing conditions.
- » Growth may be retarded, and leaves may be chlorotic.
- » Death of seedling in nursery bed and non-survival of transplants in main field when seedlings are infected.
- » Significant reduction in survival of plants and drastic reduction in flowering and fruit yield.

6.1.3. Interaction with other pathogens: Bell pepper is prone to many soil-borne diseases among which the bacterial wilt (*Ralstonia solanacearum*) in combination with root-knot nematode (*M. incognita*) takes heavy toll every year all over the world (Naik, 2004).

6.1.4. Management

6.1.4.1. Physical methods: Rabbing nursery beds with castor husk @ 7 kg/m^2 or soil solarization with 25 μm LLDPE clear film for 15 days during summer proved beneficial.

Burning of paddy husk or saw dust on infested nursery proved to be reasonably effective.

Experiments in Sicily (Cartia and Greco, 1987; Cartia *et al.*, 1988, 1989), aimed at the control of soil-borne pathogens of pepper in greenhouse culture, yield and fruit size from solarized soil averaged 12.2 and 2.3 times, respectively, over those in controls.

6.1.4.2. Cultural methods: Several cultural methods like selection of nematode-free nursery sites, destruction of infested roots after crop harvest, crop rotation with non-hosts (maize, wheat, sorghum) or antagonistic crops (marigold, mustard, sesame) or intercropping with marigold, onion and garlic, flooding, fallowing, deep summer plowing and harrowing either alone or in combination proved to be reasonably effective and economical to check multiplication of root-knot nematodes on capsicum.

i) Summer plowing: Plowing not only leads to disturbance and instability in nematode communities but also causes mortality by exposing them to solar heat and desiccation. Generally, 2 to 3 summer plowings each at 10 days interval during April-May (40-46 °C) have been recorded to reduce 96% *M. javanica* population while fallowing during the same period registered 45.5% reduction. Additional use of plastic sheets for covering soil either in nursery beds or in fields further enhances nematode reduction. Such an approach also helps in reducing the intensity of weeds, fungi and bacteria in the soil.

ii) Crop rotation: Khan *et al.* (1975) found that rotation of chilli with bottle gourd, marigold, and spinach or fallowing of field reduced *M. incognita* population on chilli. Growing chilli along with marigold gave the highest degree of reduction in root-knot nematode population on chilli roots, followed by onion, garlic and asparagus (Trivedi and Tiagi, 1984).

iii) Intercropping: Intercropping with marigold, onion, and garlic, proved to be reasonably effective and economical to check multiplication of root-knot nematodes on capsicum.

iv) Organic amendments: Khan and Ghadipur (2004) reported that inhibition in the reproduction of *M. incognita* and 21.3% increased fruit yield was obtained by in row application of fly ash at 0.6 kg/m^2.

Soil application of decaffeinated tea leaves effectively reduced root galling on chilli plant (Trivedi and Tiagi, 1984).

v) Biofumigation: Biofumigation experiments have been carried out in Murcia (Spain), in a greenhouse affected by a population of *M. incognita*, race 1. The fumigant effect of sheep manure plus 25% chicken manure was analyzed after one, two and three years of treatment. Doses were decreased every year from the initial dose of 100 t/ha to 75 t/ha in the 2nd year and to 60 t/ha in the 3rd year. Results were compared to a methyl bromide (MB) (98: 2) treatment at a dose of 30 g/m^2 and to the control plots. Biofumigation is as effective in controlling *M. incognita* as MB, and saprophagous nematodes increased after biofumigation, while nematode fauna biodiversity in the MB treatments was null. The average indexes of *M. incognita* root galling at the end of the crop in the month of August are as follows, according to treatments: 3rd year of biofumigation (1.6), 2nd year (1.9), 1st year (4.5), MB treatment (1.2) and control group (6.8). There are no significant differences between biofumigation from the 2nd (11.5 kg m^2) and 3rd (11.7 kg m^2) years and MB (11.5 kg m^2) treatments in sweet pepper production.

Biofumigation with fresh sheep manure (if it is done from July to October) is promising to be an efficient disinfectant although the parameters recorded (plant growth, yield, and pathogen incidence) were not at the same level of control (Table 6.1). An improvement of the results can be obtained if biofumigation is repeated for the following seasons. This practice is a good solution for soil disinfection in organic agriculture (Bello *et al.*, 1997).

Table 6.1. Effects of biofumigation on 5 plastic greenhouses (Season 1998/99).

Treatments	Plants height (cm)	Commercial yield (kg/m^2)	% plants with *Meloidogyne*
MB 60 g/m^2	163.8 a	9.7 a	8.2 a
Biofumigation and solarization (4 weeks)	135.3 b	8.7 b	39.0 b
Untreated	122.4 c	4.3 c	64.8 c

No significant differences between values with the same letter (P> 0.05)

Biofumigation experiments were initiated using local organic amendments (sheep and chicken manure) to look for non-chemical alternatives equally efficient as MB to manage *M. incognita* infecting bell pepper. Biofumigation with manure was not effective during the first year. When applied for a second year the control of *Meloidogyne* was random but was improved when repeated applications were used. The development and production of the plants were comparable with MB treatment in the second year. Physical and chemical properties of the soil were improved and lixiviates were not produced (Guerrero *et al.*, 2008).

6.1.4.3. Chemical methods: Carbofuran at 3 per cent was effective against root-knot nematodes infecting chilli (Kandasamy and Sivakumar, 1981).

Dichlofenthion was found effective against *M. incognita* in chillies (Alam *et al.*, 1973, Saxena *et al.*, 1974).

Carbofuran, ethoprophos and fenamiphos each at 1 to 2 kg a.i. per ha were found effective in reducing the root-knot nematode population and in increasing fruit yields of chillies (Handa and Mathur, 1981).

Bare root dip of chilli seedlings in 0.05% Carbosulfan solution for 1 hour was found effective. Bare-root dip treatment of capsicum seedlings in 0.1% carbosulfan/monocrotophos for 6 hr. eliminates root-knot infection.

Application of Carbofuran 0.2 g/m^2 to chilli nursery, improved plant stand and increased the seedling weight by controlling *M. incognita*.

The demonstration trials on chilli showed that nursery application of Carbofuran @ 0.3 g a.i./m^2 and root dip with Carbosulfan 500 ppm significantly reduced *M. incognita.*

6.1.4.4. Biological methods

i) Antagonistic bacteria: Field application of a nematicidal *Bacillus thuringiensis* strain to capsicum in Puerto Rico reduced galling in roots due to *M. incognita* and increased yield significantly (Zuckermann *et al.*, 1993).

Nursery treatment with either *Trichoderma viride* formulation (FYM enriched) @ 2.5 g/m^2 or *Pseudomonas fluorescens* formulation @ 20 g/m^2.

Bio-agent *Pseudomonas fluorescens* was evaluated under field conditions for its efficacy against root-knot nematode *M. incognita* infecting capsicum. Seed treatment with *P. fluorescens* alone was effective (Naik, 2004).

Avermectins (secondary metabolites of *Streptomyces avermitilis*) effectively reduced root-knot nematode infection in chilli under field conditions (Parvatha Reddy and Nagesh, 2002).

Burkholderia cepacia decreased egg hatch of *M. incognita* and mobility of second stage juveniles (Meyer *et al.*, 2000).

ii) Antagonistic fungi: Application of 'Royal 350' (*Arthrobotrys irregularis* cultured on oat seed medium) at 140 g/m^2 a month before transplantation of capsicum resulted in good protection against root-knot nematodes. Among the egg parasites, efficacy of *Purpureocellium lilacinum* (commercially formulated as 'Biocon' in Philippines) has been found to be comparatively higher in suppressing the population of *Meloidogyne* spp. on capsicum.

Pandey and Trivedi (1992) reported a significant reduction in number of galls, eggs/egg mass, hatching and final soil population of *M. incognita* infecting chillies in the presence of *P. lilacinum*.

Vicente and Acosta (1992) reported that soil application of *P. lilacinum* one week before planting gave effective control of *M. incognita* and *Rotylenchulus reniformis*. Significantly more and heavier fruits were obtained from *P. lilacinum* treated plots than control. A similar trend was observed in the percentage reduction in population.

Bio-agent *Pochonia chlamydosporia* and *Pseudomonas fluorescens* were evaluated under field conditions for their efficacy against root-knot nematode *M. incognita* infecting capsicum. Treatment of the nursery bed with the formulation of *P. chlamydosporia* at 50 g/m^2 was significantly effective in reducing the galling index, number of nematodes in roots and soil, increasing the percent parasitization of eggs by bioagent and also yields. Seed treatment with *P. fluorescens* alone and the nursery bed treatment with *P. chlamydosporia* alone were effective. The main purpose of these studies was to produce capsicum seedlings that are colonized by *P. chlamydosporia* before transplanting so that they could carry the bio-agent to the main field. During this process, the field soil would be enriched with the propagules of bioagent component of the management in two or three seasons (Naik, 2004).

6.1.4.5. Host resistance: Recommended resistant varieties for root-knot nematode on Chilli include BC-21-2, Pusa Jwala, Agnirekha, Suryamukhi, Utkal Ava (BC- 14-2), CA- 2057, CA-2123, TC-52-18, CA-2123, C-26-4-5, and BR-31 (Anon, 2015).

Chilli cvs. Pusa Jwala, Wonder Hot, Teja, Utkal Abha, Utkal Rashmi were found resistant to root-knot nematodes (Patnaik *et al.*, 2004).

Capsicum cvs. Mississipi-68, Santanka, Anaheim Chile and Italian Pickling are reported to be resistant to root-knot nematodes (Hare, 1951). Black Indica, Naharia and Pant C1 were reported to be resistant to *M. incognita*, while All Big as moderately resistant. California Wonder and Naharia were resistant to *M. javanica*, while Early California Wonder as moderately resistant. Naharia was reported to be resistant to *M. arenaria*.

Bell pepper grafted on Graffito rootstock of *Capsicum annuum* had good compatibility, do not affect scion growth or production, and is resistant to *Phytophthora capsici* and root-knot nematodes (Morra and Billoto, 2006).

Sweet pepper (*Capsicum annuum*) cultivars possessing the *N* gene, which controls resistance to RKNs (*M. incognita*, *M. arenaria*, and *M. javanica*), have been effective as rootstocks to control RKNs in pepper (Table 6.2) (Oka *et al.*, 2004; Thies and Fery, 1998, 2000).

Table 6.2. Rootstock for root-knot resistance in sweet pepper

Scion/rootstock	RGI (1-10 scale)	Fruit yield (kg/6m^2) 20 plants
Celia non-grafted	7.8a	20.2b
Celia grafted on AR-96023 (*C. annuum*)	0.7b	43.8a

6.1.4.6. Integrated methods

i) Physical and cultural methods: Integrating summer plowing with soil solarization with polythene mulching effectively reduced *M. javanica* population in brinjal (Jain and Gupta, 1991) (Table 6.3).

Table 6.3. Integrated management of root-knot nematodes by summer plowing/ fallowing and nursery bed treatment in capsicum.

Treatment	Reduction in nematode population (%)		Yield in kg/m^2	
	Hisar	Vellayani	Hisar	Vellayani
Plowing + Fallowing for 15 days	87.00	66.20	28.98	---
Plowing + Covering with polythene sheet for 15 days	94.90	77.00	18.84	78.00
No plowing + No covering	77.60	---	---	---

ii) Physical, chemicals and botanicals: For effective management of root-knot nematode in chilli, nursery should be covered with polythene sheet during May/June for 15 days after giving light irrigation. At the time of sowing, application of Carbofuran 3G @ 10 g/m^2 in the nursery prior to sowing was found effective. Also, before transplanting, application of Neem cake @ 500 kg/ha in the main field proved beneficial.

iii) Physical and chemicals: Effective management of root-knot nematode (*Meloidogyne* spp.) in chilli was obtained by solarization of soil under turf for two weeks during peak summer, soil incorporation of Carbofuran @ 3.3 g/m^2 before sowing followed by bare root dip treatment of transplantable seedlings with Carbosulfan @ 500 ppm at transplanting in the main field.

iv) Physical and botanicals/bioagents/chemicals: The effect of solarization, and its combinations with dazomet (400 kg/ha), chicken manure (10 tons/ha) or straw (500 kg/ha), on soil-borne nematodes and crops were demonstrated at various sites. Combinations of solarization with basamid (400 kg/ha), chicken manure (10 tons/ha) or *Trichoderma* were the most effective applications for the management of root-knot nematodes infecting bell pepper. MB and solarization + straw (500 kg/ha) were partially effective. However, it can be said that all treatments effectively controlled nematodes (Table 6.4). The galling index overall averages of six pepper plastic houses were 0.10, 0.89, 0.98, 1.92, 2.04, and 5.95 for solarization + *Trichoderma*, solarization + manure, solarization + dazomet, MB, solarization + straw, and check, respectively.

Table 6.4. Effects of treatments on the nematode population as percentage of non-treated check during the growing season of sweet pepper.

Date	Treatments and control rates (%)				
	Solarization + Basamid	Solarization + *Trichoderma*	Solarization + Manure	Solarization + Straw	Methyl bromide
03-10-2000	100.0	100.0	100.0	100.0	100.0
19-10-2000	98.8	100.0	100.0	100.0	100.0
07-11-2000	100.0	100.0	100.0	100.0	100.0
28-11-2000	89.9	100.0	80.0	100.0	97.8
19-12-2000	100.0	100.0	100.0	100.0	100.0
09-01-2001	96.3	99.2	79.9	99.3	100.0

30-01-2001	99.7	100.0	100.0	100.0	99.3
20-02-2001	100.0	100.0	100.0	100.0	97.0
13-03-2001	100.0	99.3	99.9	98.1	97.9
03-04-2001	79.7	98.7	96.2	94.3	91.6
25-04-2001	99.5	95.5	99.8	70.9	90.4
16-05-2001	72.9	88.8	94.4	93.2	87.4
06-06-2001	74.6	91.4	95.8	99.6	59.4
23-07-2001	40.8	59.3	84.9	0.0	0.0

100 is complete control and 0 shows no control.

v) Biofumigation and solarization: Biofumigation combined with soil solarization in a greenhouse in Spain provided levels of *M. incognita* control in bell pepper, similar to levels achieved using methyl bromide (Bello *et al.*, 2004).

The best results were obtained when biofumigation with solarization was applied at the end of August or the beginning of September. Sweet pepper yield using fresh sheep manure with soybean flour, or fresh sheep manure with chicken manure treatments, was similar to sweet pepper yield when MB was used. The reiteration of biofumigation with solarization over two or more years led to an improvement in pathogen and weed control, higher plants, an increase in yield and improvements in the soil physical properties, with a higher level of macro- and micronutrients as well as electrical conductivity.

vi) Two bioagents: Enhancement of the yield of the bell pepper crop and reduction in the number of nematodes in the roots and soil and root-galling index (*M. incognita*) under field conditions was achieved by integration of *P. fluorescens* and *P. chlamydosporia* in the nursery bed (Naik, 2004).

Naik (2004) reported that combination of *T. harzianum* along with *P. fluorescens* increased bell pepper yield (3.320 kg/3m^2 plot) followed by combination of *T. harzianum* and *P. chlamydosporia* (2.960 kg/3m^2). The least gall index was present where the combination of *T. harzianum* and *P. fluorescens* was used (1.5) followed by *T. harzianum* + *P. chlamydosporia* (1.7) and *P. fluorescens* + *P. chlamydosporia* (1.7).

Integrated management of *M. incognita* infecting capsicum was achieved by seed treatment with *P. fluorescens* at 50g/kg seed combined with nursery bed treatment with *P. chlamydosporia* at 50g/m^2. The above treatment was significantly effective in increasing the seedling growth and root colonization with bioagents and in reducing root galling (Table 6.5) (Rao *et al.*, 2004).

Table 6.5. Effect of integration of bioagents on plant growth and management of *Meloidogyne incognita* infecting capsicum in nursery

Treatment/Dose	Seedling weight (g)	No. of galls/10 seedlings	Root colonization of *P. chlamydosporia* (cfu/g)	Root colonization of *P. fluorescens* (cfu/g)
P. chlamydosporia - Nursery treat (25 g/m^2)	3.6	62	15,896	---
P. chlamydosporia - Nursery treat (50 g/m^2)	3.8	59	17,459	---

P. fluorescens - seed treat. (50g/kg)	4.0	55	---	12,563
P. fluorescens - seed treat. (50g/kg) + *P. chlamydosporia* - Nursery treat. (25 g/m²)	4.5	51	16,256	12,249
P. fluorescens - seed treat. (50g/kg) + *P. chlamydosporia* - Nursery treat. (50 g/m²)	4.4	53	16,789	11,897
Control	3.2	82	---	---
CD (P = 0.05)	0.34	7.96	959.72	729.65

Capsicum seedlings raised in nursery treated with *P. fluorescens* at 50 g/kg seed (seed treatment) combined with nursery bed treatment with *P. chlamydosporia* at 50g/m² when transplanted in field, significantly reduced nematode population both in soil and roots, root galling (4.4 as compared to 8.1 in control) and increased root colonization by bioagents, propagule density in soil, parasitization of eggs and fruit yield (5.5 kg/6m² as compared to 4.0 kg/6m² in control) (Table 6.6) (Rao *et al.*, 2004).

Table 6.6. Effect of integration of bioagents on plant growth and management of *Meloidogyne incognita* infecting capsicum under field conditions

Treatment*	Root-knot index (1-10 scale)	Yield (kg/6 m²)	% eggs parasitized by *P. chlamydosporia*	Root colonization (cfu/g)		Propagule density (cfu/g soil)	
				Pc	*Pf*	*Pc*	*Pf*
T1	6.3	4.4	41.00	22,456	---	19,569	---
T2	5.6	4.7	45.00	25,789	---	22,845	---
T3	6.0	5.1	---	---	22,568	---	15,987
T4	4.8	5.3	40.00	21,679	21,789	19,234	16,843
T5	4.4	5.5	42.67	24,587	21,567	23,221	15,359
Control	8.1	4.0	---	---	---	---	---
CD 5%	0.45	0.28	3.78	1246.76	1089.86	1178.32	873.65

*T1 - *P. chlamydosporia* - Nursery treat (25 g/m²), T2 - *P. chlamydosporia* - Nursery treat (50 g/m²), T3 - *P. fluorescens* - seed treat (50g/kg), T4 - *P. fluorescens* - seed treat (50g/kg) + *P. chlamydosporia* - Nursery treat (25 g/m²), T5 - *P. fluorescens* - seed treat (50g/kg) + *P. chlamydosporia* - Nursery treat (50 g/m²).

Integration of *P. lilacinum* with *P. chlamydosporia* in nursery was found to be most effective in increasing the plant growth of capsicum seedlings and reducing the nematode population both in soil and roots by 59 and 72%, respectively. When capsicum seedlings raised in nursery treated with *P. lilacinum* + *P. chlamydosporia* were transplanted in the main field, there was reduction in root gall index by 51% and increase in fruit yield by 44% compared to control (Naik, 2004) (Table 6.7).

Table 6.7. Effect of integration of bioagents for the management of *Meloidogyne incognita* and fruit yield of capsicum.

Treatment	Dose/m^2	Root-knot index (1-10 scale)	Yield in kg/2 m^2
P. chlamydosporia	50 g (10^6 cfu/g)	6.20	2.58
T. harzianum	50 g (10^6 cfu/g)	7.20	2.36
P. lilacinum	50 g (10^6 cfu/g)	6.70	2.61
P. chlamydosporia + *P. lilacinum*	25 g + 25 g	4.35	3.28
T. harzianum + *P. lilacinum*	25 g + 25 g	4.75	2.86
P. chlamydosporia + *T. harzianum*	25 g + 25 g	5.35	2.07
P. chlamydosporia + *T. harzianum* + *P. lilacinum*	17 g + 17 g + 17 g	5.10	2.05
Carbofuran	25 g	7.80	2.30
Control	---	8.75	2.04
CD (P = 0.05)	---	0.44	0.77

vii) Bioagents and botanicals: In capsicum, the seedling stand was good where the combination of Neem based *P. fluorescens* and *T. harzianum* weres used (4.4), followed by seed treatment with *P. fluorescens* + soil application of *T. harzianum* (Naik, 2004).

Capsicum seedling treatment with *P. fluorescens* and soil application of neem cake enriched with *P. fluorescens* was found to be significantly effective in reducing the population of nematodes. This treatment reduced root-populations of *R. reniformis* by 69% and *M. incognita* by 68%. The yield of capsicum was significantly increased by 19% (Table 6.8). In general, the growth of capsicum plants treated with *P. fluorescens* was better (Rao, 2011).

Table 6.8. Effect of neem cake enriched with *P. fluorescens* and vermi-compost on root population densities of *M. incognita* and *R. reniformis*, PGPR root-colonization and yield of capsicum (mean of 5 replicates)

Treatments	No. of *M. incognita*/ 10 g root	No. of *R. reniformis*/ 10 g root	Root colonization of *P. fluorescens* (CFU/g ×10^6)	Increase in yield of capsicum (%)
T1	22.54	19.83	1.7	6
T2	15.28	13.65	4.0	14
T3	16.43	15.93	2.3	12
T4	12.32	10.07	6.4	19
T5	14.13	14.95	3.6	16
T6	22.17	19.83	-	5
T7	26.74	24.38	-	4
T8	18.72	17.55	2.5	8
T9	19.86	18.53	2.0	5

T10	38.20	32.50	-	-
CD (P = 0.05)	3.24	2.68	0.45	-

T1-seedlings of capsicum treated with *P. fluorescens* (CFU 2×108/ml); T2 - untreated seedlings sown in the plots mixed with neem cake enriched with *P. fluorescens* 20 g/m^2; T3 - untreated seedlings sown in the plots mixed with vermicompost enriched with *P. fluorescens* 100 g/m^2; T4- seedlings treated with *P. fluorescens* sown in plots mixed with neem cake enriched with *P. fluorescens*; T5- seedlings treated with *P. fluorescens* sown in plots mixed with vermicompost enriched with *P. fluorescens*; T6- Untreated seedlings transplanted in plots mixed with neem cake alone 20g/m^2; T7- Untreated seedlings transplanted in plots mixed with vermicompost alone 20g/m^2; T8- seedlings treated with *P. fluorescens* were transplanted in plots mixed with neem cake alone 20g/m^2; T9- seedlings treated with *P. fluorescens* transplanted in plots mixed with vermicompost alone 20g/m^2; T10 - untreated seedlings transplanted in the plots without any treatment, served as control.

viii) Botanicals and bioagents/chemicals: Soil application of poultry manure at 10 t/ha a week prior to transplanting + *P. lilacinum* (10^9/conidia/g) at 25 kg/ha spore dust with carrier at the time of transplanting; or neem cake at 2 t/ha a week prior to transplanting + *P. lilacinum* (10^9 conidia/g) at 25 kg/ha spore dust with carrier at the time of transplanting; or carbofuran at 2 kg/ha in 2 equal splits (one at the time of transplanting and the other after 75 days) + *P. lilacinum* (10^9 conidia/g) at 25 kg/ha spore dust with carrier at the time of transplanting considerably reduced root galling and also gave higher capsicum fruit yield over control (Vyas *et al.*, 2009) (Table 6.9).

Table 6.9. Effect of different organic amendments, bioagent and nematicide for the management of *Meloidogyne incognita* infecting capsicum

Treatment	Gall index (0-5 scale)	Yield (tons/ha)	BCR
P. lilacinum at 25 kg/ha + Poultry manure at 10 t/ha	2.0	42.088	18.5
P. lilacinum at 25 kg/ha + Mustard cake at 2 t/ha	2.0	44.801	7.7
P. lilacinum at 25 kg/ha + Neem cake at 2 t/ha	1.8	41.745	5.8
P. lilacinum at 25 kg/ha + Carbofuran at 2 kg a.i. /ha	1.9	41.498	13.6
P. lilacinum at 25 kg/ha	2.5	39.941	77.6
Poultry manure at 10 t/ha	2.3	40.000	18.5
Mustard cake at 2 t/ha	1.9	41.810	6.7
Neem cake at 2 t/ha	1.9	39.954	5.2
Carbofuran at 2 kg a.i. /ha	2.0	40.530	14.9
Control	3.9	28.107	---

ix) Botanicals and chemicals: In greenhouse and field studies, Neem cake @ 50 g/m^2 + Carbofuran 3G @ 10 g/m^2 application in the pots as well as in the field was most effective in reducing nematode population as well as increasing the plant growth and yield in chilli (Thakur, 2016).

6.2. INTERACTION OF ROOT-KNOT NEMATODE WITH BACTERIAL WILT

6.2.1. Interaction

Capsicum is prone to many soil-borne diseases among which the bacterial wilt (*Ralstonia solanacearum*) in combination with root-knot nematode (*Meloidogyne incognita*) takes heavy toll every year all over the world (Naik *et al.*, 2003). The disease ratings of bacterial wilt in treated plants (inoculated with *M. incognita* one month later with *R. solanacearum*; inoculated with *R. solanacearum* and 2 weeks later with the nematodes; or inoculated with nematodes and *R. solanacearum* simultaneously) were higher than ratings in plants inoculated with bacteria only. It is concluded that a complex infection of root-knot nematodes and *R. solanacearum* increased disease severity and reduced the resistance of bacterial wilt-resistant chilli plants.

6.2.2. Management

6.2.2.1. Integrated methods

i) Bioagents and botanicals: Naik *et al.* (2003) found that treatment of the nursery beds with Neem based formulations of *T. harzianum* + *P. fluorescens* followed by seed treatment with talc based *P. fluorescens* + nursery bed treatment with Neem based *P. fluorescens* enhanced the seedling stand.

Enhancement of the bell pepper crop yield by 37% and significant reduction in bacterial wilt disease complex to the extent of 70% was obtained by integration of Neem based formulations of *P. fluorescens* and *P. chlamydosporia/T. harzianum* at 40 g/m^2 in nursery beds and transplanting these seedlings in the main field (Table 6.10) (Rao *et al.*, 2002). Combinations of the bioagents did not affect the colonization of the individual bioagents on the roots and hence the transplants carried the bioagents to the main field which controlled the wilt disease complex.

Table 6.10. Effect of integration of neem-based bioagents on the growth of transplants and management of disease complex and yield of capsicum.

Treatment	Seedling weight (g)	Root-knot index (1-10)	Disease index (1-9)	Yield in kg/4 m^2
Seed treat. with *P. fluorescens*	421	5.6	6.4	4.3
Seed treat. with neem-based *P. fluorescens*	428	5.2	6.7	4.7
Nursery treat. with *P. fluorescens*	435	4.6	5.4	4.8
Nursery treat. with *P. chlamydosporia*	364	4.4	7.3	3.2
Nursery treat. with *T. harzianum*	374	4.8	7.0	3.4
Nursery treat. with *P. fluorescens* + *P. chlamydosporia*	463	4.1	5.2	4.0
Nursery treat. with *P. fluorescens* + *T. harzianum*	493	3.8	3.5	5.1
Control	340	8.7	8.2	2.6
CD at 5%	27.20	0.49	0.38	0.25

ii) Two bioagents: Combined application of neem-based formulations of *P. fluorescens* and *P. chlamydosporia/T. harzianum* at 40 g/m^2 in nursery beds and transplanting these seedlings in the main field resulted in significant reduction in disease index and root-knot index in capsicum to the tune of 70% and increased the crop yield by 37% (Rao *et al.*, 2002) (Table 6.11). Combinations of the bioagents did not affect the colonization of the individual bioagents on the roots and hence the transplants carried the bioagents to the main field. This has resulted in the effective management of the pathogens involved in the disease complex.

Table 6.11. Effect of integration of neem-based bioagents on plant growth and management of root-knot and bacterial wilt disease complex and yield of capsicum.

Treatment	Seedling weight (g)	Root-knot index (1-10)	Disease index (1-9)	Yield in kg/4 m^2
Seed treatment with *P. fluorescens*	421	5.6	6.4	4.3
Seed treatment with neem-based *P. fluorescens*	428	5.2	6.7	4.7
Nursery treatment with *P. fluorescens*	435	4.6	5.4	4.8
Nursery treatment with *P. chlamydosporia*	364	4.4	7.3	3.2
Nursery treatment with *T. harzianum*	374	4.8	7.0	3.4
Nursery treatment with *P. fluorescens* + *P. chlamydosporia*	463	4.1	5.2	4.0
Nursery treatment with *P. fluorescens* + *T. harzianum*	493	3.8	3.5	5.1
Control	340	8.7	8.2	2.6
CD at 5%	27.20	0.49	0.38	0.25

The seedling stand was good when the nursery beds were combinedly treated with neem-based formulations of *T. harzianum* + *P. fluorescens* followed by seed treatment with talc-based *P. fluorescens* + nursery bed treatment with neem-based *P. fluorescens* (Table 6.12) (Naik *et al.*, 2003).

Table 6.12. Effect of integration of bioagents on nursery stand of capsicum seedlings

Treatment	Nursery stand of seedlings*
Seed treatment with talc-based *P. fluorescens*	3.4
Nursery bed treatment with neem-based *P. fluorescens*	3.2
Seed treatment with talc-based *P. fluorescens* + Nursery bed treatment with neem-based *P. fluorescens*	4.2
Nursery bed treatment with *T. harzianum* + *P. fluorescens*	4.4
Control	3.0

* 3 – Poor, 4 – Good, 5 – Very good

Integration of seed treatment with *P. fluorescens* and nursery bed treatment with *P. chlamydosporia* at 25 or 50 g/m^2 increased plant height, weight of seedlings and root colonization with bioagents; and reduced root galling in bell pepper (Table 6.13) (Naik *et al.*, 2003).

Table 6.13. Effect of integration of bioagents on plant growth, root galling and root colonization by bioagents in capsicum under nursery conditions

Treatment*	Plant height (cm)	Seedling weight (g)	No. of galls/10 seedlings	*P. chlamydosporia* root colonization (cfu/g)	*P. fluorescens* root colonization (cfu/g)
T_1	12.58	3.6	62	15,896	---
T_2	14.63	3.8	59	17,459	---
T_3	15.87	4.0	55	---	12,563
T_4	18.54	4.5	51	16,256	12,349
T_5	17.42	4.4	53	16,789	11,897
T_6	11.23	3.2	82	---	---
CD (P = 0.05)	1.67	0.34	7.96	959.72	729.65

*T_{1-} Nursery bed treatment *P. chlamydosporia* at 25 g/m², T_{2-} Nursery bed treatment *P. chlamydosporia* at 50 g/m², T_{3-} Seed treatment with *P. fluorescens*, T_{4-} T_1 + T_3, T_{5-} T_2 + T_3, T_{6-} Control.

The seedlings raised in nursery beds treated with *P. chlamydosporia* at 25 or 50 g/m² + seed treatment with *P. fluorescens* when transplanted in the main field gave least root galling and increased fruit yield, colonization of roots with the bioagents, and parasitization of eggs by *P. chlamydosporia* (Table 6.14) (Naik *et al.*, 2003).

Table 6.14. Effect of integration of bioagents on fruit yield, root galling and root colonization by bioagents in capsicum under field conditions

Treatment*	Root-knot index	Fruit yield (kg/6m²)	*P. chlamydosporia* root colonization (cfu/g)	*P. fluorescens* root colonization (cfu/g)	% eggs parasitized by *P. chlamydosporia*
T1	6.3	4.4	22,456	---	41.00
T2	5.6	4.7	25,789	---	45.00
T3	6.0	5.1	---	22,568	---
T4	4.8	5.3	21,679	21,789	40.00
T5	4.4	5.5	24,587	21,567	42.67
T6	8.1	4.0	---	---	---
CD (P 0.05)	0.45	0.28	1246.76	1089.86	3.78

*T1 - Nursery bed treatment *P. chlamydosporia* at 25 g/m², T2 - Nursery bed treatment *P. chlamydosporia* at 50 g/m², T3 - Seed treatment with *P. fluorescens*, T4 - T1 + T3, T5 - T2 + T3, T6 - Control.

Naik (2004) reported that enhancement of bell pepper fruit yield (3.320 kg/3m² plot) was achieved by integration of *T. harzianum* along with *P. fluorescens* followed by integration of *T. harzianum* and *P. chlamydosporia* (2.960 kg/3m^2). Integration of *T. harzianum* and *P. fluorescens*, followed by *T. harzianum* + *P. chlamydosporia*, and *P. fluorescens* + *P. chlamydosporia* gave the least root galling of 1.5, 1.7 and 1.7, respectively. Integration of *T. harzianum* + *P. fluorescens* followed by *T. harzianum* + *P. chlamydosporia* showed the percent bacterial wilt disease (*R. solanacearum)* incidence of 16.90 and 17.20, respectively (Table 6.15).

Table 6.15. Effect of integration of bioagents on root galling, wilt disease incidence and fruit yield of capsicum

Treatment	No. of fruits / plant	Fruit yield (kg/3m^2)	Root-knot index	Wilt disease incidence (%)
Neem + wheat bran formulation (8:2) of *T. harzianum*	31.2	2.30	1.90	23.8 (29.20)
Liquid broth formulation of *P. fluorescens*	34.2	2.40	1.75	21.4 (27.56)
Neem + pongamia + wheat bran formulation (4:4:2) of *P. chlamydosporia*	32.8	2.04	1.92	26.7 (31.11)
T. harzianum + *P. fluorescens*	46.2	3.32	1.50	16.9 (24.20)
T. harzianum + *P. chlamydosporia*	40.6	2.96	1.70	17.2 (24.50)
P. fluorescens + *P. chlamydosporia*	40.0	2.48	1.70	17.8 (24.95)
Neem + wheat bran	27.2	1.93	2.25	30.5 (33.52)
Neem + pongamia + wheat bran	29.8	2.00	2.00	36.4 (37.11)
Control	26.2	1.78	3.80	96.0 (78.46)
CD (P = 0.05)	7.38	0.41	0.32	3.28

Naik *et al.* (2003) reported that soil application of organically developed *Purpureocillium lilacinum* (50 g/m^2 of formulated product containing 10^6 spores/g) along with pure culture of *Bacillus pumilis* (10^8 cfu/ml) to the nursery beds of bell pepper increased plant growth, root galling and root colonization by the bioagents (Table 6.16).

Table 6.16. Effect of integration of bioagents on plant growth and management of disease complex in bell pepper in nursery beds

Treatment*	Seedling length (cm)	Seedling weight (g)	No. of galls /10 plants	Root colonization (cfu/g)	
				P. lilacinum	*B. pumilis*
T1	12.63	3.8	52	13,342	---
T2	15.53	4.0	50	16,539	---
T3	15.17	4.2	55	---	14,431

T4	18.43	4.4	44	12,984	13,964
T5	19.25	4.6	40	16,839	14,291
T6	10.53	3.3	76	---	---
CD (P = 0.05)	1.29	0.37	6.28	1087.95	976.75

*T1 – Nursery bed treatment with *P. lilacinum* (25 g/ m^2), T2 - Nursery bed treatment with *P. lilacinum* (50 g/ m^2), T3 - Nursery bed treatment with *B. pumilis*, T4 – T1+ T3, T5 – T2 + T3, T6 – Control.

Transplanting of such biological control agents (*Purpureocillium lilacinum*) colonized plants in the main field resulted in significant reduction of root galling and wilt incidence, and increased root colonization, propagule density in soil, egg parasitization and fruit yield (Table 6.17) (Naik *et al.*, 2003).

Table 6.17. Effect of integration of bioagents on root galling, wilt incidence, root colonization, propagule density and yield of bell pepper under field conditions

Treatment*	**Root-knot index**	**Mort-ality %**	**Yieldkg/ 6m^2**	**Root colonization (cfu/g)**		**Propagule density (cfu/g soil)**	
				P. lilacinum	***B. pumilis***	***P. lilacinum***	***B. pumilis***
T1	6.5	68.90	4.8	30,653	---	21,346	---
T2	5.3	52.45	5.4	34,289	---	23,467	---
T3	7.0	34.62	7.5	---	25,875	---	17,321
T4	4.3	25.86	8.2	31,764	21,459	20,457	15,347
T5	4.1	17.69	8.5	34,762	20,764	22,689	14,569
T6	8.4	91.45	0.5	---	---	---	---
CD(P=0.05)	0.58	8.94	0.35	1176.52	1087.54	1234.87	897.34

*T1 – Nursery bed treatment with *P. lilacinus* (25 g/ m^2), T2 - Nursery bed treatment with *P. lilacinus* (50 g/ m^2), T3 - Nursery bed treatment with *B. pumilis*, T4 – T1+ T3, T5 – T2 + T3, T6 – Control.

iii) Bioagents, botanicals and AMF: The efficacy of *P. fluorescens* and *Glomus mosseae*, against root-knot and reniform nematodes (*Meloidogyne incognita* and *Rotylenchulus reniformis*) were evaluated to produce nematode free seedlings of capsicum.

Seed treatment with talc-based formulation of *P. fluorescens* was able to increase the growth of the seedlings of capsicum. Seed treatment with talc-based formulation of *P. fluorescens* and talc based or neem formulation of *P. fluorescens* with coco-peat treatment reduced root galling due to *M. incognita.* This resulted in production of a greater number of nematode free seedlings of capsicum. There was a significant increase in the production of nematode free capsicum seedlings in *G. mosseae* with coco-peat treatment + seed treatment with talc-based formulation of *P. fluorescens* + neem-based formulation of *P. fluorescens* with coco-peat treatment. There appears to be synergistic interaction among *P. fluorescens, G. mosseae* and neem cake and complementing each other resulting in additive effect on plant height. Neem cake helped in the increased root colonization of *P. fluorescens* and *G. mossease* in capsicum (Table 6.18). When these seedlings were transplanted in the main field, there was significant reduction in the infestation of *M. incognita, M. javanica* (root-knot nematodes), *Rotylenchulus reniformis* (reniform nematode) on capsicum (Gavaskar *et al.*, 2011).

Table 6.18. Effect of application of *P. fluorescens, G. mossease* and coco peat on growth of tomato seedlings, infestation of *M. incognita* and root-colonization with bioagents (mean of 5 replicates)

Treatments	Seedling height (cm)	% nematode free seedlings	Root colonization of *P. fluorescens* (cfu/ g×10^6)	Root colonization of *G. mosseae* (%)
T1	19.04	87	1.56	-
T2	20.52	90	3.42	-
T3	21.60	91	4.82	-
T4	21.28	85	0.00	27
T5	23.85	92	2.25	-
T6	21.35	90	4.94	-
T7	22.86	92	6.39	-
T8	24.27	94	5.68	34
T9	25.16	97	8.26	42
T10	22.37	93	4.66	37
T11	24.46	94	23.76	40
T12	17.26	86	15.20	-
CD (P = 0.05)	2.66	-	1.72	-

T1 - seed treatment with talc based formulation of *P. fluorescens*, T2 – talc-based formulation of *P. fluorescens* with coco-peat treatment, T3 – neem-based formulation of *P. fluorescens* with coco-peat treatment, T4 - *G. mosseae* with coco-peat treatment, T5 - seed treatment with talc-based formulation of *P. fluorescens* + *G. mosseae* with coco-peat treatment, T6 - seed treatment with talc-based formulation of *P. fluorescens* + talc-based formulation of *P. fluorescens* with coco-peat treatment, T7 - seed treatment with talc-based formulation of *P. fluorescens* + neem-based formulation of *P. fluorescens* with coco-peat treatment, T8 - seed treatment with talc-based formulation of *P. fluorescens* + talc-based formulation of *P. fluorescens* with coco-peat treatment + *G. mosseae* with coco-peat treatment, T9 - seed treatment with talc-based formulation of *P. fluorescens* + neem-based formulation of *P. fluorescens* with coco-peat treatment + *G. mosseae* with coco-peat treatment, T10 – talc-based formulation of *P. fluorescens* with coco-peat treatment + *G. mosseae* with coco-peat treatment, T11 – neem-based formulation of *P. fluorescens* with coco-peat treatment + *G. mosseae* with coco-peat treatment, T12 - control without seed and coco-peat treatment.

6.3. INTERACTION OF ROOT-KNOT NEMATODE, *Meloidogyne javanica* WITH *Fusarium oxysporum*

6.3.1. Interaction

The disease complex involving *Meloidogyne javanica* and *Fusarium oxysporum/F. solani* inflict severe losses to chilli crop (Haseeb, 2003). The simultaneous occurrence of both root-knot nematode and Fusarium wilt disease caused enhanced disease development and chilli yield loss.

6.3.2. Management

6.3.2.1. Integrated methods

i) Bioagents and botanicals: Integration of *T. harzianum* at 50 kg/ha with neem seed powder at 250 kg/ha in nursery beds was found to be highly effective in increasing the number of germinated seedlings (87/0.5 m^2 as compared to 39/0.5 m^2 in control) and fresh weight of seedlings/bed (146.9g as compared to 61g in control). The above treatment was also effective in reducing root galling due to *M. incognita* (0.7 RKI as compared to 3.5 RKI in control) and percent root infection by *F. solani* (5% as compared to 41.5% in control) in nursery beds (Table 6.19) (Kumar *et al.*, 2009).

Table 6.19. Effect of biocontrol agents and organic amendments against *M. incognita-F. solani* disease complex in chilli

Treatment	No. of seedlings emerged/bed	Fresh total wt. of seedlings/ bed (g)	Root-knot index	% root infection
Untreated control	39	61.0	3.5	41.5
Trichoderma harzianum	67	110.3	1.5	15.0
Aspergillus niger	62	100.0	2.3	21.5
Purpureocellium lilacinum	60	97.5	1.5	15.0
Pseudomonas fluorescens	65	105.7	1.9	17.3
Neem seed powder	64	105.0	1.3	15.5
Farm yard manure	45	71.9	3.3	35.5
T. harzianum + Neem seed powder	87	146.9	0.7	5.0
A. niger + Neem seed powder	77	128.7	1.0	13.0
P. lilacinus + Neem seed powder	74	123.0	0.9	13.5
P. fluorescens + Neem seed powder	81	136.7	0.7	5.5
T. harzianum + Farmyard manure	80	134.5	1.2	7.7
A. niger + Farmyard manure	71	115.3	2.0	15.5
P. lilacinus + Farmyard manure	70	113.0	1.3	23.5
P. fluorescens + Farmyard manure	75	124.9	1.5	13.0
Carbofuran	72	117.0	0.6	25.5
CD (P = 0.05)	3.53	7.91	0.06	0.65

ii) Two bioagents: *Pseudomonas aeruginosa* and *Purpureocellium lilacinum* when used together significantly reduced infection of the disease complex on chilli (Perveen *et al.*, 1998). Use of *P. aeruginosa* and *P. lilacinum* used alone or together significantly ($P<0.05$) increased plant height and reduced infection of *M. javanica* and root infecting fungi viz., *Macrophomina phaseolina, Rhizoctonia solani, F. solani* and *F. oxysporum* on chilli. *P. aeruginosa was* more effective than *P. lilacinum* in reducing the *M. javanica* infection (Table 6.20).

Table 6.20. Effect of *Pseudomonas aeruginosa* and *Purpureocellium lilacinum* on plant height and control of root rot disease complex in chilli

Treatment	Plant height (cm)	Root-knot index	Infection %			
			M. phaseolina	*R. solani*	*F. solani*	*F. oxysporum*
Control	10.5	3.3	31	19	75	37
P. lilacinum (*PL*)	14.5	2.9	19	0	56	44
P. aeruginosa (*PA*)	16.5	2.5	12	6	75	21
PL + PA	14.7	2.1	12	25	69	12
CD (P = 0.05)	2.2	0.34	6.1	6.1	6.1	6.1

Bare root-dip treatment with *Pseudomonas aeruginosa* along with or without *T. harzianum*, *T. koningii* and *T. hamatum* significantly controlled infection of roots by *F. solani* and *M. javanica* on chilli. Combined use of *T. harzianum* along with *P. aeruginosa* caused the greatest reduction in root galling by *M. javanica* (Siddiqui *et al.*, 1999).

6.4. INTERACTION OF ROOT-KNOT NEMATODE WITH FOOT AND ROOT ROT

6.4.1. Interaction

The root-knot nematode, *Meloidogyne incognita* interacts with foot and root rot, *Phytophthora capsici* in causing a root rot disease complex. Kim *et al.* (1989) reported that the disease incidence was raised to 54% when hot pepper was inoculated with *Meloidogyne* sp. and *Phytophthora* sp.

6.4.2. Management

Capsicum seed treatment with consortia of *P. fluorescens* (3 strains) at the rate of 10 g/kg and subsequent application of consortia formulation of *P. fluorescens* at the rate of 15 g/m^2, significantly reduced *M. incognita* in capsicum roots by 74%; reduced the incidence of *Phytophthora* foot and root rot and increased the yield by 26%. Benefit: cost ratio (calculated for the additional cost of the bio-pesticides and additional returns accrued by the application of the bio-pesticide) was 3.8 (Rao *et al.*, 2009).

6.5. INTERACTION OF ROOT-KNOT NEMATODE WITH DAMPING-OFF

6.5.1. Interaction

Pythium aphanidermatum and *Rhizoctonia solani* were both found to interact with *M. incognita* on chilli, causing some loss of nematode resistance in two cultivars tested (Hasan, 1985a, b), whereas the interaction of *P. debaryanum* with this nematode appeared to be due to physiological response of the plants to nematode infection, making the roots more susceptible to invasion by the fungus (Brodie and Cooper, 1964).

6.5.2. Management

6.5.2.1. Integrated methods: Biological control of *P. aphanidermatum*-*M. incognita* disease complex in chilli with organic amendments (FYM and neem cake); antagonistic organisms *T. viride* and *T.*

harzianum (against *P. aphanidermatum*); and *Paecilomyces lilacinus* (against *M. incognita*) was reported by Karthikeyan *et al.* (1999).

6.6. CONCLUSION

Plant parasitic nematodes constitute one of the major limiting factors for cultivation of chilli and bell pepper. Several nematode species are associated with chilli and bell pepper and few of them negatively impact yield. Severe yield losses have been reported in most regions where chilli and bell pepper are grown. To select suitable nematode management practices, correct diagnosis of the causal agent is of prime importance. Overall, these nematode problems in chilli and bell pepper are better managed when integrated management practices are used, i.e. regulatory methods (plant quarantine and seed certification), cultural practices (crop rotation, cover crops), genetic control, and ultimately by the use of nematicides. Therefore, for a sustainable cropping of chilli and bell pepper; growers, extension services, and researchers must consider these nematodes holistically, the impact they cause and whether these management practices are economically, environmentally, and technically sound.

Integrated nematode management could be tackled with the intelligent planning of nematode-suppressive crop sequences, summer plowing, organic manuring, clean cultivation, adjusting sowing time, water and irrigation management and sensible use of nematicides. The increasing concerns on the ill effects of chemical pesticides on the environment have driven recent research interests on the use of several alternative strategies like botanicals, biopesticides and cropping system research for management of nematodes. Some of the successes have been obtained for managing plant parasitic nematodes with neem-based formulations, fungal formulation of *Purpureocillium lilacinum* and *Trichoderma* spp. and bacterial formulations of *Pasteuria penetrans* and *Pseudomonas fluorescens*. Inconsistent efficacy of the biological control agents in different agro-climatic regions is the main reason for not adopting them widely under field scale. Integration of more than one option can be explored based on their compatibility, economic viability and availability to the farmers. The major challenge ahead for the greater benefit of the farmers is the need for development of holistic approaches for managing field problems including nematodes induced disease complexes with other soil-borne pathogens. There is a need for development of economic, eco-friendly, and sustainable bio-intensive integrated nematode management strategies using all the available components.

Chapter - 7

FRENCH BEAN, *Phaseolus vulgaris*

7.1. INTRODUCTION

French bean (*Phaseolus vulgaris*), also known as common, haricot, kidney, string, salad bean, runner bean or snap bean, is an important leguminous vegetable crop. It is consumed as tender pods, shelled green beans and dry beans. It is a nutritious vegetable which contains proteins (1.7g), calcium (50 mg), phosphorus (28 mg), iron (1.7 mg), carotene (132 mg), Thiamine (0.08 mg), Riboflavin (0.06 mg) and vitamin C (24 mg/100 g of edible pods). It is largely grown in hilly areas of Himachal Pradesh, Jammu and Kashmir and north-eastern states during summer, winter and autumn crop in parts of Uttar Pradesh, Maharashtra, Karnataka and Andhra Pradesh.

Root-knot and reniform nematodes are the major limiting factors in successful production of French bean crop.

7.2. ROOT-KNOT NEMATODES, *Meloidogyne incognita, M. javanica*

M. incognita and *M. javanica* appear to be the most common root-knot species of French beans and have been reported causing damage in tropics and subtropics. The root-knot nematode, *M. arenaria* was recorded on beans for the first time from Bihar, India (Lal and Ansari, 1960).

7.2.1. Crop losses

M. incognita was responsible for 18.77 % avoidable yield loss in French bean cv. contender (Mohapatra, 2015). Crop losses have been estimated by various workers varying from 19.38 to 43.48% loss in pod yield of French bean (Das, 1994; Parvatha Reddy and Singh, 1981), while *M. javanica* caused 30-40% loss in yield (Sharma *et al.*, 2002).

7.2.2. Symptoms

Above-ground symptoms of severe root-knot infestation include patches of chlorotic, stunted, necrotic, or wilted plants. Earlier wilting was noticed when infested plants were under temperature or moisture stress compared to other plants.

The characteristic under-ground symptom of infected plants is the formation of root galls stimulated by feeding of nematodes. Root-galling and nematode reproduction harms the plant because they both take away nutrients and photosynthates from the developing top growth and harvestable bean crop. Gall size on roots of French beans is variable. Large egg-masses are present on galls. Severely galled roots can be short and thick.

7.2.3. Interaction with other pathogens

M. incognita in association with soil-borne fungus such as *F. solani* was responsible for root rot disease complex in French beans. In the presence of nematode, the root rot due to *F. solani* increased considerably (Singh *et al.*, 1981).

M. incognita in association with soil-borne fungus such as *R. solani* was responsible for root-rot disease complex in French beans. In the presence of nematode, the root rot due to *R. solani* increased considerably (Parvatha Reddy *et al.*, 1979).

7.2.4. Management

7.2.4.1. Physical methods: Rabbing by rice husk or sawdust 10-15 days before sowing is effective in reducing root-knot nematode infestation in French beans (Ahmed and Choudhury, 2004c).

Root-knot nematodes can be effectively controlled by a 4-8 weeks solarization, if the land will remain non-cropped during summer. Data in Table 7.1 show that solarization gave good control of nematodes. The Reproduction factor (RF) in solarized treatment was 0.47, while in non-solarized plots, reduction was 0.98.

Table 7.1. Effect of solarization on the control of nematodes on broad bean

Treatment	Average number of nematodes per 200 g soil*		Reproduction factor (RF) RF = Pf/Pi
	(Pi)	(Pf)	
Bean - solarized	85	40	0.47
Bean - non-solarized	95	93	0.98

* Pi = Initial nematode population; Pf = Final nematode population.

7.2.4.2. Cultural methods

i) Crop rotation: Cropping sequence of French bean-*ahu* rice-sesame is effective in controlling *M. incognita* in French beans (Ahmed and Choudhury, 2004a).

The root-knot nematode damage in bean crop was reduced by rotation with certain cover/ green manure crops like crotalaria, hairy indigo, joint vetch, pigeon pea, and velvet bean.

Rotation of root-knot nematode resistant processing tomatoes with common beans was successful in preventing root-knot damage. This rotation worked because root-knot nematodes were unable to reproduce on the resistant tomatoes, thereby reducing their numbers to below damaging levels. A similar reduction in root-knot nematodes can be achieved by planting beans after a crop of root-knot nematode-resistant cotton.

Growing of chilli or groundnut or finger millet in rotation with French beans significantly reduced the root-knot nematode population (Ramappa, 1988).

ii) Intercropping: Reduction in root-knot nematode, *M. incoginita* reproduction was obtained when French bean was intercropped with Chenopodium, bell pepper, maize, marigold, and sesame. Maize, marigold, and sesame did not allow multiplication of the nematode and reduced the population by 0.26, 0.61 and 0.68 times, respectively.

Intercropping of French bean with finger millet, chilli and groundnut helps in the management of *M. incognita* Race 2 (Ramappa, 1988).

iii) Time of planting: Early planting during the end of winter or early in the spring (temperatures below 15 °C) would help bean plants to escape from nematode infection and establish well. Yield would increase because the large root system can withstand the damage caused by delayed nematode invasion. Moreover, the beans would be harvested by the end of spring or early in summer, thus limiting the number of generations produced (often to only one) and overall population densities.

iv) Organic amendments: Ahmed and Choudhury (2004b) reported that the reduction in root galling (64.4 compared to 122.9 in control), number of egg masses (16.0 compared to 44.4 in control), final nematode population (253 compared to 437 in control), and in increasing plant growth, and yield (3.375 t/ha compared to 0.715 t/ha in control) by spot application of neem cake at 16.6 g/spot 15 days before sowing.

7.2.4.3. Chemical methods: Parvatha Reddy (1984) reported that the effective management of the root-knot nematode, *M. incognita* infecting French bean and enhancement of pod yield was obtained by seed treatment with Carbofuran and Fenamiphos each at 1 percent concentration. Seed treatment with Oxamyl at 3-10% prevented development of *M. incognita*.

Root-knot nematodes on French beans can be effectively managed by soil application of Carbofuran 3G @ 1 kg a.i. /ha (Choudhury *et al.*, 2011). Soil application of Carbofuran 3G @ 2 kg a.i. /ha was found to be most effective in reducing the root galls and egg masses by 76.50% and increasing the pod yield by 50.81% with benefit: cost ratio of 10.74 (Mhase, 2012).

Effective management of the population of root-knot nematode and French bean pod yield enhancement to the extent of 100 to 112 per cent was achieved by soil incorporation of Fenamiphos, Ethoprophos, and Carbofuran each at 1 to 2 kg a.i. per ha.

7.2.4.4. Host resistance: French bean cvs. Banat, Blue Lake Stringless, Bountiful Flat, Brown Beauty, Cambridge Countess, Gallaroy, Kenya-3, Pinto W5-114, Seafarer, and Sutton's Masterpiece were reported to be resistant to *M. incognita* (Singh *et al.*, 1981). French bean cv Contender showed moderately resistant reaction to *M. javanica*.

Wyatt *et al.* (1983) released the first snap bean cv. Nemasnap resistant to *M. incognita*. In Kenya, the cvs. Kahuti, Red Haricot, Rono, Saginaw and Kiburn were resistant to *M. incognita* and *M. javanica* (Ngundo, 1977). French bean cvs. Kibbu, Manoa Wonder, Red Haricot, Rono and Saginaw were reported to be resistant to both *M. incognita* and *M. javanica*.

7.2.4.5. Integrated management

i) Bioagents and botanicals: The combination of *Bacillus subtilis* (isolate K194) inoculum and cow manure led to a 54% reduction in numbers of root-knot nematodes, compared to the untreated control. Consequently, damage by root-knot nematodes produced galls with galling indices 1.6 and 4.5, respectively in plots treated with the combination *(B. subtilis* and cow manure) and the untreated control, respectively. Compared to the other treatments, combining *B. subtilis* and organic amendments resulted in the highest nematode diversity. It can therefore be concluded that the root-knot nematodes associated with common beans can be maintained at levels below economic threshold using *B. subtilis* combined with cow manure, an integration which also demonstrated conservation of the nematode diversity (Wepuhkhulu *et al.*, 2011).

7.3. RENIFORM NEMATODE, *ROTYLENCHULUS RENIFORMIS*

7.3.1. Symptoms

General symptoms include reduced root systems, leaf chlorosis, overall stunting of host plants, and reduced yields and plant longevity. When viewed under a dissecting microscope, the female nematodes and their egg masses are seen on the plant roots.

7.3.2. Interaction with other pathogens

Vadhera *et al.* (1995) demonstrated that the incidence of root rot was highest in simultaneous inoculation of *R. reniformis* and *F. solani* on beans. The combination treatment resulted in maximum reduction in plant growth parameters.

7.3.3. Management

7.3.3.1. Cultural methods: Effective management of the reniform nematode in French beans was obtained with soil application of Karanj and Neem oil cakes at 2 t/ha.

7.3.3.2. Chemical methods: Satisfactory nematode control has been obtained with integration of six foliar sprays of Oxamyl at 0.56 kg a.i. /ha along with soil application of 2.24 kg a.i. /ha of the same chemical. Furrow application of 2.5 kg a.i. /ha of Carbofuran was also found effective.

7.3.3.3. Biological methods: Sitaramiah and Sikora (1981) noted 41% fewer *R. reniformis* penetrated *Glomus mosseae* colonized bush bean roots after 8 days of inoculation.

7.4. INTERACTION OF ROOT-KNOT NEMATODEWITH ROOT ROT

7.4.1. Interaction

Effects of *Meloidogyne incognita* and *Macrophomina phaseolina* on root-rot of two cultivars of French bean were examined in a greenhouse experiment. Severity of *Macrophomina* root-rot increased by 54.5, 94.6, and 9.6% when both pathogens were introduced simultaneously, the nematode first by 2 weeks, and the fungus first by 2 weeks, respectively. Nematode infection and reproduction was adversely affected when the fungus was introduced first. Cultivar "Harvester" was more tolerant to both pathogens and less susceptible to the nematode than "Romano Italian" (Al-Hazmi, 1985).

7.4.2. Management

7.4.2.1. Cultural methods: The beneficial effects of neem cake were observed not only against *M. incognita* and the fungus *M. phaseolina* individually, but also when both the pathogens formed a disease complex on French beans.

7.5. INTERACTION OF ROOT-KNOT NEMATODE WITH *RHIZOBIUM*

7.5.1. Interaction

Infection by root-knot nematodes is known to suppress nodulation and hence nitrogen fixation in beans. *Meloidogyne incognita* infection caused significant ($P \leq 0.05$) reductions in nodulation. In a second pot experiment, bean cv. GLP-24 was inoculated with *Rhizobium leguminosarum* bv. *phaseoli* alone and in various combinations with *M. incognita*. Both nodulation and the dinitrogen fixation processes were adversely affected especially in plants where nematode inoculation preceded rhizobial inoculation (Kimenju *et al.*, 1999).

7.5.2. Effect of *Meloidogyne* infection on nodulation of bean genotypes

Numbers of nodules on bean plants infected with *M. incognita* were significantly ($P \leq 0.05$) lower than in control, except for lines E1 and M28 (Table 7.2). Reduction in nodulation ranged from 35.4% in line NOB to 100% in line E4. Numbers of nodules in plants inoculated with *M. incognita* were significantly ($P \leq 0.05$) different among the bean genotypes. For instance, no nodules were observed on the roots of bean line E4 inoculated with nematodes as compared to more than 111 nodules on roots of bean line NOB. *Meloidogyne* infection caused significant ($P \leq 0.05$) reduction in proportions of effective nodules in the bean genotypes tested, except for bean cv. GLP-24 (Kimenju *et al.*, 1999).

Table 7.2. Numbers of nodules and percentage of effective nodules in *Meloidogyne*-infected and non-infected bean plants

Bean genotype	Nodule numbers		Effective nodules (%)	
	Inoculated[1]	Control[2]	Inoculated	Control
GLP-24	90.0	166.8	70.0	98.0*
GLP-1004	20.0	58.6	14.0	50.0
NOB	111.2	172.2	72.0	90.0
KK8	70.2	132.2	36.0	78.0
KK14	88.6	161.8	38.0	88.0
KK15	77.4	157.6	40.0	80.0
KK22	38.0	124.6	32.0	62.0
E1	42.6	92.8	20.0	74.0
E3	19.4	121.4	8.0	40.0
E4	0	54.0	0	76.0
M14	1.4	30.2	0	68.0
M24	45.4	99.8	22.0	62.0
M26	41.6	92.4	10.0	58.0
M28	25.4	39.2*	20.0	70.0
M29	27.6	59.4	46.0	74.0
M30	9.2	28.0	8.0	38.0
L31	32.2	116.6	36.0	70.0
L32	32.4	107.8	34.0	66.0
L40	4.6	40.0	12.0	44.0
L45	19.0	46.2	18.0	50.0
CD ($P \leq 0.05$)	19.9	24.7	15.2	12.6

[1]Inoculated = Nematode-infected, [2]Control = Nematode-free.

Data are means of five replications.

7.5.3. Effect of time of *M. incognita* inoculation on bean nodulation

The number of nodules was significantly ($P \leq 0.05$) higher in bean cv. GLP-24 plants inoculated with *rhizobia* alone than in plants inoculated with combinations of *rhizobia* and *M. incognita.* Plants inoculated with nematodes at emergence and with *rhizobia* 10 days after emergence (DAE) had the lowest number of nodules on their roots. The percentage of effective nodules was significantly ($P \leq 0.05$) higher in bean plants inoculated with *rhizobia* alone than those inoculated with *rhizobia* and *M. incognita* (Table 7.3). Plants inoculated with nematodes at emergence had lower ($P \leq 0.05$) proportions of functional nodules than those inoculated with nematodes 10 DAE (Kimenju *et al.,* 1999).

Table 7.3. Effect of *Meloidogyne* infection on bean nodulation and functioning of nodules in bean cv. GLP-24 plants[1]

Treatments	Nodule number/ plant	Effective nodules (%)
Rhizobium alone at emergence	96.5 a[2]	83.0 a
Nematode + Rhizobium (both at emergence)	74.4 b	48.8 c
Nematode (at emergence) + Rhizobium (10 DAE)	54.2 c	47.4 c
Rhizobium (at emergence) + Nematodes (10 DAE)	78.2 b	66.0 b

DAE = Days after emergence
[1]Data are means of 10 replications.

[2]Means followed by the same letter along the columns are not significantly ($P \leq 0.05$) different.

7.6. LESION NEMATODE, *Pratylenchus scribneri*

7.6.1. Symptoms

Above-ground symptoms appear patchy with unthrifty and/ or yellowed plants, not unlike the symptoms of water or nitrogen deficiency. Below-ground early infection symptoms show brown root discoloration, with fewer and shorter root branches. Bean plants severely infected by root-lesion nematodes have a reduced root system and, depending upon the cv., may exhibit brown or black lesions on the roots. plants may show chlorosis, stunting, necrosis of leaf margins, and wilting during periods of moisture stress. These lesions result from penetration of and feeding activities of nematodes in epidermal and cortical cells. In severe cases, plants are easily pulled from the ground as a result of root destruction.

7.6.2. Management

7.6.2.1. Biological methods: An endospore forming bacterial parasite, *Pasteuria thornei* was shown to reduce the numbers of *Pratylenchus scribneri* recovered from the soil and roots of beans (Mankau, 1980). Small amounts of soil infested with spores of this bacterium could be used to transmit the organism to uninfested sites.

7.7. CONCLUSION

So far, limited importance has been placed on the management of French bean nematodes, while much attention has been devoted to the use of different nematicides. But as eco-friendly approaches become more popular to maintain a pollution-free environment, integrated pest management practices for control of nematodes on French beans are gaining priority, and the use of bio-agents has recently become an important aspect in control of nematode diseases.

Research on the impact of plant-parasitic nematodes and the development of management strategies to limit their damage on French beans need to be pursued. Various gaps in terms of the generation of knowledge need to be addressed concerning the distribution and impact of nematode pests on French beans. Initiatives to reduce population levels of plant-parasitic nematodes in French beans should be made available to the farmers which are practical and cost-effective to ensure sustainable crop production.

There is a need for collection of information on the distribution and yield losses due to major nematode species on a regional basis. Research on cropping patterns and systems is another important aspect that will help in devising nematode disease management programs. Only when these research results are acquired can we then go forward in this field and expect to gain the desired results.

Chapter - 8

COWPEA, *Vigna unguiculata*

8.1. INTRODUCTION

Cowpea (*Vigna unguiculata*) also known as southern pea and black-eye pea, is one of the important vegetable crops. It is cultivated for its long, green or purplish pods to be cooked as vegetable or for dry seeds used as pulse. Its foliage is used as fodder or green manure. In India, cowpea is grown almost throughout the country.

The major limiting factors in successful production of cowpea crop include root-knot, reniform and pigeon pea cyst nematodes.

8.2. ROOT-KNOT NEMATODES, *Meloidogyne incognita, M. javanica*

8.2.1. Crop losses

Parvatha Reddy and Singh (1981) reported that *M. incognita* was responsible for 28.60% loss in pod yield of cowpea.

8.2.2. Symptoms

- » Root-knot infected field showed stunting of plants, yellowing of leaves, and patchy growth.
- » Heavy galling on roots.
- » Poor growth of seedlings in nursery.
- » Yield is considerably reduced.

8.2.3. Interaction with other pathogens

The presence of heavy infestations of *M. javanica* on a cowpea cv. tolerant to wilt incited by *Fusarium oxysporum* f. sp. *tracheiphilum* caused increased wilting when compared to the susceptible cultivar.

8.2.4. Management

8.2.4.1. Cultural methods

i)Summer plowing: Summer plowing was found effective for the management of root-knot nematodes and increased the cowpea pod yield over non-plowed field.

ii) Organic amendments: The root-knot nematodes associated with cowpea can be effectively managed by the application of neem or eupatorium leaves at 15 t/ha two weeks before sowing. Neem cake incorporation in the previous crop, caused reduction in density of all nematodes in the soil on the following cowpea crop (Jain and Hasan, 1986). Cocoa pod husks incorporated at 6 t/ha caused 28% reductions in galling and 6.7% increases in yield (Egunjobi, 1985).

iii) Crop rotation: The crop rotation sequence recommended for the management of root-knot nematode includes cowpea - graminaceous crops/*Crotalaria spectabilis*. Populations of root-knot decreased greatly when compared to fallowed plots when *C. spectabilis* was grown as a weed free cover crop. One crop of rice was sufficient to effectively reduce root-knot nematode infestations in succeeding susceptible cowpea crop which is superior to rotation with non-host crops.

iv) Intercropping: *M. javanica* populations were lower when cowpea and maize were grown under mixed rather than under sole crop cropping systems.

v) Mulching: Mulching with cowpea foliage was also highly effective in suppressing root-knot nematode populations.

8.2.4.2. Chemical methods: Seed treatment of cowpea with Carbofuran @ 3% w/w gave 35% reduction in gall index and 68% increase in yield over control.

Seed treatment with fenamiphos and carbofuran each at 1 percent concentration was found effective in controlling the root-knot nematode (*M. incognita*) infecting cowpea and in increasing their pod yields (Parvatha Reddy, 1984). Carbosulfon at 1,000 ppm used as seed treatment reduced juvenile penetration in cowpea plants. Posse and monocrotophos at 1,000 ppm reduced root-knot index (2.2 and 2.8, respectively) as compared to untreated check (3.5) when cowpea seeds were dipped in these chemicals for 12 hours.

Carbofuran, ethoprophos and fenamiphos each at 1 to 2 kg a.i. per ha were found effective in reducing the root-knot nematode population and in increasing pod yields of cowpea (40 to 50 per cent) (Parvatha Reddy, 1985a; Rao and Singh, 1978). An increase in seed yield to the tune of 61.8% was reported with application of fenamiphos at 2.5 kg a.i. /ha (Jain and Hasan, 1984).

Seed soaking of cowpea cv. CO 1 for 12 hr. in Monocrotophos at 0.1% gave a reduction of 92% of juvenile and adult nematode population in roots.

Soaking of seeds of cowpea cv. CO 2 for 12 hr. with 0.1% solutions of Phosphamidon and Methyl demeton caused higher shoot weight of 46.20 and 44.57 g/plant and lower root- knot index of 2.59 and 2.69, respectively as against the shoot weight of 36.85 g/plant and the root-knot index of 5.00 in the untreated control.

8.2.4.3. Biological methods

i) Antagonistic bacteria: *Pasteuria penetrans* has shown its potential against *M. incognita* (Sharma and Swarup, 1988).

ii) Antagonistic fungi: Cowpea seed treatment with *Purpureocillium lilacinum* spores was reported to be effective in reducing the root-knot infestation level (Midha, 1985). Application of certain organic materials, viz. Subabool leaves, Neem cake or Saw dust prior to the application of *P. lilacinum* not only enhanced the growth and multiplication of the fungus resulting into high percentage of egg mass infection but also effectively reduced the soil incidence in cowpea (Hasan, 1988). Application of *P. lilacinum* at 15 kg/ha along with certain organic materials effectively reduced the root-knot nematode incidence in cowpea (Hasan and Jain, 1992).

P. lilacinum has shown its potential against *M. incognita* (Sharma and Swarup, 1988). *P. lilacinum* reduces the damage caused by *M. incognita* (Khan and Hussain, 1988).

iii) Arbuscular mycorrhizal fungi: Soil application of AMF increased plant growth and resulted in fewer nematodes and galls on the roots of cowpea (Jain and Hasan, 1988; Jain and Sethi, 1988).

8.2.4.4. Host resistance: For the effective management of root-knot nematode, the farmers of middle Gujarat growing vegetables are advised to grow root-knot resistant varieties i.e., cowpea (Anand Vegetable Cowpea-1) in *kharif* and tomato (Hisar Lalit) in *rabi* for three years.

Cowpea genotypes viz., 82-1-B (Gujarat-1) and C-152 tested against *M. incognita* recorded resistant reaction.

8.2.4.5. Integrated methods

i) Botanicals and bioagents: Hasan and Jain (1992) reported that soil application of *P. lilacinus* cultured on sorghum seeds together with certain organic matter effectively reduced the incidence of *M. incognita* and increased the crop yield of cowpea.

ii) Cultural and chemical: Summer plowing along with seed treatment with carbosulfon 3% w/w or seed soaking in monocrotophos at 0.1% for 6 hr. gave effective control of root-knot nematodes and increased the cowpea yield.

iii) AMF and botanicals: Combined application of *G. fasciculatum* and Achook, a neem product was very effective in reducing root-knot nematode population in cowpea (Jain and Hasan, 1995).

Combined application of *G. fasciculatum* and Achook, a Neem product was very effective in reducing root-knot nematode population in cowpea (Jain and Hasan, 1995).

Reduction in *M. incognita* population was obtained in cowpea by combined soil application of chopped leaves of *Prosopis fuliflora* along with *G. fasciculatum* that increased the spore production and root colonization of *G. fasciculatum*.

iv) Physical and chemicals: Solarization of nematode infested field + seed dressing with Carbosulfan 25 DS @ 3% a.i. (w/w) in cowpea proved effective for management of root-knot (*Meloidogyne* spp.),

Summer plowing along with seed treatment with Carbosulfon 3% w/w or seed soaking in Monocrotophos at 0.1% for 6 hr. gave effective control of root-knot nematodes and increased the cowpea yield.

8.3. RENIFORM NEMATODE, *Rotylenchulus reniformis*

The emergence of cowpea seedlings was delayed by 7 days and seedling stand was reduced to the tune of 6 to 11% due to *R. reniformis* at one nematode/g of soil (Nanjappa *et al.*, 1978).

8.3.1. Crop losses

R. reniformis was responsible for 13.20 to 32.00% loss in pod yield of cowpea (Palanisamy and Sivakumar, 1981; Hasan and Jain, 1998).

8.3.2. Symptoms

The reniform nematode *R. reniformis* is known to attack and cause growth reduction of cowpea. This nematode is most likely to cause or contribute to yield losses on cultivated cowpea. The symptoms include leaf chlorosis, delayed emergence of seedlings, stunted growth, wilting, and collapse of plant.

Infestation of *R. reniformis* causes growth reduction and adversely affects the germination of and seedling emergence. The emergence of cowpea seedlings was delayed by 7 days and seedling stand was reduced to the tune of 6 to 11% due to *R. reniformis* at one nematode/g of soil (Nanjappa *et al.*, 1978). Sud *et al.* (1984) reported that 20 larvae of *R. reniformis* per gram of soil were more damaging on 7-day-old cowpea seedlings than older ones.

8.3.3. Physiological races

The nematode has two races based on their ability to parasitize cowpea, castor, or cotton with 'race A' reproducing on all three hosts and 'race B' only on cowpea (Dasgupta and Seshadri, 1971).

8.3.4. Management

8.3.4.1. Physical methods: Solarization was considered as an effective method for reducing the nematode densities to a depth of 15 cm. Enhanced cowpea yield and significant reduction of soil population of the reniform nematode up to 15 cm depth was achieved by soil solarization under field conditions.

Interesting results have also been reported using soil solarization to control the reniform nematode, *R. reniformis*. Heald and Thomas (1983) observed that the development of cowpea in plots solarized for six weeks was greatly enhanced as was the subsequent yield of lettuce. Nematode densities at the harvest of the cowpea crop were 47 specimens/100 g soil in solarized plots and 14,400 in the controls. Satisfactory control was also similarly obtained by McSorley and Parrado (1986), even though nematode population densities increased during the following tomato crop.

8.3.4.2. Cultural methods: Summer plowing was found effective for the management of reniform nematodes and increased the cowpea yield over non-plowed field.

Soil application of Neem seed kernel powder at 2.5 t/ha was most effective in reducing the number of females/plant (46.2 compared to 80.7 in check), number of egg masses/plant (26.0 compared to 45.5 in check), number of eggs and larvae/egg mass (46.2 compared to 80.7 in check) and final soil population/200 ml soil (143.5 compared to 831.0 in check) of the reniform nematode (Ram and Baheti, 2004).

Summer plowing was found effective for the management of reniform nematodes and increased the cowpea yield over non-plowed field.

The reniform nematodes associated with cowpea can be effectively managed by the application of Neem or Eupatorium leaves at 15 t/ha two weeks before sowing.

The reniform nematode densities were suppressed when cowpea was grown intercropped with maize.

8.3.4.3. Biological methods

i) Antagonistic fungi: *P. lilacinum* reduces the damage caused by *R. reniformis* (Khan and Hussain, 1988).

ii) Arbuscular mycorrhizal fungi: *Glomus fasciculatum* significantly reduced juvenile penetration and development of *R. reniformis* on mycorrhizal cowpea roots. Development of gelatinous matrix was delayed, and fewer egg masses were produced on the plants (Deepthi, 1993).

8.3.4.4. Host resistance: Cowpea cvs. V-16 and Pusa Phalguni are reported to be resistant to *R. reniformis* (Thaker and Patel, 1984).

8.3.4.5. Integrated methods

i) Physical and chemicals: Summer plowing along with seed treatment with Carbosulfan 3% w/w or seed soaking in Monocrotophos at 0.1% for 6 hr. gave effective control of reniform nematodes and increased the cowpea yield.

ii) Bioagents and botanicals: Soil application of Neem cake @ 100 g/m^2 + *Trichoderma viride* @ 2.5 kg/ha is effective against reniform nematode and enhancing yield of cowpea.

8.4. PIGEON PEA CYST NEMATODE, *Heterodera cajani*

The cyst nematode has been found associated with cowpea in a number of regions of India.

8.4.1. Symptoms

The cyst nematode retarded emergence of leaves and reduced the number of flowering buds, flowers, growing pods and yield.

8.4.2. Interaction with other pathogens

If *R. Solani* is added to cowpea before *H. cajani*, nematode reproduction was severely depressed (Walia and Gupta, 1986a, 1986b) and only when the fungus was added two weeks after the nematode there was an adverse effect on plant growth (Walia and Gupta, 1986a).

8.4.3. Management

8.4.3.1. Cultural methods: Cowpea rotated with paddy rice may be less affected by the cyst nematode, because of the negative effect of flooding on nematode densities.

8.4.3.2. Biological methods

i) Antagonistic bacteria: *Pasteuria penetrans* has shown its potential against *H. cajani* (Sharma and Swarup, 1988).

The inability of *H. cajani* (race 8) to multiply to expected level in culture pots revealed that the second stage juveniles were encumbered with the endospores of *Pasteuria penetrans* (Walia *et al.*, 1990).

Application of *Pseudomonas fluorescens* at 2.5 kg/ha at sowing significantly decreased *H. cajani* population in roots at 45 DAS by 18.32% and final nematode population by 29.55% and thereby gave 35.93% and 32.46% higher pod and grain yield, respectively.

ii) Antagonistic fungi: *P. lilacinum* has shown its potential against *H. cajani* (Sharma and Swarup, 1988).

iii) Arbuscular mycorrhizal fungi: Soil application of *Glomus fasciculatum* and *G. epigaeus* improved plant growth of cowpea and reduced larval penetration and cyst production of *H. cajani* (Jain and Sethi, 1987; 1988a).

8.4.3.3. Host resistance: Cowpea cv. CO 6 was found resistant to the pigeon pea cyst nematode. The cowpea cv. Barsati Mutant has been reported to be tolerant to the cyst nematode (Sharma and Sethi, 1976).

8.5. INTERACTION OF ROOT-KNOT AND RENIFORM NEMATODE WITH ROOT ROT

8.5.1. Interaction

The root-knot, *Meloidogyne incognita* and the reniform, *Rotylenchulus reniformis* nematodes interacts with root rot fungus, *Rhizoctonia solani* in causing root rot disease complex in cowpea.

8.5.2. Management

a) Antagonistic fungi: *Purpureocellium lilacinum* reduces the damage caused by *M. incognita, R. reniformis* and *R. solani* disease complex in cowpea (Khan and Husain, 1988).

8.6. INTERACTION OF ROOT-KNOT NEMATODES WITH ROOT ROT

8.6.1. Interaction

Synergistic interactions were reported between *M. javanica* and *R. solani* in cowpea. Inoculation of cowpea with *M. incognita* and *R. solani* led to breakdown of resistance to both organisms (Khan and Husain, 1989), and the greatest decrease in plant dry weight occurred if the nematode was inoculated two weeks before the fungus (Varshney *et al.*, 1987).

8.6.2. Management

8.6.2.1. Biological methods: Khan and Husain (1988, 1990) reported that application of *P. lilacinum* at 2g mycelium/pot reduced damage to cowpea caused by *M. incognita, R. reniformis* and *R. solani* disease complex in cowpea.

The root-knot nematode reproduction parameters such as number of galls, egg masses and final soil nematode population under *M. incognita* alone and *M. incognita* + *R. solani* complex, were also reduced maximally and significantly in *Trichoderma viride* and *Gliocladium virens* application at 10 g/kg soil. Similarly, under complex of both the pathogens, *T. viride* and *G. virens* enhanced all the plant growth parameters of cowpea significantly (Verma *et al.*, 2010).

8.7. INTERACTION OF ROOT-KNOT NEMATODE WITH STEM AND ROOT ROT

8.7.1. Interaction

The root-knot nematode, *Meloidogyne incognita* in association with stem and root rot fungus, *Macrophomina phaseolina* is responsible for root rot disease complex.

8.7.2. Management

8.7.2.1. Integrated methods: Devi and Goswami (1992) demonstrated that *G. fasciculatum* in cowpea together with mustard cake helped in reducing the disease severity caused by *M. incognita* and *M. phaseolina*. They observed that the pre-establishment of AMF checked the entry of *M. incognita* larvae as also colonization of pathogenic fungus.

8.8. INTERACTION OF ROOT-KNOT NEMATODE WITH FUSARIUM WILT

8.8.1. Interaction

The wilt fungus, *F. oxysporum* and root-knot nematode, *M. incognita* co-infect cowpea. Histopathological studies revealed that in nematode + fungus inoculated cowpea roots, conidia of *F. oxysporum* could be observed in the cortex as well as xylem vessels adjacent to the giant cells but not inside the giant cells induced by *M. incognita* (Singh *et al.*, 2007).

8.8.2. Management

8.8.2.1. Integrated methods: The minimum gall diameter index was reported in reduced dose of both neem cake and carbofuran (Singh *et al.*, 2007).

8.9. CONCLUSION

So far, limited importance has been placed on the management of cowpea nematodes, while much attention has been devoted to the use of different nematicides. But as eco-friendly approaches become more popular to maintain a pollution-free environment, integrated pest management practices for control of nematodes on cowpea are gaining priority, and the use of bio-agents has recently become an important aspect in control of nematode diseases.

Research on the impact of plant-parasitic nematodes and the development of management strategies to limit their damage on cowpea need to be pursued. Various gaps in terms of the generation of knowledge need to be addressed concerning the distribution and impact of nematode pests on cowpea. Initiatives to reduce population levels of plant-parasitic nematodes in cowpea should be made available to the farmers which are practical and cost-effective to ensure sustainable crop production.

There is a need for collection of information on the distribution and yield losses due to major nematode species on a regional basis. Research on cropping patterns and systems is another important aspect that will help in devising nematode disease management programs. Only when these research results are acquired can we then go forward in this field and expect to gain the desired results.

Chapter - 9

PEA, *Pisum sativum*

9.1. INTRODUCTION

Pea is an important vegetable crop widely grown throughout the world. As a cool season crop, it is extensively grown in temperate zone; but restricted to cooler altitudes in the tropics and winter season in the sub-tropics. A rich source of proteins (25%), amino acids and sugars (12%), green peas are an all-time favorite vegetable. It is rich in vitamins A, B, C and minerals. It is being cultivated in 0.276 million hectares producing 1.972 million tons of peas with an average yield of 18.9 tons per hectare. Garden pea is grown in Uttar Pradesh, Himachal Pradesh, West Bengal, Jammu and Kashmir, Punjab, and Haryana.

Root-knot and reniform nematodes are the important limiting factors in successful cultivation of pea crop.

9.2. ROOT-KNOT NEMATODES, *Meloidogyne* spp.

9.2.1. Crop losses: Pea pod yield loss of 20.00 - 50.61% was attributed to the root-knot nematode, *Meloidogyne incognita* (Sharma, 1989; Upadhayay and Dwivedi, 1987; Parvatha Reddy, 1985b).

9.2.2. Symptoms: The infested fields show patches in which pea plants are stunted, chlorotic and have few flowers which produce small and often empty pods. The root systems are reduced in size, exhibit large galls and poor nodulation. Senescence also tends to occur earlier.

9.2.3. Interaction with other pathogens: The root-knot nematode, *Meloidogyne incognita* along with wilt fungus, *Fusarium oxysporum* f. sp. *pisi* causes wilting of pea plant that leads to death (Abuzar, 2012). Maximum reduction in plant height was observed in nematode inoculated one week prior to fungus. Both the organisms suppressed the plant growth characters in general and plant height in particular. Maximum nematode population was found to be in nematode alone and nematode followed by fungus inoculation. Fifty days after inoculation of Fusarium alone, Fusarium prior to nematode, and nematode prior to Fusarium resulted in wilt index of 3.5, 4.0 and 4.2, respectively.

9.2.4. Management

9.2.4.1. Cultural methods: Intercropping with marigold or mustard with pea reduced the damage due to root-knot nematodes.

Peas escape nematode attack, in the subtropics when sowing is postponed to mid-autumn, when temperatures drop.

9.2.4.2. Chemical methods

i) Seed treatment: Bhagawati and Phukan (1990) found that Carbofuran at 3% w/w as seed treatment was effective in reducing galls and egg masses in roots of pea and increased yields. Seed treatment with Fenamiphos and Carbofuran both at 1 per cent concentration was effective in controlling the root-knot nematode (*M. incognita*) infecting peas and in increasing their pod yields (Parvatha Reddy, 1984).

ii) Main field treatment: Enhancement of pea pod yields to the extent of 58-85 per cent and reduction in the root-knot nematode population occurred following soil incorporation of Carbofuran, Ethoprophos, and Fenamiphos each at 1 to 2 kg a.i. per ha (Parvatha Reddy, 1985a).

Decrease in the root-knot nematode, *M. incognita* in pea was achieved by drill application of Carbofuran at 1.5 kg a.i. /ha below the seed level.

9.2.4.3. Biological methods

i) Antagonistic bacteria: *Bacillus thuringiensis* reduced rot-knot infection (*M. incognita*) both in naturally infested and sterilized soils by 69.4 and 74.7%, respectively (Chahal and Chahal, 1991) (Tables 9.1 and 9.2).

Table 9.1. Effect of *Bacillus thuringiensis* on *Meloidogyne incognita* in pea in naturally infested soil.

Treatment	Dry shoot wt. (g)	Dry root wt. (g)	Galls/plant	Inhibition %
Control	2.3	2.34	85	---
B. thuringiensis	4.4	3.10	26	69.4

Table 9.2. Effect of *Bacillus thuringiensis* on *Meloidogyne incognita* in pea in sterilized soil.

Treatment	Dry shoot wt. (g)	Dry root wt. (g)	Galls/plant	Inhibition %
Control	3.4	2.85	95	---
B. thuringiensis	4.9	3.25	24	74.7

9.2.4.4. Host resistance: Pea cv. Wando is reported to be highly resistant to *M. incognita* and moderately resistant to *M. javanica*.

Pea selection 'Pant P-42' showed highest tolerance among all the selections tested and can be recommended for field infested with both root-knot and Fusarium wilt pathogens (Abuzar, 2012).

9.2.4.5. Integrated methods

i) Bioagents and botanicals: Integration of *Trichoderma harzianum, T. viride, and T. viiride* fungi with basil, caraway, and fennel wastes provided satisfactory control of the root-knot nematode, *M. incognita* in peas. The integrated treatments of plant wastes and biocontrol agents not only did reduce suppressive effect of the nematodes, but also caused an additional increase in growth and pod production criteria of pea (El-Nagdi *et al.*, 2019).

9.3. ROOT LESION NEMATODE, *Pratylenchus penetrans*

9.3.1. Symptoms

Large, circular patches of stunted plants are often associated with root lesion nematode-infested areas in pea fields. However, large areas of a field can also be infested. Pea roots infested with nematodes usually have a single primary root with no secondary roots. Several soilborne root rotting pathogens are usually associated with diseased roots because of wounds caused by the feeding nematodes enabling fungi/bacteria to colonize roots. Infected plants may develop chlorotic leaves that discolor from the base towards the top of the plant. A study showed that 150 nematodes/100 cm^3 soil can cause ~50% yield reduction in beans (Inglis, 2001).

9.3.2. Biology/epidemiology

Root lesion nematodes have a wide host range, including many crops grown in rotation with pea. Roots of infested plants are usually dark from secondary infection by root-rotting fungi and oxidation of phenolic compounds. Nematodes generally reside within the top 2 feet of soil and can be moved by farm equipment and water. Root debris moved around the field can also disseminate nematodes (Inglis, 2001).

9.3.3. Management

Crop rotation is not highly effective for managing root lesion nematodes due to the wide host range of the nematode, although some crops are less susceptible than others. Incorporation of white mustard, rapeseed, perennial ryegrass, and other green manure crops into the soil has been used to manage root lesion nematodes. Soil fumigation (e.g. methyl isothiocynate) can be effective for managing nematodes; however, this practice is not economically feasible for pea crops (Inglis, 2001).

9.4. RENIFORM NEMATODE, *Rotylenchulus reniformis*

9.4.1. Crop losses

Dalal and Vats (1998) reported that *R. reniformis* was responsible for 15.80% loss in pod yield of peas.

9.4.2. Symptoms

The reniform nematode causes root necrosis resulting in severe root pruning and subsequent dwarfing of plants. Fibrous or feeder roots are mostly attacked which may reduce the absorption ability and other physiological functions of the plant. This may lead to stunting, yellowing, and wilting.

9.4.3. Interaction with other pathogens

Vats and Dalal (1997) reported maximum and early wilting in pea due to prior inoculation of *R. reniformis* followed by *F. o.* f. sp. *pisi*.

The interaction of *Rotylenchulus reniformis* and *F. oxysporum* f. sp. *pisi* race 3 and its importance in "early yellowing" disease of peas was investigated by Labruyere *et al.* (1959). Neither pest alone causes severe root rot in peas but together they cause an extensive decay of the root cortex which is the chief symptom in roots of peas with "early yellowing".

The reduction in the number of bacterial root nodules was more pronounced with R. reniformis infection as compared to leguminous species inoculated with *Rhizobium* species alone (Montasser *et al.*, 2016).

9.4.4. Management

Intercropping with marigold or mustard with pea reduced the damage of reniform nematodes.

9.5. CYST NEMATODE, *Heterodera goettingiana*

The cyst nematode, *Heterodera goettingiana*, is an aggressive parasite that can be quite damaging to a pea crop (Inglis, 1998).

9.5.1. Crop losses

The threshold of economic damage is five or more eggs per gram of soil. Yield loss may be 20% at five eggs per gram; complete crop failure can occur at 20 or more eggs per gram. In Washington State, yield losses caused by *Heterodera goettingiana* were estimated to be 5 to 10% in green pea during 1994 (Koenning *et al.*, 1999).

9.5.2. Symptoms

Infected peas are stunted with yellowing leaves, and leaves tend to point up. Foliage symptoms often are not evident until bloom, when yellowing spreads rapidly from the bottom of the plant to the top. Sometimes the color is a very bright yellow. Invaded plants lack fibrous roots and nitrogen-fixing nodules. Tiny (less than 0.5 mm) cream-color, lemon-shaped cysts can be found embedded in root tissue, but once cysts mature and turn dark brown, they detach easily from root systems and are difficult to detect. Cysts are the dead bodies of females and may contain up to 300 eggs. Initial damage to the field is restricted to small distinct areas, but with further cropping to pea, the areas expand.

Root invasion is promoted by cool soil temperatures (about 55 °F). Thus peas planted early can be affected more severely than late plantings.

9.5.3. Host range

Pea and faba bean are the main economic hosts. Certain vetches also can be hosts.

9.5.4. Survival and spread

Cysts are very persistent in the soil, surviving more than 12 years, and can be spread by moving contaminated soil.

9.5.5. Management

9.5.5.1. Cultural methods: Crop rotation effectively reduces soil populations because the nematode has a limited host range. Rotate 4 years out of host crops unless the field is infested at an economically damaging level; then, an even longer break is needed. Avoid planting fava bean in fields with this nematode. Plant later when soil temperature has warmed.

9.5.5.2. Biological methods: MeloCon WG at 2 to 4 lb/A for nematode suppression.

9.6. CONCLUSION

So far, limited importance has been placed on the management of pea nematodes, while much attention has been devoted to the use of different nematicides. But as eco-friendly approaches become more popular to maintain a pollution-free environment, integrated pest management practices for control of nematodes on peas are gaining priority, and the use of bio-agents has recently become an important aspect in control of nematode diseases.

Research on the impact of plant-parasitic nematodes and the development of management strategies to limit their damage on peas need to be pursued. Various gaps in terms of the generation of knowledge need to be addressed concerning the distribution and impact of nematode pests on peas. Initiatives to reduce population levels of plant-parasitic nematodes in leguminous vegetable crops should be made available to the farmers which are practical and cost-effective to ensure sustainable crop production.

There is a need for collection of information on the distribution and yield losses due to major nematode species on a regional basis. Research on cropping patterns and systems is another important aspect that will help in devising nematode disease management programs. Only when these research results are acquired can we then go forward in this field and expect to gain the desired results.

Chapter - 10

PIGEON PEA, *Cajanus cajan*

Based on surveys and crop loss estimation trials, it is evident that *Heterodera cajani*, *Rotylenchulus reniformis*, and *Meloidogyne* spp. are important nematode pests of pigeon pea in India.

10.1. PIGEON PEA CYST NEMATODE, *Heterodera cajani*

10.1.1. Geographic distribution

Heterodera cajani has been reported from Egypt, India and Pakistan. It is widespread in sandy-loam soils in northern India and in black cotton soils in western and southern India (Reddy *et al.*, 1993; Varaprasad *et al.*, 1997).

10.1.2. Crop losses

Yield losses of over 30% and 28% reduction in plant growth have been reported (Sharma *et al.*, 1993).

10.1.3. Symptoms

The symptoms of nematode injury include stunting, reduced leaf lamina size and yellowing on cotyledonary leaves (Gaur and Singh, 1977). Flowers and pods are reduced in size and number and the root system may also be poorly developed. The most important characteristic symptom is the presence of cysts on the root surface. Identification of "pearly root" caused by the presence of white females is a useful symptom of *H. cajani* infestation in pigeon pea at the vegetative stage. The cyst nematode retarded emergence of leaves and reduced the number of flowering buds, flowers, growing pods and yield.

It has been found to reduce bacterial nodulation which in turn affects the crop yield. Significant reduction in growth parameters was recorded at 100 larvae and above per kg of soil (Sharma and Nene, 1985; Zaki and Bhatti, 1986a). The nematode also affects the macro and micro-nutrients of pigeonpea (Zaki and Bhatti, 1986b). Under field conditions, the reduction in plant growth and yield of pigeonpea were 27.6 and 30.1%, respectively (Saxena and Reddy, 1987).

10.1.4. Host range

Besides *Cajanus cajan* (pigeon pea), the host range of cyst nematode include *Cyamopsis tetragonoloba* (cluster bean), *Lablab purpureus* (hyachinth bean), *Vigna aconitifolia* (aconite-leaved kidney beans),

Vigna radiata (bean, mung), *Vigna mungo* (black gram), *Phaseolus vulgaris* (common bean), *Pisum sativum* (pea), *Sesamum indicum* (sesame) and *Vigna unguiculata* (cowpea) (Sharma and Nene, 1985).

10.1.5. Survival

Eggs within cysts are able to withstand extremes of desiccation; a maximum reduction of about 30% of viable eggs was recorded after 3 weeks exposure to 0% relative humidity (RH), which is similar to the reduction found in dry soil after storage of between 2 and 12 months. Cysts stored in moist soil for up to 12 months gave a greater percentage hatch in cowpea root diffusate (CRD) than those from air dried soil but the actual number of J2 emerging per cyst was lower. Eggs hatched during the first 4 months of storage in moist soil but only during the first 2 months in air dried soil (Gaur *et al.*, 1996).

10.1.6. Management

10.1.6.1. Physical methods: The population of all parasitic nematodes of pigeonpea was markedly affected by soil solarization. Included in this population were *H. cajani* cysts, eggs, and larvae; *Rotylenchulus reniformis*; *Helicotylenchus retusus*; *Pratylenchus* spp; *Tylenchorhynchus* spp., and *Heterodera* spp. Soil solarization with irrigation was most effective in reducing populations of cysts, eggs, and larvae of *H. cajani.*

10.1.6.2. Chemical methods: Fensulfothion, when evaluated as a seed treatment for pigeon pea, reduced nematode populations, with high concentrations and long soaking period being the most effective (Zaki and Bhatti, 1986a).

10.1.6.3. Biological methods

i) Antagonistic fungi: *Cladosporium herbarum* was more rapid in parasitizing the cysts of *H. cajani* by parasitizing 97% of the cysts in 96 hr. while *Botryodiplodia theobromae* parasitized only 49% of the cysts in the same time. It was noted that young cysts were more prone to infection by *C. herbarum* than matured cysts. Upon infection, the cyst walls became fragile and the cyst color changed from brown to black. On some matured infected cysts, fungal mycelium was seen on the cyst wall, but the young cyst walls did not have any such growth on them. This observation suggests that some nematoxin may be involved in the parasitization process. Thus, *C. herbarum* offers a great potential to pigeon pea growers in the management of pigeon pea cyst nematode (Kannan and Lingaraju, 2002).

ii) Antagonistic bacteria: Application of *Pseudomonas fluorescens* at 2.5 kg/ha at sowing significantly decreased *H. cajani* population in roots at 45 DAS by 18.32% and final nematode population by 29.55% and thereby gave 35.93% and 32.46% higher pod and grain yield, respectively.

Juveniles of *H. cajani* have been found to be naturally infected by the bacterium *Pasteuria penetrans*; a 27% reduction in the number of juveniles in roots and the maximum reduction in cyst population were obtained when spore-infected soil was incubated at 30 °C before application to cowpea seedlings infested with *H. cajani* (Bhattacharya and Swarup, 1989).

10.1.6.4. Integrated management

i) Two bioagents: Integration of *T. harzianum* at 5 kg/ha along with *P. chlamydosporia* at 2 kg/ha significantly increased the plant height and pigeon pea yield (0.56 kg/plant as compared to 0.40 kg/plant in control) and reduced the seedling mortality, eggs per cyst, parasitization of cysts by the bioagents and cyst nematode population (Table 10.1) (Kumar and Prabhu, 2008) .

Table 10.1. Effect of bioagents on plant growth and management of *Heterodera cajani* infecting pigeon pea

Treatment	Seedling mortality	Plant height 30 DAS (cm)	Yield (kg/ plant)	eggs/ cyst	Parasitized cysts at harvest (%)	Final cyst popn.
T. harzianum - 5 kg/ha	13.21	18.97	0.529	36.8	---	69.3
P. chlamydosporia-2 kg/ha	13.67	17.50	0.545	30.1	52.4	60.9
T. harzianum - 5 kg/ha + *P. chlamydosporia*-2 kg/ha	9.20	24.06	0.560	24.0	46.7	52.7
Carbofuran – 2 kg a.i./ha	2.16	27.11	0.634	18.4	---	45.5
Control	18.40	16.35	0.404	45.1	---	148.3
CD (P = 0.05)	1.14	2.09	0.0053	3.45	4.51	7.52

ii) Bioagents and botanicals: Combined treatment of 10% neem seed kernel powder + *Trichoderma viride* at 10 g/kg seeds was found to reduce *H. cajani* in pigeon pea by 58% and increased yield by 32%. Application of neem cake at 100 kg/ha + *Trichoderma viride* at 2.5 kg/ha was also found to reduce *H. cajani* by 62% and increased pigeon pea yield by 34%.

Seed treatment with neem seed kernel powder at 10% w/w + *Pseudomonas fluorescens* at 10 g/ kg seed was highly effective which recorded 30.11% reduction in nematode population and 29.10% increase in yield over control. Soil application of neem cake at 10 g/m^2 + *P. fluorescens* at 2.5 kg/ha was also highly effective which recorded 31.63% % reduction in nematode population and 17.88% increase in yield over control.

iii) AMF, botanicals and physical: Integration of soil solarization (transparent polythene sheet of 400 gauge thickness for a period of 4 weeks) with AM fungi at 100 kg/ha provided good plant growth and yield of pigeon pea by effectively controlling the detrimental effects of *H. cajani*. Maximum reduction in nematode population was observed in soil solarization + soil application of neem seed powder at 50 kg/ha + AMF, and soil solarization + seed treatment with neem seed powder at 10% w/w + AMF (Nageswari and Mishra, 2005).

10.2. RENIFORM NEMATODE, *Rotylenchulus reniformis*

10.2.1. Symptoms

Feeding causes damage to the plant resulting in stunting and generally poor plant growth. The reniform nematode does not typically cause complete plant death. However, they reduce the productivity of the crop. Populations can be quite uniform in their distribution across a field, making detection of early plant symptoms difficult.

Pigeon pea proved to be the preferred host. Their roots harbored the greatest number of nematodes of all stages. A total of 15,901 parasitic females in different stages of development were seen attached to 1 g. of roots and there were 406,000 juveniles, preparasitic females, and males per 300 cc. of soil.

10.2.2. Management

10.2.2.1. Cultural methods: Crop rotation with non-host crops such as wheat or sorghum reduce base populations. Long fallows can help also to break the life cycle; however, it is important to control any weeds.

10.2.2.2. Host resistance: Two promising short duration pigeon pea breeding lines (ICPL 83024 and ICPL 85045) and selections from medium-duration lines (ICPL 83057, 85068, 85073, 89050, 89051, and 90097) have been identified as tolerant to the reniform nematode (Sharma *et al.*, 2005).

10.3. INTERACTION OF ROOT-KNOT NEMATODE WITH FUSARIUM WILT

10.3.1. Interaction

The wilt disease complex caused by *Fusarium udum* in association with *Meloidogyne incognita* has been reported as the most severe constraint in the cultivation of pigeon pea. Among plant-parasitic nematodes, *M. incognita* is considered most serious threat to its cultivation. Similarly, among fungi, *Fusarium* wilt is one of the most destructive diseases of pigeon pea. When both these pathogens attack the crop, intensity of damage is increased by several folds. Inoculation with *F. udum* and *M. incognita* together significantly increased wilt severity in pigeon pea seedlings compared with inoculation of the fungus alone.

Simultaneous or sequential inoculation of *M. incognita* and *F. udum* increased the severity of the disease. Highest reduction in plant height, fresh/dry weight was observed in plants inoculated with *M. incognita* and *F. udum* simultaneously followed by *M. incognita* prior and *F. udum* 7 days later and *F. udum* prior and *M. incognita* 7 days later, respectively. Reproduction of *M. incognita* was enhanced in the presence of *F. udum* but percent root colonization by *F. udum* was suppressed in the presence of *M. incognita*. Highest final nematode population and gall index of *M. incognita* were observed in simultaneous inoculation of *M. incognita* and *F. udum* and lowest in *F. udum* prior and *M. incognita* 7 days later, while highest percent root colonization was found in *F. udum* prior and *M. incognita* 7 days later followed by *M. incognita* and *F. udum* simultaneously and *M. incognita* prior and *F. udum* 7 days later, respectively (Table 10.2) (Perveen *et al.*, 1988).

Table 10.2. Effect of *Meloidogyne incognita* and *Fusarium udum* on plant growth and root galling and disease index in pigeon pea

Treatment	Plant height (cm)	Dry plant wt. (g)	Gall index	Disease index
Control	33.9	1.03	---	---
M. incognita	32.6	0.73	1.75	0.00
F. udum	32.6	0.81	---	45.32
M. incognita + *F. udum* simultaneouly	23.6	0.38	2.15	38.21
M. incognita (pre) + *F. udum* (post)	27.3	0.57	2.75	32.14
F. udum (pre) + *M. incognita* (post)	30.6	0.65	1.25	42.33
CD (P = 0.05)	1.3	0.031	0.143	0.273

10.3.2. Management

10.3.2.1. Cultural methods: Plots planted to sorghum, safflower and chickpea in the preceding year contained fewer *H. cajani* eggs and juveniles than did plots previously planted to pigeon pea, cowpea or *Vigna radiata*. Continuous cropping of sorghum in the rainy season and safflower in the post-rainy season markedly reduced the population density of *H. cajani*.

Intercropping sorghum with a tolerant pigeon pea cultivar could be effective in increasing the productivity of traditional production systems in *H. cajani* infested regions (Sharma *et al.*, 1996b).

10.3.2.2. Chemical methods: Carbofuran and phorate reduced disease severity both in plants inoculated with both organisms and in plants inoculated with fungus alone. Greatest reduction occurred at the highest dose (20 mg a.i. /kg soil) (Hasan, 1989).

10.3.2.3. Biological methods

i) Antagonistic fungi: In a farmers' field trial conducted at Arania, Bulandshahr district of Uttar Pradesh during 2001; wilting of plants was not seen in plots treated with carbofuran, neem seed powder, neem jivan, *T. harzianum* and *Aspergillus niger*, 1-5% wilting was observed in plots treated with *P. lilacinum* (Table 10.3) (Haseeb and Shukla, 2005).

Table 10.3. Effect of seed treatment with biopesticides, chemicals and botanicals on wilt and parasitic nematodes infecting pigeon pea under field conditions during 2001.

Treatment	Grain yield (kg)/100 m^2	% wilting of plants	% root infection by *F. udum*	Nematodes /g of root	Cysts & larvae/100 ml soil
Carbofuran	27.57	0.0	10.0	15	20
Neem seed powder (soil application)	22.66	0.0	5.0	18	22
Neem seed powder (seed treatment)	19.66	0.0	5.0	17	26
T. harzianum	18.66	0.0	5.0	20	27
Neem jivan	17.33	0.0	10.0	24	27
A. niger	16.49	0.0	10.0	26	30
P. lilacinum	15.83	2.0	15.0	30	36
Dimethoate	15.83	5.0	20.0	31	32
Latex of Calotropis	12.66	5.0	25.0	30	33
Control	8.83	33.0	65.0	44	37
CD (P = 0.05)	1.58	1.8	2.7	2.9	2.1

Pigeon pea seed treatment with *Trichoderma harzianum* and *Pochonia chlamydosporia* gave maximum control of the disease complex in which wilt severity decreased by 79-81%, and root galling by 44-49%, and egg mass production by 33-35% (Khan *et al.*, 2010).

10.3.2.4. Host resistance: Use of varieties resistant to Fusarium wilt is one of the best options to avoid the losses caused by the disease complex. However, sometimes the varieties developed as resistant to Fusarium wilt when grown under farmers' field conditions become susceptible to wilt due to the presence of root-knot nematodes.

Among 10 pigeon pea accessions (identified as resistant to Fusarium wilt) evaluated against combined infection of *M. javanica* and *F. udum* under pot culture conditions, increase in wilting was observed in five accessions namely, ICP 8859, AWR 74/15, KPL 44, ICPL 89049 and ICPL 12745. Whereas in other 5 accessions (KPL 43, PI 397430, BWR 370, GPS 33 and ICPL 89048), wilting was not influenced much in the presence of *M. javanica* (Singh *et al.*, 2004).

The plant height, fresh and dry shoot and root weights were significantly lower in combined inoculation compared to either nematode or fungus alone. The root-knot index varied from 3.0 to

4.5 in these accessions in both treatments having nematode alone or nematode and fungus together. The lowest root-knot index was observed in KPL 43 (1.5) and GPS 33 (1.75) (Table 10.4) (Singh *et al.*, 2004).

Table 10.4. Effect of combined inoculation of *M. javanica* and *F. udum* on plant growth and disease complex in different accessions of pigeon pea

Pigeon pea accessions	Plant height (cm)	Fresh shoot weight (g)	Fresh root weight (g)	Gall index	% Wilt incidence
KPL 43	25.9	4.94	5.12	2.45	0
KPL 44	17.4	2.50	3.67	1.97	33
AWR 74/15	18.3	1.92	2.55	1.56	60
ICP 8859	14.7	1.52	2.33	3.39	50
ICPL 89049	24.3	2.97	4.23	2.17	50
PI 397430	24.8	3.73	3.57	3.06	0
BWR 370	28.4	4.18	4.92	3.17	0
GPS 33	32.0	3.93	5.75	2.25	0
ICPL 89048	28.3	3.65	4.17	3.25	0
ICP 12745	19.4	2.67	3.19	2.50	50
CD (P = 0.05)	2.40	0.43	NS	1.60	---

Five accessions of pigeon pea which were resistant to *F. udum* (KPL 43, PI 397430, BWR 370, GPS 33 and ICPL 89048) were also found resistant to combined inoculation of *M. javanica* and *F. udum*, which indicated that wilt resistance in these accessions was not influenced by the presence of *M. javanica* (Singh *et al.*, 2004).

10.3.2.5. Integrated methods

i) Bioagents, botanicals and chemicals: In the trial conducted in the field during 2002, no wilting was observed in plots treated with neem seed powder + dimethoate/*T. harzianum*/*P. lilacinus*/latex, and neem seed powder and *T. harzianum* alone. Neem seed powder + *T. harzianum* was found to be the most effective treatment in increasing the yield and suppressing the pathogens, followed by carbofuran, neem seed powder + *P. lilacinus*, neem seed powder + dimethoate and neem seed powder + latex. Data regarding *F. udum* infection in roots gathered at the pre-flowering stage (90 DAS) indicated that all the treatments maintained significant protection of the roots as compared to control (Table 10.5) (Haseeb and Shukla, 2005).

Table 10.5. Effect of seed treatment with bioagents, chemicals and botanicals on wilt and parasitic nematodes infecting pigeon pea under field conditions during 2002.

Treatment	Grain yield (kg) /100 m^2	% wilting of plants	% root infection by *F. udum*	Nematodes /g of root	Cysts & larvae/100 ml soil
Carbofuran	23.47	5.0	35	9	27
Neem seed powder (soil application)	19.55	0.0	20	23	45
Neem seed powder (seed treatment)	18.33	0.0	20	30	43
T. harzianum	15.46	0.0	25	31	44
P. lilacinum	12.24	5.0	35	36	55
Dimethoate	11.75	10.0	50	39	58
Latex of Calotropis	11.34	10.0	50	39	59
NSP (soil) + Dimethoate	21.05	0.0	15	16	36
NSP (soil) + *T. harzianum*	23.35	0.0	15	10	28
NSP (soil) + *P. lilacinum*	21.85	0.0	20	14	34
NSP (soil) + latex	20.75	0.0	15	18	43
Control	5.35	45.0	75	51	67
CD (P = 0.05)	**1.22**	2.8	**3.9**	**0.68**	**3.1**

ii) Bioagents and AMF: Combinations of biocontrol agents and arbuscular mycorrhizal fungi [*Bacillus subtilis, Bradyrhizobium japonicum* and *Glomus fasciculatum* (Siddiqui and Mahmood, 1995a); *Purpureocelliums lilacinum, Pochonia chlamydosporia* and *Gigaspora margarita* (Siddiqui and Mahmood, 1995b); and *Trichoderma harzianum, V. chlamydosporium* and *Glomus mosseae* (Siddiqui and Mahmood, 1996).] have been used to manage the wilt disease complex of pigeon pea caused by *H. cajani* and *Fusarium udum* (*Gibberella indica*).

iii) AMF and botanicals: The treatment constituting FYM, karanj oilseed cake and arbuscular mycorrhizal fungus, *Glomus fasciculatum* reduced the disease incidence caused by root knot nematode, *M. incognita* and root wilt fungus, *F. udum* on pigeon pea to a great extent with the most promising improvement in plant growth parameters (Goswami *et al.*, 2007).

iv) Bioagents and AMF: Combined application of *T. harzianum, P. chlamydosporia* and *G. mosseae* enhanced the activity against *H. cajani – F. udum* wilt disease complex in pigeon pea.

Chapter - 11

CHICKPEA, *Cicer arietinum*

Nematode species causing significant economic damage in chickpea include root-knot nematodes (*M. incognita*, and *M. javanica*), cyst nematode (*Heterodera cajani*), and root lesion nematode (*Pratylenchus thornei*).

11.1. ROOT-KNOT NEMATODES, *Meloidogyne* spp.

M. arenaria, M. incognita, and *M. javanica* are the principal root-knot nematodes infecting chickpeas in the Indian subcontinent.

11.1.1. Crop losses

The root-knot nematodes, *M. incognita* and *M. javanica*, cause yield losses of 19 to 40% to chickpea in India (Ali and Sharma, 2003) with thresholds for damage for these species varying from 200 to 2000 eggs and/or juveniles per liter soil at the time of sowing (Sharma *et al.*, 1992). Chickpea yield losses of up to 60% were caused by severe M. incognita infections in India (Ali, 1995)

11.1.2. Symptoms

Uneven patches of chickpea plant growth are commonly associated with soils infested with root-knot nematodes. Heavily infected plants show stunted growth, less branching, and leaves with pale green to yellow color. Root galling is the typical symptom of parasitism by *Meloidogyne* spp. in susceptible crops. *M. arenaria, M. incognita*, and *M. javanica* produce large galls on chickpea. Deformation and blockage of vascular tissues at feeding sites limit translocation of water and nutrients and suppress plant growth and reduce seed yield (Perry and Moens, 2006).

11.1.3. Management

11.1.3.1. Cultural methods: Crops such as ground nut and winter cereals are not suitable hosts for *M. arenaria, M. incognita*, and *M. javanica* (Sikora *et al.*, 2005) and could therefore be included in rotation with chickpea.

11.1.3.2. Biological methods: Soil application of *Trichoderma viride* showed significantly higher plant growth parameters over control. The gall development and final nematode population of *Meloidogyne incognita* decreased with the increasing level of *T viride* in pots as well as under field conditions Maximum soil fungi were decreased with increasing level of *T viride* and maximum root nodules and minimum root-knot nematodes were observed in 4000 spores of *T viride*.

11.1.3.3. Host resistance: Three chickpea cultivars (N 31, N 59, and ICCC 42) and a promising chickpea breeding line (ICCV 90043) have been identified as tolerant to the root-knot nematode.

11.2. PIGEON PEA CYST NEMATODE, *Heterodera cajani*

Heterodera cajani was reported infecting chickpea in several localities of India (Ali, 1995).

11.2.1. Symptoms

Soil infestations by cyst-forming nematodes are first limited to small circular areas that may further extend to include the entire field. Symptoms on infected chickpeas are more apparent at the flowering stage. Infected plants are stunted, with pale green leaves at an early stage of growth which later turn chlorotic. A reduction in the number of flowers and pods is observed in infected plants, which contain small or no seeds if heavily infected (Greco *et al.*, 1988). Infected roots are poorly developed, lacking nitrogen-fixation nodules and showing cysts of adult nematode females as the sole diagnostic signs (Greco *et al.*, 1992).

The second stage juveniles attack the roots and develop to produce lemon shaped white females protruding out of root. The female deposits many eggs in a large gelatinous egg mass but retains some in body, which after death turns into a brown cyst. Infected chickpea roots are characterized by the visible swollen adult females protruding from the root surface. The lemon shaped cysts change from white to brown as females mature.

11.2.2. Life cycle

The nematode completes the life cycle in 3-4 weeks and has several generations in a single crop season.

11.2.3. Management

11.2.3.1. Physical methods: The population of all parasitic nematodes of chickpea was markedly affected by soil solarization. Included in this population were *H. cajani* cysts, eggs, and larvae; *Rotylenchulus reniformis*; *Helicotylenchus retusus*; *Pratylenchus* spp.; *Tylenchorhynchus* spp.; and *Heterodera* spp. Solarization with irrigation was most effective in reducing populations of cysts, eggs, and larvae of *H. cajani*.

11.2.3.2. Cultural methods: Summer plowing of the field and crop rotation for 2-3 years with cereals may reduce nematode population densities.

11.2.3.3. Chemical methods: Application of carbofuran or phorate @ 1-2 kg a.i. /ha may also reduce crop damage.

11.3. LESION NEMATODE, *Pratylenchus thornei*

In India, the world's largest producer and consumer of chickpea, *P. thornei* is emerging as a serious threat to chickpea production, with high populations reported in Madhya Pradesh (Baghel and Singh, 2013), Rajasthan (Ali and Sharma, 2003), Maharashtra (Varaprasad *et al.*, 1997), and Uttar Pradesh (Sebastian and Gupta, 1995).

11.3.1. Crop losses

Yield losses of 25% were obtained in chickpea fields with 11,600 *P. thornei*/kg of soil at planting (Thompson *et al.*, 2000; Reen *et al.*, 2014). A damage threshold as low as 31 nematode per liter of soil was reported for *P. thornei* by Di Vito *et al.* (1992) in field conditions in Syria, with 2000 nematodes per liter at planting resulting in yield losses up to 58%. Chickpea yield being reduced to about 40% with a nematode population at planting of 2 nematodes per cm^3 of soil (Di Vito *et al*, 1992).

11.3.2. Symptoms

Under field conditions, root-lesion nematode damage to chickpea occurs in patches, with affected plants showing stunting, leaf chlorosis, root necrosis, and reduction of root growth and shoot weight because of destruction of cortical parenchyma. Root-lesion nematodes are migratory endoparasites that cause extensive damage to cortical cells in the pathway of migration and during feeding (Castillo *et al.*, 1998). Infection by *P. thornei* is characterized by dark brown to black lesions on chickpea roots.

11.3.3. Host range

P. thornei has a wide host range, including highly susceptible crops such as grain legumes (e.g., chickpea, faba bean, lentil, pea, and vetch) and small grain cereals; and moderately susceptible crops such as cauliflower, lettuce, and potato. Spinach and sugar beet are resistant to *P. thornei* (Greeco *et al.*, 1988).

11.3.4. Survival

In the absence of a host crop, *Pratylenchus* species survive by means of several strategies, but particularly on roots of weeds or in the soil in a quiescent state (anhydrobiosis). The ability of *Pratylenchus* species to undergo anhydrobiosis is the main reason that these nematodes are difficult to eradicate from infested fields (Castillo and Vovlas, 2007). Anhydrobiosis of *P. thornei* was observed in Israel and southern Spain, where nematodes survived the dry fallow season in soil or within root debris in an anhydrobiotic state until reactivated by fall rains (Talavera and Valor, 2000).

11.3.5. Management

11.3.5.1. Physical methods: Daily (2 p.m.) soil temperatures of soil solarized plots were above 45 °C and reached a maximum of 50 °C at 5 cm depth. They exceeded 40 °C at 15 cm depth (max. 46 °C) and were in the range 33-35 °C at 30 cm depth. Temperatures higher than 42-43 °C are known to kill nematodes after a few hours' exposure. Temperatures above 35 °C can also injure nematodes if prolonged for several days. Therefore, according to the observed temperatures, good nematode control would have occurred in the top 20 cm soil. This assumption was confirmed by the data on percentage of *P. thornei* that survived in the soil at the end of the solarization periods (Table 15.67). Only 11.6 percent of the nematodes survived after two weeks of solarization and nearly 100 percent control was achieved by treatment for eight weeks. However, by harvest time, nematode populations had increased in all solarized plots.

Nematode numbers remained significantly lower than the control only in plots solarized for at least four weeks. This was reflected in the nematode infestation of the roots, which was significantly less than the controls in all plots solarized for more than two weeks (Table 11.1).

Table 11.1. Effect of soil solarization on the numbers of *Pratylenchus thornei* extracted from soil and roots of chickpea.

Treatment	Nematodes/50g soil			Nematodes /5g roots
	Before mulching	After mulching	At harvest	
Control (non-solarized)	259	237	57.5	2860
2 weeks solarized	248	29	43.7	2519
4 weeks solarized	297	11	19.7	1377
6 weeks solarized	219	5	24.7	1347
8 weeks solarized	240	1	22.5	1273
CD ($P \geq 0.05$)	N.S.	53.1	27.2	863.2

The total weight of the aerial plant parts increased significantly (28-34 percent) in all plots solarized for six to eight weeks (Table 11.2). However, although grain yield increased in all solarized plots, significant increases (23-29 percent) were only observed in those solarized for six to eight weeks.

Table 11.2. Effect of soil solarization on the yield of chickpea grown in a field infested by *Pratylenchus thornei*

Treatment	Yield of chickpea (kg/ha)	
	Straw and grain	Grain
Control (non-solarized)	2563	1389
2 weeks solarized	2889	1516
4 weeks solarized	2825	1488
6 weeks solarized	3286	1710
8 weeks solarized	3444	1798
CD ($P \geq 0.05$)	476.5	237.3

The results clearly show that soil solarization is effective for the control of *P. thornei* on chickpeas. This method of control has also shown promise against several soil-borne plant pathogens, including *Orobanche* spp., which could become another constraint to the production of this pulse crop.

11.4. RENIFORM NEMATODE, *Rotylenchulus reniformis*

Rotylenchulus reniformis is the most economically important species of reniform nematodes and is also an important pathogen of chickpea in India (Ali, 1995). Infections of chickpea by *R. reniformis* have been associated with chickpea decline. High soil infestations with *R. reniformis* occur in irrigated chickpea fields, whereas population densities of this nematode are lower in rain fed areas (Ali, 1995).

11.4.1. Crop losses

The reniform nematode is responsible for up to 80% yield reduction occurring at 10 nematodes per g of soil (Mahapatra and Pahdi, 1986).

11.4.2. Symptoms

Infected plants are distributed in patches and show pale green leaves, stunted growth, and early senescence. The damage threshold for *R. reniformis* in greenhouse grown chickpeas was estimated at 0.5 to 1.0 nematodes per g of soil ((Mahapatra and Pahdi, 1986).

11.4.3. Host range

R. reniformis has a wide host range, including legumes (common bean, black gram, chickpea, pigeon pea, and soybean), cereals (maize, rice, sorghum, and wheat), vegetables (cucumber, eggplant, potato, and tomato), fruits (banana, citrus, grapevine, and tropical fruits), and oilseed crops (castor, groundnut, and sunflower).

11.4.4. Management

The use of soil solarization was effective in reducing the population density of *R. reniformis.*

Altering planting dates have had some success; for example, advancing of planting date of paddy rice (*Oryza sativa*) in India significantly reduced soil population densities of *R. reniformis* (Haidar *et al.*, 2001).

11.5. CONCLUSION

Attacks by root-knot, root-lesion, and cyst-forming nematodes can significantly limit chickpea production worldwide, particularly in subsistence agriculture. Although the extent of crop yield loss is mainly determined by the nematode population density in soil, other factors in the agro-ecosystem (e.g., climate, soil conditions, cultivar susceptibility, *etc.*) may also contribute to significant yield reductions. Accurate and quick diagnoses of the nematode species and pathogenic variants are a prerequisite for the proper management of nematodes in chickpea. Management of nematodes that attack chickpea must be based on strategies that integrate the use of available host resistance or tolerance in chickpea cultivars together with crop rotation and cultural modifications that can be practiced by farmers (e.g., time of planting). Practical and efficient integrated management of plant-parasitic nematodes in chickpea requires that high-yielding cultivars with appropriate marketable traits and resistance or tolerance to nematodes be made available to the farmers. Additional studies are also needed to improve cultural management practices. Finally, further studies are needed to unravel mechanisms underlying the interactions between *Meloidogyne* species and races of *F. oxysporum* f. sp. *ciceris*. Such information would foster appropriate use of the wilt-resistant cultivars in chickpea fields infested with the two pathogens.

Chapter - 12

GREEN GRAM, *Vigna radiata*

12.1. INTRODUCTION

In India, the total coverage area under mung bean was about 4.07 million ha with a production of 1.9 million tons and the productivity was 467 kg/ha. More than 80 per cent of mung bean production comes from 10 states namely Rajasthan, Madhya Pradesh, Maharashtra, Bihar, Karnataka, Tamil Nadu, Gujarat, Andhra Pradesh, Odisha, and Telangana. In Orissa, the total area coverage and production of mung bean during 2017-18 was 0.22 million ha (which is 5.42% of total mung bean cultivated area in India) and 0.08 million ton (which is 4.22% of total mung bean production in India), respectively (Anon, 2018).

The root-knot nematodes (*Meloidogyne incognita* and *M. javanica*), the lesion nematode (*Pratylenchus thornei*), and the reniform nematode (*Rotylenchulus reniformis*) are the major nematode problems on green gram.

12.2. ROOT KNOT NEMATODES, *Meloidogyne* spp.

The root-knot nematodes (*Meloidogyne incognita* and *M. javanica*) are considered as one of the most destructive pests of green gram limiting its production.

12.2.1. Geographical distribution

The root-knot nematodes are distributed in Africa, Asia, and Australia.

12.2.2. Crop losses

Root-knot nematode is responsible for 29% yield loss in green gram worth Rs. 2,001 million in India (Walia and Chakraborty, 2018). Root-knot nematodes can reduce mung bean yield by 13 to 16% (Khan *et al.*, 2002). The major damage is caused by root-knot nematode (*Meloidogyne incognita*), with 29-31% yield loss to green gram in India (Mishra and Chakrabarti, 2021). In mung bean, avoidable yield losses under field conditions due to *M. javanica* ranged from 42.1 to 93.4 percent using carbofuran at 1.2 and 4.0 kg a.i. /ha (Gupta and Verma, 1990).

12.2.3. Symptoms

Symptoms of root-knot nematode infection on green gram plants include root galling, wilting of plants during moisture stress, and reduced vigor. Galls are produced mostly on root tips and on their vicinity, but all along the roots the beaded appearance of multiple galls (the result of coalescing of adjacent galls) is also common. In severely infected plant, the root system is reduced, and the rootlets are almost completely absent. A severe infestation by root-knot nematode finally results in poor pod formation with deformed seeds that are less developed, and smaller in size and weight. *M. incognita* deteriorated the growth and caused significant yield losses in mung bean.

Nematode infection predisposes plants to fungal and bacterial root pathogens. The younger mung bean plants were more susceptible to nematodes than older plants. The nodulation on mung bean roots which had been infected by *Meloidogyne* spp. decreased significantly compared to non-infected plants.

12.2.4. Life cycle

M. incognita and *M. javanica* complete their life cycle in 27–32 days at an optimum temperature of 25–30 °C. During the cropping season, more than one life cycle is completed by the nematode. Fertilized females lay on average 200–500 eggs in a gelatinous matrix.

12.2.5. Spread

Primary spread is by means of egg masses in infected plant debris and soil or collateral and other hosts like Solanaceous, Malvaceous and Leguminous plants that act as sources of inoculum. Secondary spread is through autonomous second stage juveniles that may also be water dispersed.

12.2.6. Interaction with other pathogens

The root-knot nematode interacts and aggravates the severity of Fusarium wilt.

12.2.7. Management

12.2.7.1. Cultural methods: The farmers of middle Gujarat (AES III) growing green gram during *kharif* season in root-knot nematode infested soil are advised to adopt crop rotation with cabbage in *rabi* and cluster bean (vegetable purpose) in summer for two years to manage root-knot nematodes.

Cropping sequence green gram/black gram-*toria*-paddy is effective against root-knot nematode.

Intercropping of green gram with sesame (2:1) or application of Neem cake/Neem based organic amendment @1 t/ha were found effective.

Wani and Bhat (2012) reported that soil amended with Nimin coated urea (Neem – based product with Neem-triterpenes) was found effective against *M. incognita* and improved plant-growth characteristics and chlorophyll content of plant leaves.

Tagetes spp. are other plants which can be successfully used as inter-crops or alternative crosp in the nematode infested field as their root exudates are effective against root-knot nematodes and this method of management may prove economical to farmers around cities as this is a cash crop.

Soil application of neem cake @ 2 tons/ ha significantly increased the shoot length, fresh and dry weight of shoot of green gram and reduced root-knot nematode infection (Sumita and Das, 2014).

12.2.7.2. Chemical methods: Seed treatment with Carbosulfan 25DS @ 3% a.i. w/w before sowing is recommended for the control of root-knot nematode.

Seed soaking in Carbosulfan 25EC @ 0.01% for 6 hr. is effective (Choudhury *et al.*, 2011).

Soil application of carbofuran @ 1-2 kg a.i. /ha in root-knot nematode infested soil and increased the yield of chickpea.

The yield per plot and plant height were maximum when seed treatment with carbosulfan (Posse 25 ST) at 1% a.i. (w/w) + soil application of phorate at 1 kg a.i. /ha were combined (Sumita and Das, 1990).

Seed soaking treatment of Emamectin benzoate @ 1.0 g/lit and Carbosulfan @ 2.0 ml/lit for 6 hours proved to be effective in improving the plant growth parameters and reducing root-knot disease (*M. incognita*) and final nematode population (Patel and Patel, 2017).

12.2.7.3. Biological methods: Seed treatment with *Trichoderma viride* @ 5 g/kg seed was found effective.

Seed treatment with *Bacillus subtilis* @ (133.33 x 10^6 cells /seed), *B. thuringiensis* @ (70 x 10^6 cells /seed) and *B. cereus* @ (74.66 x 10^6 cells /seed) and soil drenching with *B. subtilis* (8.83 x 10^9 cells/ml), *B. thuringiensis* (2.1 x 10^9 cells/ml) and *B. cereus* (9.3 x 10^9 cells/ml) showed a significant increase in seed germination, shoot length, shoot weight, root length and root weight and a significant reduction in root galling in mung bean (Dawar *et al.*, 2008).

Among the different bio-agents, Biofor-pf (*Psudomonas fluorescens*) @ 100 kg/ ha exhibited the best result in reducing the galls, egg masses and final soil nematode population (*M. incognita*) (Sumita and Das, 2014).

12.2.7.4. Host resistance: Growing of green gram varieties (T-44, L-62, K-851, PDM-14, ML 1265 and COGG-912) resistant to root-knot nematode (*M. incognita*).

12.2.7.5. Integrated methods: Seed treatment with NSKP @ 5 g/kg seed + *Trichoderma viride* (2 x 10^6 cfu/g) @ 5 g/kg seed is effective in reducing the infestation of *M. incognita* in green gram (Choudhury *et al.*, 2011).

Sowing of Carbosulfan 25DS @ 3% treated seed in deep summer plowed field was found effective for the management of root knot nematode in green gram.

Soil application of Neem cake @ 100 kg/ha along with *Trichoderma viride* (2 x 10^8 spores/g) @ 2.5 kg/ha at the time of sowing was found effective to manage root-knot nematodes and to increase in green gram yield.

12.3. INTERACTION OF ROOT-KNOT NEMATODE WITH FUSARIUM WILT

12.3.1. Interaction

Nematode-fungus disease complex particularly of *Meloidogyne incognita* and *Fusarium oxysporum* poses a great problem to the cultivation of green gram by inflicting severe yield losses (Mahapatra and Swain, 2001).

12.3.2. Management

The present study was carried out to determine the comparative efficacy of chemicals, neem seed powder and microbial antagonists against *M. incognita* - *F. oxysporum* disease complex on green gram cv ML-1108 (Haseeb *et al.*, 2005). Data (Table 14.49) shows that the effect of various treatments on all the growth and yield parameters viz. shoot length, fresh weights of shoot and root, seed weights were mostly significant ($P \leq 0.05$) as compared to untreated inoculated plants. Maximum increase in all the growth and yield parameters was found in carbofuran treated plants followed by neem seed powder, *T. harzianum*, bavistin and *P. fluorescens* treated plants as compared to untreated inoculated plants (Haseeb *et al.*, 2005).

Table 14.49. Effect of carbofuran, bavistin, neem seed powder, *Trichoderma harzianum* and *Pseudomonas fluorescens* against *Meloidogyne incognita* and *Fusarium oxysporum* disease complex on green gram cv. ML-1108

Treatments	Fresh weight (g)		Shoot height (cm)	Seed weight (g)
	Root	Shoot		
Nematode + fungus (Control)	22.1 (28.3)	4.8 (36.2)	18.6 (32.3)	5.34 (37.1)
Nematode + fungus + carbofuran	29.2 (5.2)	7.0 (6.8)	25.9 (5.8)	7.88 (7.2)
Nematode + fungus + neem seed powder	27.8 (9.7)	6.7 (10.2)	24.8 (9.8)	7.62 (10.4)
Nematode + fungus + *T. harzianum*	26.3 (14.6)	6.4 (15.1)	22.7 (17.3)	7.12 (16.2)
Nematode + fungus + bavistin	24.1 (21.8)	6.0 (20.1)	22.1 (19.7)	6.77 (20.4)
Nematode + fungus + *P. fluorescens*	23.8 (22.6)	5.4 (22.8)	21.9 (20.5)	6.66 (21.7)
CD (P = 0.05)	2.5	0.6	2.0	0.70

Values in parenthesis represent percent reduction over uninoculated control.

Data presented in Table 14.50 shows that the greatest suppression of nematode reproduction i.e. reproduction factor (Rf) and root-knot index was achieved by application of carbofuran (Rf =0.46, RKI =0.25) followed by neem seed powder (Rf =0.99, RKI =0.50), *T. harzianum* (Rf =1.27, RKI =1.50), *P. fluorescens* (Rf =1.38, RKI =1.75) and bavistin (Rf =1.50, RKI =2.25) as compared to controls (Rf =1.74, RKI =3.00). Bavistin was highly effective in suppression of taproot colonization by fungus (8% root colonized) followed by neem seed powder (30%), *T. harzianum* (35%), *P. fluorescens* (65%) and carbofuran (75%) (Haseeb *et al.*, 2005).

Table 14.50. Effect of carbofuran, bavistin, neem seed powder, *Trichoderma harzianum* and *Pseudomonas fluorescens* on root-knot development, reproduction of *Meloidogyne incognita* and infection of *Fusarium oxysporum* on green gram cv. ML-1108

Treatments	Final nematode population (Pf)			Reproduction factor Rf = Pf / Pi	Root-knot index	Disease index (%)
	Root	Soil	Total			
Nematode + fungus (Control)	1488	2000	3488	1.74	3.00	95
Nematode + fungus + carbofuran	420	500	920	0.46	0.25	75
Nematode + fungus + neem seed powder	1474	500	1974	0.99	0.50	30
Nematode + fungus + *T. harzianum*	1536	1000	2536	1.27	1.50	35
Nematode + fungus + bavistin	1508	1500	3008	1.50	2.25	8
Nematode + fungus + *P. fluorescens*	1512	1250	2762	1.38	1.75	65
CD (P = 0.05)	144.1	121.9	258.4	0.11	0.14	6.0

Pi: Initial population.

12.4. RENIFORM NEMATODE, *Rotylenchulus reniformis*

12.4.1. Crop losses

A 13% yield loss was attributed to the reniform nematode (Dayal and Sharma, 2007).

12.4.2. Symptoms

General symptoms include reduced root systems, leaf chlorosis, overall stunting of host plants, and reduced yields and plant longevity. Root lesion nematodes reduce development of lateral roots, which decreases the ability of plants to extract water and nutrients. Female nematodes and their eggs are often visible when plant roots are viewed under a dissecting microscope. Seedlings can be killed outright, and survivors may be stunted and produce less grain.

12.4.3. Management

12.4.3.1. Cultural methods: Mahua (*Madhuca longifolia*) seed kernel at 20% w/w was highly efficient, resulting in maximum plant growth characters (improving shoot length and weight, root length and weight, and number of nodules per plant) and minimum nematode reproduction (decreasing nematode reproduction, i.e. number of females per plant, number of egg masses per plant, number of eggs and larvae per egg mass, and soil population) (Dayal and Sharma, 2007).

Sorghum is an excellent crop to include in rotations where *P. thornei* is present because it is moderately resistant.

12.4.3.2. Biological methods: Integration of seed treatment with 5 g/kg each of *Trichoderma viride*, *Pseudomonas fluorensens* or 10 g Neem seed kernel powder was found effective for management of reniform nematode in green gram.

12.4.3.3. Host resistance: Grow green gram varieties (Pusa-103, ML-62, ML-80, and PDM-14) resistant to reniform nematode.

12.5. ROOT-LESION NEMATODE, *Pratylenchus thornei*

Root-lesion nematodes (RLN) (*P. thornei*) can reduce the yield of mung bean, and large populations will remain in the soil causing a major issue in the cropping system.

12.5.1. Symptoms

Above-ground symptoms of RLN attack on mung bean include poor establishment, stunting, yellowing of lower leaves, and poor tillering. Infected plants may wilt prematurely in dry periods and at the end of the season. Primary and secondary roots may show a general browning and discoloration. Roots may develop small lesions (up to 2 mm long) or black striping. Mung bean yield is reduced when *P. thornei* is present, because of damaged and ineffective root-systems.

12.5.2 Biology

Mung bean varieties are susceptible to infection by the RLN *P. thornei* but are resistant to *P. neglectus*. When a susceptible crop is grown, *P. thornei* will build up to high levels and populations will persist in the soil to infect other susceptible crops such as wheat and chickpeas. Resistant crops such as sorghum and sunflowers do not allow RLN reproduction, and populations of the nematodes will not increase. RLN are found on a wide variety of soil types.

12.5.3. Spread

RLN appears to be spread in soil moved by surface water, vehicles, and farm machinery.

12.5.4. Management

Numbers of RLN do not reduce quickly in the absence of a host crop (or weeds), so growing resistant and/or tolerant crop species or non-host crops is a major part of management. Crops such as sorghum and sunflower are moderately resistant to *P. thornei* and can be used in rotation to reduce the RLN population.

12.6. CONCLUSION

The most effective management tactics for nematode control on mung bean are disease-resistant cultivars (Hassan and Devi, 2004, Singh and Prasad, 2016) and botanical seed dressings, such as neem and oilseed cakes (Dayal and Sharma, 2007). Growing sorghum, small grains, grasses, or leaving a clean summer fallow between crops can reduce the nematode populations and should be routinely incorporated into the cropping sequence (Stirling *et al.*, 2006, Mohler and Johnson, 2009).

Chapter - 13

CLUSTER BEAN, *Cyamopsis tetragonoloba*

13.1. INTRODUCTION

India accounts for about 80% of the world cluster bean (guar) production. Cluster bean is mainly grown in India (Rajasthan, Haryana, Gujarat, Punjab), Pakisthan, Sudan and USA. In Rajasthan, cluster bean is mainly grown in Bikaner, Sri Ganganagar, Udaipur, Jalore, Sikar, Churu and Jodhpur districts. Total area under the crop in India was 4.26 million hectare with annual production of 2.42 million tons and productivity of 567 kg ha^{-1} (Anon, 2019).

Cluster bean (*Cyamopsis tetragonoloba*) is also popularly called as 'guar' in India. Its tender pods being edible and are used in preparation of curry. The green pods are rich in vitamins A and C. Since it is draught tolerant, it can be grown in summer. It is also grown as a forage and green manure crop. Its seeds are reported to be good cattle feed.

Several plant parasitic nematodes viz., *Meloidogyne incognita, M. javanica, Rotylenchulus reniformis, Helicotylenchus incises, Pratylenchus delattrei, Tylenchorhynchus capitatus, Xiphinema* sp., *Tylenchus* sp., *Criconema* sp. and *Aphelenchus* sp. have been found associated with cluster bean in India and abroad (Bishnoi and Yadav, 1989; Shaukat *et al.*, 2010). Amongst nematodes, root-knot nematode (*Meloidogyne incognita*) is reported to cause significant losses on cluster bean (Yadav *et al.*, 2009). Root-knot nematode and *Fusarium* wilt disease complex is the major limiting factor in successful production of cluster bean crop.

13.2. ROOT-KNOT NEMATODE, *Meloidogyne incognita*

13.2.1. Crop losses

The avoidable loss due to root-knot nematode, *M. incognita* was recorded at 42.88 per cent in cluster bean (Nama and Baheti, 2021).

13.2.2. Symptoms

Plants affected by *M. incognita* show symptoms of water and nutrient stress, yellowing, wilting, and stunting in the above-ground parts. Under-ground symptoms include typical root galling. At high levels of infestation, plant mortality may occur.

13.2.3. Management

The maximum increase in plant growth characters and significant reduction in root-knot nematode (*M. incognita*) population was obtained with the application of neem cake at 400 kg/ha followed by castor cake at 400 kg/ha and neem cake at 200 kg/ha over untreated check (Baheti *et al.*, 2018).

Seed treatment with with acephate at 2 per cent gave maximum increase in shoot length, root length, shoot weight, root weight and number of nodules; and significant reduction in nematode population viz., galls/plant, egg masses/plant and final larval population/100 cc soil (Baheti *et al.*, 2017).

Soil application of phorate at 2 kg per hectare significantly decreased root-knot index and final nematode population over untreated check. Phorate at 2 kg per hectare in root-knot infested field (*Meloidogyne incognita*) significantly enhanced the yield of cluster bean (11.94 q/ha) over untreated check (6.82 q/ha). Application of chemical increased 75.07 per cent yield of cluster bean infected with *M. incognota* (Table 13.1) (Nama and Baheti, 2021).

Table 13.1. Effect of phorate on root-knot index (*Meloidogyne incognita*) and yield of cluster bean

Treatment	Root-knot index	Final nematode population
Phorate 2 kg a.i. /ha	3.4	695.80
Untreated check	4.8	1276.33

Effects of organic amendments and bio-agents in combination significantly increase the yield of cluster bean and reduced nematode populations as compared to applied alone. The highest yield and lowest nematode population were found in neem cake + *Pseudomonas fluorescens* (Kumar *et al.*, 2020).

13.3. INTERACTION OF ROOT-KNOT NEMATODE WITH ROOT ROT

13.3.1. Interaction

The association of the root-knot nematode, *Meloidogyne javanica* with root infecting fungi (*Macrophomina phaseolina, Rhizoctonia solani, Fusarium solani, F. oxysporum*) produces greater losses than either pathogen alone.

13.3.2. Management

13.3.2.1. Integrated methods: *Pseudomonas aeruginosa* and *Purpureocellium lilacinum* used alone or together significantly ($P<0.05$) reduced infection of root knot nematode *M. javanica* and root infecting fungi viz., *M. phaseolina, R. solani, F. solani* and *F. oxysporum* on cluster bean. *Pseudomonas aeruginosa* was more effective than *P. lilacinum* in reducing the root-knot nematode, *M. javanica* infection. Combined use of *P. lilacinum* and *P. aeruginosa* was more effective in reducing the infection of root-knot nematode and *F. solani* on cluster bean than either used alone. Use of *P. aeruginosa* and *P. lilacinum* significantly ($P<0.05$) increased plant height of cluster bean (Table 13.2) (Perveen *et al.*, 1998).

Table 13.2. Effect of *P. aeruginosa* and *P. lilacinum* on plant height and control of root rot disease complex in cluster bean

Treatment	Plant height (cm)	Root-knot index	Infection %			
			M. phaseolina	*R. solani*	*F. solani*	*F. oxysporum*
Control	24.5	4.1	31	0	75	81
P. lilacinum (PL)	28.0	3.5	6	0	62	81
P. aeruginosa (PA)	27.7	2.0	19	0	44	50
PL + PA	32.0	1.4	6	0	37	44
CD (P = 0.05)	2.2	0.34	6.1	6.1	6.1	6.1

13.4. RENIFORM NEMATODE, *Rotylenchulus reniformis*

13.4.1. Symptoms

Infected plants show stunted growth and early senescence. Such plants in field are distributed in patches. Leaves of infected plants become pale green and show uneven growth especially at a young stage. Only females infect plant roots. Affected plants have a reduced number of *Rhizobium* nodules. Roots at the point of infection appear bulgy and dirty because of the protruding posterior half of the nematode, attached egg masses, and adhering soil particles and organic matter.

13.4.2. Management

Soil application of Aak (*Calotropis procera*), Lantana (*Lantana camara*) and Water hyacinth (*Eichhornia crassipes*) at 15% were found most effective over others in reducing *R. reniformis* population and in enhancing the plant growth characters (Rajvaniya *et al.*, 2021).

Crop rotation with mustard, oats, onion, sugarcane, and sun hemp can effectively reduce the nematode populations (Caswell *et al.*, 1991; Robinson *et al.*, 1997). Clean fallowing and deep summer plowing are effective practices that can significantly reduce population density of *R. reniformis*.

The application of systemic dithiocarbamate nematicide (carbofuran) at sowing after weeding, deep plowing, and following for 1 month effectively reduced *R. reniformis* to a satisfactory level (Sivakumar *et al.*, 1973).

Chapter - 14

CABBAGE, *Brassica oleracea* var. *capitata* and CAULIFLOWER, *B. oleracea* var. *botrytis*

Stunt, cyst, and root-knot nematodes are the major limiting problems in successful cultivation of cabbage and cauliflower crops.

14.1. STUNT NEMATODE, *Tylenchorhynchus brassicae*

Siddiqi *et al.* (1972) encountered *T. brassicae* as the most prevalent species around the roots of cabbage and cauliflower in Uttar Pradesh.

14.1.1. Symptoms

Cabbage field infested with stunt nematode show patchy appearance. *T. brassicae* caused reduction in fresh weight of cabbage and cauliflower plants and suppressed root growth, resulting in reduced water absorption.

14.1.2. Host range

The most efficient hosts of the stunt nematode include cabbage, cauliflower, and tomato, although all crucifers are susceptible. The multiplication rate on cabbage (cv. Suttons Earliest), cauliflower (cv. Patna) and Knol-khol (cv. Purple Vienna) was 13, 18 and 29, respectively. Common weeds such as *Chenopodium album, Trianthema monogyna,* and *Portulaca oleracea* were found to be hosts of *T. brassicae.*

14.1.3. Life cycle

The optimum soil temperature (30 °C) and soil moisture (25-30%) for reproduction of *T. brassicae* was same for both cabbage and cauliflower. Further, high temperature (35-40 °C) and low moisture are highly unfavorable for survival of *T. brassicae* in the absence of host. However, at moderate temperature and moisture conditions, the nematode can survive even up to 240 days.

14.1.4. Interaction with other pathogens

The stunt nematode, *T. brassicae* interacts with the root rot fungus, *Rhizoctonia solani* and adversely affects the emergence of crucifer seedlings when occur together (Khan *et al.*, 1971).

14.1.5. Histopathology and histochemistry

The histopathology of cabbage and cauliflower roots shows that *T. brassicae* penetrated throughout the intercellular region of cortex. The nematodes were mostly situated 1-3 cell layer deep in the cortex and were parallel to the longitudinal axis of the roots.

Alam *et al.* (1976) reported that the roots of cabbage and cauliflower infected with 10,000 *T. brassicae* enhanced total soluble phenols, ortho-dihydroxy phenols, free amino acids and proteins, but not in total carbohydrates. Total phenols, ortho-dihydroxy phenols and free amino acids increased with increasing seedling age at inoculation (Khan *et al.*, 1980).

14.1.6. Management

14.1.6.1. Cultural methods

i) Crop rotation: Siddiqi *et al.* (1973) reported that rotation of cabbage and cauliflower with wheat resulted in sudden decline in population of *T. brassicae*. The mustard (*rabi*), radish (summer), sesame (*kharif*) cake sequence considerably reduced the population of *Tylenchorhynchus* spp. (Haque and Gaur, 1985).

ii) Intercropping: Interculture of seedlings of marigold, neem, and Persian lilac reduced multiplication rate of *T. brassicae* and improved plant growth. Intercropping of cabbage and cauliflower with neem reduced the multiplication rate of *T. brassicae* and improved the plant health (Siddiqi and Saxena, 1987a, b).

iii) Botanicals: Seed dressing with latex of *Calotropis gigantea, C. procera, Euphorbia milii, E. nerifolia* and *E. tirucalli* significantly inhibited *T. brassicae* multiplication on cabbage and cauliflower and improved plant growth (Siddiqi and Alam, 1989a).

Bare-root dip of cabbage and cauliflower seedlings in leaf extracts of Neem and Persian lilac (Siddiqi and Alam, 1989b) reduced multiplication rate of *T. brassicae* and improved plant growth.

Amendment of soil with oil cakes of Neem, Groundnut, Mustard, and Castor (equivalent to 100 kg N/ha) was effective in reducing the population of *T. brassicae* around the roots of cabbage and cauliflower (Siddiqi *et al.*, 1976). Neem cake was the best.

Enhanced crop growth and reduction in reproduction of *T. brassicae* was obtained by soil incorporation of Neem cake (Siddiqi and Saxena, 1987a, b).

iv) Biofumigation: Soil amendment with chopped shoot parts of jack, *Ficus carica, F. elastica, F. glomerata, Ipomea fistulosa, Nerium indicum,* and *Tabernaemontana coronaria* gave good control of *T. brassicae* on cabbage and cauliflower (Siddiqi *et al.*, 1987).

14.1.6.2. Chemical methods: Carbofuran, Ethoprophos, and Fenamiphos each at 1 to 2 kg a.i. per ha were found effective in reducing the root-knot nematode population and in increasing flower yields of cauliflower.

Carbofuran at 1.5 kg a.i. per ha was effective in reducing *T. brassicae* and *T. dubius* population and in increasing yields of cabbage (Varma *et al.*, 1978).

14.1.6.3. Host resistance: *Brassica rapa* cv. Sarson (yellow mustard) was found to be least susceptible to *T. brassicae* and could be grown on infected soil in preference to *B. nigra* (black mustard) (Khan *et al.*, 1986).

14.2. CYST NEMATODE, *Heterodera cruciferae*

14.2.1. Symptoms

The first field symptom is the appearance of small, conspicuous areas where plants are stunted. Later in the season, these areas are marked by absence of plants and presence of weed patches. In severely infested fields, areas of decline usually can be identified by excessive leaf wilting during hot weather. Affected areas increase in size with continuous cropping of host crops.

Infected plants are much smaller than normal ones. Frequently, they have a dense system of secondary roots, called a hairy-root condition. Examination with a hand lens usually shows numerous small, white, bead-like structures on roots of infected plants. These structures are lemon-shaped and later become cysts containing eggs that develop into future generations of nematodes.

Heterodera cruciferae only infects crucifers. Nematode significantly reduces the size of heads. As the disease progresses, leaves may wilt or curl, especially during hot weather, tap root remains small, roots are discolored and branched profusely, produce loose, and small heads. Invasion of infected roots by fungi is common. A characteristic sign of this pathogen is the appearance of lemon-shaped cysts on the root surface, which are white, tan or reddish in color. Plants often die prematurely.

14.2.2. Host range

In addition to sugar beets, other hosts are mangel-wurzel, table beet, cabbage, cauliflower, Brussels sprouts, broccoli, rape, turnip, rutabaga, and radish. Weed hosts are dock, knotweed, lambs' quarters, mustard, nightshade, purslane, saltbush, and red root. Any of these that survive from year to year may act as reservoirs for the pests, counteracting much of the value of crop rotation.

14.2.3 Favorable conditions

Loamy soils favor disease development.

14.2.4. Survival and spread

- The means of survival is through cysts which overwinter in soil. After planting of the crop, eggs from cysts hatch and the second stage larvae infect the roots.
- The means of nematode spread include infected planting stock, farm implements, and contaminated soil. The nematode spread through rain and irrigation water.

14.2.5. Management

14.2.5.1. Cultural methods: Crop rotation is an effective way to reduce nematode population because the pest has a narrow host range. Crops that may be used in a short-term rotation include beans, sweet clover, corn, grains, peas, potatoes, and tomatoes. Alfalfa is suitable for a long rotation period. Length of rotation period should vary with severity of nematode infection. Slight infestations may require only a 2-year program; severe infestations require a 5- to 6-year interval between host crops.

14.2.5.2. Biological methods: MeloCon WG (*Purpureocellium lilacinum*) at 2.25 to 4.50 kg/ha at 4- to 6-week intervals for nematode suppression. See label for specific application types and timings. 4-hr re-entry.

Other management strategies include:

- Crop rotation with non-hosts (barley resistant, wheat and sugar beet moderately resistant) for a period of three to five years helps to reduce nematode populations.

- » Incorporation of crop residues immediately after harvest proved beneficial.
- » Eradication of weeds and volunteers helps control this disease.
- » Application of nematicides.
- » Use of resistant varieties.

17.3. ROOT-KNOT NEMATODE, *Meloidogyne incognita*

17.3.1. Symptoms

Patches of stunted plants apparent by midseason. Root-knot nematodes on crucifers induce prolific root branching and galling above the point of infection. Invasion of infected roots by fungi may occur. Above-ground symptoms include stunting, chlorosis, and wilting. Wilting occurs earlier in the day during periods of moisture stress (e.g., warm, and windy conditions). Small and unmarketable heads are produced by infected plants which survive in the growing season.

14.3.2. Favorable conditions

The most severe damage can occur in sandy soil with moderate moisture. Infection can occur at temperatures ranging from 10-35 °C. Freezing temperatures kill all life cycle stages of *Meloidogyne* species.

14.3.3. Interaction with other pathogens

Pathak and Keshari (2004) found that two weeks prior inoculation of root-knot nematode, *M. incognita* before the soil-borne fungus, *F. oxysporum* f. sp. *conglutinans* resulted in maximum wilt score (4.0) and wilt symptom index (80.0) in cauliflower plants.

14.3.4. Survival

These nematodes survive in infected root debris.

14.3.5. Management

- » Cabbage cultivar 'Oxylus' is reported to be a poor host of root-knot nematodes and recommended for cultivation by farmers.
- » Strawberry may be a suitable rotation crop in fields with root-knot nematodes because they are nonhosts to *M. incognita* and most populations of *M. javanica.*
- » Soil fumigation with Dazomet.
- » Reduction in root-knot nematode population by fallowing or flooding.
- » Crop rotation with grass crops for at least two years effectively reduces nematode populations below damaging levels.

14.4. INTERACTION OF ROOT-KNOT NEMATODE WITH CLUB ROOT

14.4.1. Interaction

The root-knot nematode, *Meloidogyne incognita* in association with club ro disease pathogen, *Plasmodiophora brassicae* causes disease complex.

14.4.2. Management

14.4.2.1. Integrated methods: PGPR strains (*Pseudomonas fluorescens, Bacillus subtilis*) combined with fungal biocontrol agents (*Trichoderma viride, T. harzianum*) were found to be effective in reducing the nematode-fungal disease complex in cabbage (Loganathan *et al.*, 2001). The bioformulation mixture of *P. fluorescens, T. viride* and chitin effectively reduced the disease complex in cabbage and cauliflower both under greenhouse and field conditions (Samiyappan, 2003) (Table 14.1).

Table 14.1. Efficacy of bioformulation mixtures against club root disease – root-knot nematode complex in cabbage under greenhouse conditions.

Treatment	Club root index	Nematode incidence	
		Population	Root-knot index
Trichoderma viride	25.99 (30.65)	129	2.66
Pseudomonas fluorescens	28.20 (32.07)	112	2.33
T. viride + *P. fluorescens*	25.33 (30.22)	114	2.33
T. viride + Chitin	25.44 (30.29)	108	2.33
P. fluorescens + Chitin	25.66 (30.43)	111	2.00
T. viride + *P. fluorescens* + Chitin	22.22 (28.12)	108	2.00
Chitin alone	31.70 (34.26)	139	3.00
Carbendazim	19.90 (26.49)	264	4.66
Carbofuran	39.90 (39.17)	106	1.66
Cabendazim + Carbofuran	15.00 (22.79)	103	1.66
Plasmodiophora brassicae alone	48.90 (44.37)	0.033	0.133
Meloidogyne incognita alone	0.03 (0.60)	280	5.00

Among different PGPR strains tested, *P. fluorescens* – Pf1 effectively reduced the root-knot index. The combined application of carbendazim and carbofuran gave least club root index both in cabbage and cauliflower under greenhouse conditions (Tables 14.2 and 14.3) (Loganathan *et al.*, 2001).

Table 14.2. Efficacy of bioformulations against club root disease-root-knot complex in cabbage and cauliflower under greenhouse conditions.

Treatment	Cabbage		Cauliflower	
	Club root index (PDI)	Root-knot index	Club root index (PDI)	Root-knot index
T. viride – TvO	35.55	2.66	36.66	2.33
T. viride – TvOL	37.77	2.33	35.55	3.00
T. viride – TvMNT7	26.66	1.33	22.22	1.00
T. viride – TvF9	36.66	2.00	37.77	1.66
T. viride – TvV4	32.77	2.33	31.66	1.33
T. harzianum	30.32	2.66	32.77	1.66
P. fluorescens – PfO2	34.71	2.00	36.66	1.33

B. subtilis – Bs1	38.88	1.66	38.32	2.00
P. fluorescens – Pf1	28.88	1.33	24.44	1.33
P. fluorescens – PfCHAO	33.88	1.66	34.70	2.66
T. viride – TvK	48.88	3.33	46.66	3.00
T. virens – Tv1	47.77	3.66	48.66	4.00
P. fluorescens – PfO1	51.11	4.33	48.32	3.66
B. subtilis – Bs2	48.88	3.66	49.99	4.33
Carbendazim	22.22	4.00	17.77	4.33
Carbofuran	33.33	3.00	28.88	2.66
Cabendazim + Carbofuran	19.99	3.00	15.55	2.33
P. brassicae	60.00	2.66	52.22	2.00
M. incognita	17.80	4.66	26.66	4.33
Control	67.77	4.66	57.72	4.66

Table 14.3. Efficacy of bioformulation mixtures against club root disease-root-knot nematode complex in cabbage under field conditions.

Treatment	**Club root index (PDI)***	**Decrease in nema. Popn. over initial (%)**	**Root-knot index**	**Yield (t/ha)**
T. viride – TvO	50.0 (45.00)	36.2	2.66	39.99
T. viride – TvOL	40.0 (39.22)	35.7	2.66	44.44
T. viride – TvMNT7	20.0 (26.56)	38.5	2.00	53.33
T. viride – TvF9	38.0 (38.05)	36.5	2.33	45.77
T. viride – TvV4	36.5 (37.16)	33.3	2.66	46.66
T. harzianum	36.2 (36.97)	32.5	3.00	48.88
P. fluorescens – PfO2	46.0 (42.70)	37.4	2.00	44.44
B. subtilis – Bs1	43.0 (40.97)	29.0	3.33	41.33
P. fluorescens – Pf1	23.0 (28.66)	46.2	1.33	59.10
P. fluorescens – PfCHAO	45.0 (42.13)	38.6	2.00	47.40
Carbendazim	18.0 (25.10)	26.4	3.66	51.11
Carbofuran	46.0(42.70)	30.2	2.00	45.77
Cabendazim + Carbofuran	11.5 (19.81)	31.1	2.00	54.66
Control	60.0 (50.79)	+52.2	4.00	33.77

*Values in parenthesis are transformed values.

14.5. CONCLUSION

The production of cruciferous vegetable crops such as cabbage and cauliflower throughout the globe is threatened by stunt nematode, *Tylenchorhynchus brassicae*; cyst nematode, *Heterodera cruciferae*; and root-knot nematode, *Meloidogyne incognita*. The root-knot nematodes interact with Fusarium wilt and the stunt nematode interacts with Rhizoctonia root rot pathogens causing the disease complexes. Cabbage cultivar 'Oxylus' is reported to be a poor host of root-knot nematodes and recommended for cultivation by farmers. Integrated nematode management strategies should be developed using crop sanitation, crop rotation, nematicide application, and biological control agents.

Chapter - 15

CUCUMBER, *Cucumis sativus*

The root-knot nematode is the most important pest on cucumber that is of some economic importance.

15.1. ROOT-KNOT NEMATODES, *Meloidogyne incognita, M. javanica*

15.1.1. Crop losses

Root-knot nematodes are responsible for annual yield loss of 12% in cucumber amounting to Rs. 110.46 million in India (Walia and Chakraborty, 2018).

15.1.2. Symptoms

The infected plants exhibit stunted plant growth, leaf yellowing and reduced yield. Large number of galls on root system and multiple galling is common. In case of heavy attacks, galls can become very large, the root system being reduced to a swollen stump without hairs. It restricts the uptake of nutrients from the root system to the foliage, resulting in a leaf yellowing and stunted plant growth. Ultimately plants wilt and die.

15.1.3. Biology and life cycle

Root-knot nematodes are soil-borne, sedentary endoparasites. After penetration the juvenile feed on conducting vessels and develops endoparasitically. Adult females are sac-like, vermiform males leave the roots and come out in soil. Females lay eggs in masses on root surface on galls, or in advanced stages inside the compound galls. During summer months, the nematode takes about 25-30 days to complete its life cycle.

15.1.4. Spread

Root-knot nematodes are disseminated through infected transplanting material, irrigation water, and agricultural implements.

15.1.5. Management

15.1.5.1. Physical methods: The most important nematode pests controlled by greenhouse steaming are root-knot nematodes attacking cucumbers.

15.1.5.2. Cultural methods: Application of Neem cake @ 30 g/plant 10 days before sowing is effective in cucumber (Anon, 2012).

In polyhouse grown cucumber, Neem cake at 320 g/m^2 proved most effective and resulted in highest yield followed by Mustard cake (Hema, 2014). The benefit: cost ratio with Neem cake was found to be 3.07 in cucumber.

The bio-fumigation approach was the most successful and demonstrated the highest vegetative, flowering, and yield characteristics values for cucumber crop production in naturally infested fields by root-knot nematodes. Cucumber cultivation after broccoli plantlets incorporation (BPI) exhibited the best horticultural traits, which reflected positively on early and total productivity. The increased yield (0.696 - 0.720 kg/plant as compared to 0.302 - 0.324 kg/plant) was gained by suppressing all nematode parameters, the number of nematode larvae, galls (), and egg masses, as well as egg hatching reduction. The highest number of effective biocides [total phenols, myrosinase activity, total glucosinolates (GSLs), and isothiocyanates (ITCs)] was released with BPI treatment (El-Remaly *et al.*, 2022).

Cucumbers planted immediately after the termination of nematode-resistant tomato cultivar Celebrity crop produced significantly higher plant fresh weight, lower percentage of galled roots and higher yields per acre than did those planted after the nematode-susceptible tomato cultivar Heatwave. The results of this study indicate that double-cropping cucumbers with a nematode-resistant tomato variety can be an effective method to improve cucumber yields in soils that have a history of root-knot nematode infestation (Hanna, 2004).

15.1.5.3. Chemical methods: Spot application of Carbofuran 3G @ 10 g/plant at sowing is effective in cucumber (Anon, 2012).

The efficacy of Metham sodium (40 ml/m^2) and Dazomet (40 g/m^2) was evaluated in polyhouse cucumber. Both the treatments proved promising. The benefit: cost ratios with Metham sodium and Dazomet were found to be 3.84 and 3.76, respectively in cucumber (Anon, 2015; Chandel *et al.*, 2014).

15.1.5.4. Biological methods

i) Antagonistic fungi: Mastievich *et al.* (1991) reported that bio-preparation of predaceous fungus *Arthrobotrys oligospora* strains, VKMF-2461D and VKMF-3062D, applied to soil gave good control of root-knot nematodes and increased yield of greenhouse cucumber cv. Estafeta-87.

Soil application of MeloCon WG (*Purpureocellium lilacinum*) at 2.25 to 4.50 kg/ha at 4- to 6-week intervals is beneficial for nematode suppression.

Maximum decline in the galls/root system was recorded with carbofuran (68%), *Trichoderma viride* ITCC-6889 (66%) and *T. harzianum* ITCC-6796 (61%). The above treatments caused around 55–62% suppression in the egg mass count. The root zone population of *M. incognita* was decreased by 87% (carbofuran) and 79–84% (*T. viride* and *T harzianum*) over control. Treatments with carbofuran, *T. viride* and *T. harzianum* significantly enhanced the plant growth of nematode infected cucumber plants (15–28%) (Mohammed and Khan, 2021).

15.1.5.5. Host resistance: Cucumbers grafted on the bur cucumber (*Sicyos angulatus*) rootstock (good compatibility with cucumbers) exhibited increased root-knot nematode resistance (Zhang *et al.*, 2006).

Cucumber cv. 'Adrian' plants grafted onto interspecific hybrid rootstock 'Strong Tosa' (*Cucurbita maxima* × *C. moschata*) had higher plant growth and total number of fruits (19.9/plant as compared to 10.1 in non-grafted plant) and yield (5.38 kg/plant as compared to 2.56 in non-grafted plant) in first season, and the same result was found in the second season also (6.96 kg/plant and more than 28.9 fruits/ plant as compared to 4.5 kg/plant and 20.4 fruits/ plant in non-grafted plant) (Ban *et al.*, 2014).

Wild cucumber rootstock, *Cucumis pustulatus* gave effective control of root-knot nematodes on cucumber and increased fruit yield (Liu *et al.* 2014) (Table 15.1).

Table 15.1. Management of root-knot nematodes in Cucurbits using wild cucumber rootstock, *Cucumis pustulatus*

Treatment	Gall index	Yield in kg/plant
Cucumber grafted on *C. pustulatus*	1	2.032
Cucumber non-grafted	5	1.234

Bur-cucumber (*Sicyos angulatus*) collected near Andong, Korea, showed good compatibility with cucumber for early summer growth and good resistance to nematodes (Lee, 1992; Lee *et al.*, 1992).

15.1.5.6. Integrated methods

i) Bioagents and botanicals: Pre-plant treatment with Purpureocillium lilacinum (20 kg/ha) in combination with organic amendments effectively reduced root-knot infection in gherkin.

Maximum reduction in gall formation in cucumber due to *M. incognita*, female numbers, egg-mass production, developmental stages and final population of juveniles in soil, was obtained by treatments like fungal bioagent, *Pochonia chlamydosporia* (10^7 clamydospores/ml) + symbiotic bacterium, *Phytorhabdus luminescens* (cell-free culture filtrates), *P. luminescens* + compost (C) and *P. chlamydosporia* + *P. luminescens* + animal compost (AC), compared with the control (Table 15.2) (Zakaria *et al.*, 2013).

Table 15.2. Management of root-knot nematodes in Cucumber using bioagents and organic amendments

Treatment	Roots	
	No. of galls/plant	Fruit yield/plant
P. chlamydosporia (10^7) + Compost (C)	72^{bc}	0.758_{d}
P. chlamydosporia + Animal compost (AC)	58^{cd}	1.196_{b}
P. luminescens (CF 100%) + Compost	69^{bcd}	0.831_{d}
P. luminescens (CF 100%) + Animal compost	53^{dc}	1.277_{ab}
P. chlamydosporia (10^7) + *P. luminescens* (CF 100%) + Compost	68^{bcd}	0.904_{d}
P. chlamydosporia (10^7) + *P. luminescens* (CF 100%) + Animal compost	43^{c}	1.415_{a}
C + *M. incognita*	74^{b}	0.612_{c}
AC + *M. incognita*	64^{bcd}	1.050_{c}
Untreated check	158^{a}	1.338_{f}
Uninoculated check (Healthy plants)	0	0.795_{d}

Numbers with the same letter within the same column are not significantly different ($P < 0.05$ using Duncan's Multiple Range Test in SAS statistical program).

ii) Bioagents and AMF: Combined inoculation of AMF and *Pseudomonas fluorescens* had positive effect on root-knot nematode control on cucumber (Jakobsen, 1999).

iii) Physical and botanicals: Enhancement of cucumber fruit yield, reduction in root galling, and second stage larvae of *M. javanica* in soil was achieved by soil amendments (using alfalfa, cauliflower, olive-cake residues, poultry manure, and tomato) or soil solarisation. Organic amendments reduced densities of *Fusarium* spp., generally increased *Aspergillus* spp., while *Trichoderma* spp. was not affected. Combinations of solarization and addition of organic amendments substantially augmented each other, particularly with poultry manure, alfalfa hay, and to a lesser extent with cauliflower and tomato residues.

The use of oil radish (*Raphanus sativus oleifera*) and arugula (*Eruca vesicaria sativa*) as biofumigation crops with plastic cover can be considered as an alternative management tool for the root-knot nematode (*M. incognita*) in greenhouse cucumber production (Haroutunian, 2013).

iv) Cultural and botanicals: In commercial greenhouse trials in Spain an integrated management system was developed, including biofumigation with sheep manure and mushroom residue and the cultivation of short-cycle vegetables acting as trap crops. Using this strategy, initial very high levels of *M. incognita* were reduced to near zero in the main susceptible cucumber crop (Bello, 1998).

v) Physical and bioagents: In a cucumber crop in glasshouse trial, the use of solarization and *Pasteuria penetrans* had an additive detrimental effect on *M. javanica* populations (Tzortzakakis and Gowen, 1994).

vi) Bioagents and chemicals: Oxamyl increased the efficacy of *P. penetrans* in trials against *M. javanica* infection of cucumber crop and the effects on nematode control were additive (Tzortzakakis and Gowen, 1994).

vii) Botanicals and chemicals: In greenhouse and field studies conducted by Tempta (2017), Neem cake @ 50 g/m^2 + Carbofuran 3G @ 10 g/m^2 application in the field was found most effective in reducing nematode population and root gall index as well as in increasing the plant growth and yield in cucumber.

The maximum reduction in numbers of juvenile 2 (J2) (66.33%), total nematode populations ($p \leq 0.05$), number of galls on the cucumber root system (23.76 %) and galling index (3.1 gall index and 77.00% reduction in gall index), and vigorous growth and high yield were observed in Nemakey (Organic) at 1.5 ml/m^2, Humic acid (15%) at 1.5 ml/m^2 and Chitosan 98 % (N- acetyl-glucosamine) at 1 g/500 ml water combination treatment 30 days after application. A significant decrease ($p \leq 0.05$) of disease severity percentage was recorded under Humic acid and Chitosan treatment (66%). The higher and significant increase in yields obtained in combination treatments (Nemakey, Humic acid and Chitosan) followed by (Nemakey and Humic acid) which recorded (6.36 g/plant) and (6.29 g/plant), respectively as compared to control (2.38 g/plant). (Table 15.3) (Ismael and Mahamood, 2020).

Table 15.3. Effect of different treatments for controlling root-knot nematode (*Meloidogyne* spp.) on number of galls in the roots of cucumber (cv. Naseem F1)

Treatment	No. of galls/ plant	RKI	Reduction in galling (%)	Yield (kg. plant^{-1})
Nemakey	50.29	3.7	51.33	4.76bc
Humic acid	50.86	3.9	50.77	4.87abc
Chitosan	54.23	3.9	47.51	4.48c
Nemakey + Humic acid	40.40	3.6	60.90	6.29ab

Nemakey + Chitosan	39.40	3.7	61.86	4.96abc
Humic acid + Chitosan	39.90	3.3	61.38	5.38abc
Nemakey + Humic acid + Chitosan	23.76	3.1	77.00	6.36a
Rugby	69.46	4.1	32.77	4.19c
Control	103.33	4.5	—	2.38d
CD (P = 0.05)	25.83	35.58	—	1.52

15.2. RENIFORM NEMATODE, *Rotylenchulus reniformis*

15.2.1. Symptoms

Above-ground symptoms are usually stunting and reduced yield. The large numbers that attack the roots weaken them and slow down growth and fruiting potential. The head of the body remains embedded in the root whereas the tail protrudes from the root surface and swells during maturation after which time, she lays eggs in a jelly-like matrix (outside the root). Only females infect plant roots.

15.2.2. Management

15.2.2.1. Biological methods: Sikora and Sitaramiah (1981) noted 36% fewer *R. reniformis* penetrated *Glomus mosseae* colonized cucumber roots after 8 days of inoculation.

15.2.2.2. Host resistance: Three cucumber varieties: Geminis 7, Poinsett 76, the hybrid Dasher II tested showed resistance to root-knot (*M. incognita*) and reniform nematodes (Echdvez-Badel, 1989).

15.2.2.3. Integrated methods

i) Bioagents and chemicals: Oxamyl increased the efficacy of *P. penetrans* in trials against *M. javanica* infection of cucumber crop and the effects on nematode control were additive (Tzortzakakis and Gowen, 1994).

15.3. TOBACCO RINGSPOT VIRUS ON CUCUMBER

15.3.1. Symptoms

A prominent symptom is the presence of small, pale circular spots on leaves. Spots may be surrounded by a yellowish halo. However, other symptoms can occur in the same field, such as a general leaf mottle, a yellow mosaic, ringspots, or necrosis.

Tiny, water-soaked spots can also occur on the fruit, and also extending into it. Older plants can be stunted, with less severe foliar symptoms.

15.3.2. Transmission

Tobacco ringspot virus (TRSV) is effectively transmitted to the roots of several species of plants by the American dagger nematode, *Xiphinema amcricanum* (Fulton, 1962; McGuire, 1964). After a 10-day acquisition feeding period, the nematodes (*X. revesi*) were transferred to healthy bait plants (*Cucumis sativus* cv. Eva, grown in a growth chamber at 25 °C) and allowed a 14-day inoculation feeding period. *X. rivesi* transmitted TRSV (Sirca *et al.*, 2007).

Other vectors of lesser importance include thrips, mites, grasshoppers, and flea beetles. Seed transmission can occur with melons but is not important with cucumbers. Plant juices on hands or tools are also infectious.

15.3.3. Management

15.3.3.1. Cultural methods: Controlling weeds may reduce disease incidence.

15.3.3.2. Chemical methods: Benomyl application inhibited feeding of *X. americanum* on cucumber, which was reflected in a marked reduction of its transmission ability of tobacco ring spot virus (McGuire and Good, 1970; Rodriguez-Kabana and King, 1977).

15.4. CONCLUSION

Nematode control measures can be conveniently divided into 2 major categories including cultural and chemical control measures. None of these measures should be relied upon exclusively for nematode management. Rather, when practicality and economics permit, each management procedure should be considered for use in conjunction with all other available measures for nematode control and used in an integrated program of nematode management.

Chapter - 16

WATERMELON, *Citrullus lanatus*

16.1. ROOT-KNOT NEMATODE, *Meloidogyne incognita*

16.1.1. Crop losses

Hasan and Jain (1998) reported that *M. incognita* was responsible for 18-33% loss in yield of watermelon. The root-knot nematode, *M. incognita* is responsible for 12% yield loss in watermelon (Singh and Kumar, 2013).

16.1.2. Symptoms

In watermelons with nematode damage, leaf chlorosis is evident, and leaves may be stunted and wilted. Yellowing of plants, stunted growth and general decline in watermelon are early symptoms. Affected areas often occur as patchy areas in a field or along a row of plants. Formation of root galls is the characteristic symptom of root-knot nematode infestation. Infested roots fail to absorb water and nutrients resulting in stunted growth, yellowing, and wilt of above ground plants.

16.1.3 Survival and spread

The primary source of inoculum includes collateral and other hosts like Solanaceous, Malvaceous and Leguminous plants and egg masses that survive in infected plant debris and soil. Secondary spread through autonomous second stage juveniles that may also be water dispersed.

16.1.4. Favorable conditions

Sandy loam light soils are favorable.

16.1.5. Management

16.1.5.1. Cultural methods

i) Fallowing: Dry fallowing (weed-free) and cultivation to bring as much soil as possible to the surface are effective in killing root-knot nematode through starvation and desiccation. To restore the soil fertility after fallowing, application of compost/FYM/ other organic amendments should be added to the soil.

ii) Organic amendments: Soil application of neem cake/pongamia cake @ 250 kg/ha or press mud @ 5 t /ha at the time of last plowing helps in reducing nematode population.

iii) Biofumigation: The nematicidal effects due to release toxic compounds after breakdown were observed in several *Brassica* species. The effect of Brassica green manure, i.e., biofumigation, for the control of root-knot nematode on muskmelon and watermelon crops was investigated. *Brassica juncea* and *Eruca sativa* green manures were compared with a commercial amendment from Brassica meals, alone or combined with E. sativa green manure, and with commercial formulations of aqueous extracts of quillay and tagetes, and the nematicide cadusaphos. In the muskmelon crop cycle *E. sativa* green manure resulted in a nematicidal effect statistically like cadusaphos and higher than *B. juncea* amendment. B juncea amendment effects were not improved by combination with pellet. Muskmelon yield recorded for chemical treatment and *B. Juncea* were 79.2 and 70.8 t/ha, respectively, and were 3 times higher than untreated control.

In the following watermelon crop the lowest population densities of *M. incognita* were recorded after the combined treatment with E. sativa green manure and pellet, and after cadusaphos, followed by E. sativa green manure alone and at a significantly higher value, by *B. juncea*. Nematicidal effects of quillay formulation, though significantly higher than untreated control, was consistently lower than biofumigation treatments. Most effective treatments for nematode control provided the highest watermelon yields, 119.0 and 85.7 t/ha for E. sativa manure plus pellet and cadusaphos, respectively (De Mastro *et al.*, 2008a).

The effectiveness of Brassica green manure biofumigation for the control of root-knot nematodes were investigated in a greenhouse experiment on a tomato - melon succession. Green manures of *Brassica juncea* and *Eruca sativa* were compared with a Brassica seed meal commercial pellet, alone or combined with *E. sativa* manure, two commercial formulations of quillay or tagetes, chemical nematicide cadusaphos and an untreated soil. In the tomato crop, green manure of E. sativa resulted in the lowest M. incognita population densities and gall formation, as its results were not significantly different from chemical treatment but were lower than *B. juncea* green manure, the pellet and quillay formulation. This last treatment was statistically similar to the untreated control. *E. sativa* green manure, either alone or combined with the pellet, provided a tomato yield two to three times higher than control, whereas *B. juncea* and cadusaphos almost doubled the yield of untreated soil.

In the following melon crop, the lowest nematode densities were recorded in soil treated with cadusaphos or the commercial pellet. Green manure of *E. sativa*, alone or combined with the pellet, resulted in a higher suppressivity than *B. juncea*, whereas a limited nematicidal activity emerged for the quillay formulation. The highest melon yield was provided by the seed meal pellet, followed by the combination of E. sativa green manure and pellet and by both single green manures (De Mastro *et al.*, 2008b).

iv) Crop rotation: Crop rotation with corn or sorghum and a weed-free fallow can reduce numbers of nematodes below damaging thresholds.

v) Cover crops: A few non-host crops such as *Mucuna pruriens* (velvet bean), *Aeschynomene americana* (joint vetch), *Tagetes* spp. (marigold), *Crotalaria juncea* (sunn hemp), and some cultivars of sorghum-sudangrass (*Sorghum bicolor* × *S. arundinaceum*) have been successfully used as cover crops for root-knot control.

vi) Green manure crops: Use of green manure and soil amendments is also beneficial. Soil incorporation of large amounts of organic matter, such as sorghum-Sudan grass green manure in combination with poultry compost, reduces populations of root-knot nematodes. Some rapeseed cultivars, such as 'Dwarf Essex' and 'Humus' also are suppressive to nematode populations.

16.1.5.2. Chemical methods: Velum Prime can be applied at 475 to 500 ml/ha through drip irrigation at 5-day interval.

16.1.5.3. Biological methods

i) Antagonistic fungi: Vicente *et al.* (1991) reported the effectiveness of soil application of *Paecilomyces lilacinus* two weeks prior to planting for the management of root-knot nematodes on watermelon under field conditions. The application of this fungus two weeks before planting was better than carbofuran or fenamiphos for controlling nematodes.

In oriental melon, root gall percentages were 32.1 (60.2% control) and 52.9% (34.5% control) in J + 200,000 and J + 100,000 *Pasteuria penetrans* endospores/1 g medium which were significantly lower than that of 80.7% in inoculated control (Table 16.1).

Table 16.19. Suppression effect of *Pasteuria penetrans* KW1 (PP) on reproduction of *Meloidogyne incognita* in oriental melon

Treatments	No. of root galls		Wt. of leaves harvested (g)	Total no. of leaves harvested	Root wt. (g/plant)
	No. of galls/root	Control effect (%)			
Control (Inoculated) MI J2 5,000/plant (N)	80.7 d	---	655 b	66 b	29.9 b
N + PP 100,000 endospores/g medium	52.9 c	34.5	717 a	81 a	47.1 ab
N + PP 200,000 endospores/g medium	32.1 b	60.2	756 a	88 ab	37.7 a
Control (Uninoculated)	0.0 a	---	732 a	76 c	11.2 ab

Data are means of 6 replicates. Root galls, leaves weight, root weight, and number of total leaves were examined 10 weeks after planting.

Means within a column followed by the same letter are not different according to Duncan's multiple-range test ($P<0.05$).

16.1.5.4. Host resistance: Watermelon grafted onto cucurbit rootstocks grown in a field infested with southern root-knot nematode, *Meloidogyne incognita* reveal that 'Emphasis' bottle gourd (*Lagenaria siceraria*) and 'Strong Tosa' squash hybrid (*Cucurbita moschata* × *C. maxima*) rootstocks exhibited severe root galling caused by root-knot nematodes; while RKVL 301 and RKVL 302 wild watermelon (*C. lanatus* var. *citroides*) rootstocks were found resistant to root-knot nematodes. Grafting watermelon varieties onto citron melon 'Ojakkyo' rootstock provides resistance to root-knot nematode. Carolina Strongback is a watermelon rootstock that is resistant to Fusarium wilt and the southern root-knot nematode. When tested in highly infested soils with fusarium and nematodes, Carolina Strongback performed well and maintained a high production level of harvested fruit.

Watermelon cv. Shehjanpuri was found resistant to *M. incognita*, while Dixie Queen was found resistant to *M. javanica.*

Wild cucumber rootstock, *Cucumis pustulatus* gave effective control of root-knot nematodes on watermelon and increased fruit yield (Liu *et al.* 2014) (Table 16.2).

Table 16.2. Management of root-knot nematodes in Cucurbits using wild cucumber rootstock, *Cucumis pustulatus*

Treatment	Gall index	Yield in kg/plant
Watermelon grafted on *C. pustulatus*	1	3.901
Watermelon non-grafted	5	1.650

Bur-cucumber (*Sicyos angulatus*) collected near Andong, Korea, showed good compatibility with watermelons for early summer growth and good resistance to nematodes (Lee, 1992; Lee *et al.*, 1992).

16.2. RENIFORM NEMATODE, *Rotylenchulus reniformis*

16.2.1. Symptoms

General symptoms include reduced root systems, leaf chlorosis, overall stunting of host plants, and reduced yields and plant longevity. Female nematodes and their eggs are often visible when plant roots are viewed under a dissecting microscope.

16.2.2. Management

16.2.2.1. Biological methods: Vincente *et al.* (1991) compared the effectiveness of the fungus (*Paecilomyces lilacinus*), a parasite of nematode eggs, with carbofuran (Furadan 10G), phenamiphos 15G at 6.73 kg a.i./ha, and phenamiphos 15G at 13.46 kg a.i./ha. There were no differences in the number of fruits (watermelon) or yield (kg/plot) among the treatments; all gave better control than untreated plots, and the fungus and carbofuran provided heavier fruit than the other two treatments and the untreated plots. The results of the nematicides are consistent with other watermelon and squash studies that found carbofuran treated plots giving higher yields, while another study found phenamiphos-treated plots giving higher yields. This study only found differences among treatments in the weight of the fruit.

16.3. INTERACTION OF ROOT-KNOT NEMATODE WITH FUSARIUM WILT

16.3.1. Interaction

On watermelon, Sumner and Johnson (1973) reported that the presence of root-knot nematodes in the soil increased wilt symptoms caused by *Fusarium oxysporum* f. sp. *niveum* (FON). The presence of *M. incognita* enhanced the susceptibility of all watermelon genotypes to Fusarium wilt. Co-inoculation with *M. incognita* led to an early development of wilt symptoms and increased disease severity. Galls were observed on roots of all nematode-inoculated plants. Whereas inoculation of the nematodes alone did not reduce plant growth, growth suppression was evident when seedlings were inoculated with both pathogens. The results indicated that *M. incognita* could enhance susceptibility of resistant watermelon genotypes to respective FON races, and host resistance alone is not sufficient for managing Fusarium wilt on watermelon in soils infested with root-knot nematodes.

16.3.2. Management

16.3.2.1. Host resistance: A few *Citrullus lanatus* var. *citroides* rootstocks such as 'Carolina Strongback' exhibited resistance to both root-knot nematode (*M. incognita*) and *Fusarium* wilt, and could be used for grafting watermelon by growers to control these two pathogens in co-infected fields.

Currently, resistance to both Fusarium wilt of watermelon and root-knot nematodes is limited to only one commercial rootstock 'RS-841' (DeRuiter Seed / Seminis Seeds, St. Louis, MO) (USDA-NIFA, 2015).

Percentage of root galling was significantly higher (P = 0.05) on the bottle gourd and squash hybrid rootstocks than on the watermelon 'Fiesta' (non-grafted) and the *C. lanatus* var. *citroides* lines.

Citrullus lanatus var. *citroides* rootstock 'Carolina Strongback' exhibited resistance to both root-knot nematode (*M. incognita*) and Fusarium wilt (*Fusarium oxysporum* f. sp. *niveum*), and could be used for grafting watermelon by growers to control these two pathogens in co-infected fields.

16.3.2.2. Integrated methods

i) Two bioagents: *Pseudomonas aeruginosa* and *Purpureocellium lilacinum* when used together significantly reduced infection of *M. javanica* and *F. solani* on watermelon. *P. aeruginosa* and *P. lilacinum* used alone or together significantly (P<0.05) reduced infection of root knot nematode *M. javanica* and root infecting fungi viz., *M. phaseolina, R. solani, F. solani* and *F. oxysporum* on watermelon. *P. aeruginosa was* more effective than *P. lilacinum* in reducing the *M. javanica* root knot nematode infection (Perveen *et al.*, 1998). Combined use of *P. lilacinum* and *P. aeruginosa* was more effective in reducing the infection of *F. solani* on watermelon than either used alone. *Use of P. aeruginosa and P. lilacinum* significantly (P<0.05) increased plant height and fresh shoot weight on watermelon (Tables 16.3 and 16.4) (Perveen *et al.*, 1998).

Table 16.3. Effect of *P. aeruginosa* and *P. lilacinum* on growth of watermelon.

Treatment	Germination %	Fresh weight of shoot (g)	Plant height (cm)
Control	60	2.5	35.1
P. lilacinum (PL)	70	3.0	37.5
P. aeruginosa (PA)	80	3.7	35.0
PL + PA	75	3.2	38.5
CD (P = 0.05)	11.1	0.55	2.2

Table 16.4. Effect of *P. aeruginosa* and *P. lilacinum* in the control of root disease complex in watermelon

Treatment	Root-knot index	Infection %			
		M. phaseolina	*R. solani*	*F. solani*	*F. oxysporum*
Control	3.8	62	44	69	12
P. lilacinum (PL)	2.8	69	12	81	0
P. aeruginosa (PA)	1.5	25	12	69	0
PL + PA	1.5	37	0	44	0
CD (P = 0.05)	0.34	6.1	6.1	6.1	6.1

16.4. CONCLUSION

Root-knot nematodes have increased in prevalence and damage severity in recent years due to the phase-out of methyl bromide. In watermelon, the pathological interaction between root-knot nematode and Fusarium wilt disease has also increased the complexity of control measures. Growers continue to suffer from economic losses despite the available management practices. Current efforts have been focused on using cultural and chemical controls as well as grafted watermelons on appropriate disease-resistant rootstocks. Using a combination of these management tactics is recommended when growers have serious problems with root-knot nematodes.

Chapter - 17

MUSK MELON, *Cucumis melo*

17.1. ROOT-KNOT NEMATODE, *Meloidogyne incognita*

17.1.1. Symptoms

Root-knot nematode (*Meloidogyne incognita*) is the most destructive plant parasitic nematode in melon growing areas of India. Plants infected with this nematode can wilt in middle of the day, especially in light sandy loam soils. Plants might become yellow and stunted. Roots of infected plants swell and produce large galls all over the roots. White female nematodes are embedded inside these galls.

As *M. incognita* larvae enter the plant root, feed and mature, the surrounding cells of the plant root increase in size and divide causing galls on the roots. The flow of nutrients and water is restricted, and plants wilt quickly when water becomes limiting. If plants are infected when young, they are often severely stunted and chlorotic. Infected vines rarely die but are generally not productive.

17.1.2. Life cycle

Second stage juveniles hatch from eggs in the soil and invade young seedlings in early spring. A second stage juvenile will establish a feeding site and go through several molting stages, swell and become a sedentary female encased within the root tissue that appears as galls. The life cycle of the RKN is about a month in optimum temperatures, 24-27.6 °C.

17.1.3. Interaction with other pathogens

Combination of fungus (*Fusarium oxysporum* f. sp. *melonis*) and nematode (*Meloidogyne javanica* 5000 eggs) caused the most decrease in growth index of susceptible melon cvs. Garmsar and Sooski.

C. pustulatus is a suitable rootstock with simultaneous resistance to root-knot nematode (*Meloidogyne incognita*) and Fusarium wilt (*Fusarium ozysporum* f. sp. *melonis*) in melon. *C. pustulatus* rootstocks are thus promising for low-input sustainable horticulture.

17.1.4. Management

17.1.4.1. Physical methods: In Japan and other Asian countries, several farmers growing successive crops, such as tomato and muskmelon susceptible to root-knot nematodes use solarization in plastic tunnels for 30 days in summer as an alternative to methyl bromide fumigation (Sano, 2002).

17.1.4.2. Cultural methods: Small grains and alfalfa can be used effectively as rotational crops.

17.1.4.3. Chemical methods: At mid-season evaluations, root galling was moderate and ranged between 2.4 in the Nimitz treatment (5 pt/A) as compared to 4.6 in the untreated control (Sidhu, 2021).

17.1.4.4. Host resistance: Melon scion cvs. 'Honey Yellow' and 'Arava' grafted onto *C. metuliferus* rootstock exhibited significantly lower galling and reduced RKN population densities in the field (Table 17.1) (Guan *et al.*, 2014). *Cucumis metuliferus* has resistance to *M. incognita* race 1 and 3. *C. metuliferus* has also been reported resistant to *M. arenaria* and *M. javanica* (Walters *et al.*, 1993). Fassuliotis (1970) found no hypersensitive reaction to infection by *M. incognita*, but the development of J2 to adult was delayed in *C. metuliferus*.

Table 17.1. Effect of rootstocks on root gall index (GI) and numbers of *Meloidogyne javanica* second-stage juveniles (J2) in melon under field conditions.

Treatment	Gall index (0-10 scale)	*Meloidogyne* density (J_2/ 100 cm^3 of soil)
Non-grafted 'Honey Yellow'	7.14 a	378.2 a
Self-grafted 'Honey Yellow'	6.70 a	515.6 a
'Honey Yellow' grafted on *C. metuliferus*	0.08 b	1.2 b
Non-grafted 'Arava'	5.20 a	200.2 a
Self-grafted 'Arava'	4.45 a	140.2 ab
'Arava' grafted on *C. metuliferus*	0.15 b	4.8 b

Wild cucumber rootstock, *Cucumis pustulatus* gave effective control of root-knot nematodes on muskmelon and increased fruit yield (Liu *et al.* 2014) (Table 17.2).

Table 17.2. Management of root-knot nematodes in muskmelon using wild cucumber rootstock, *Cucumis pustulatus*

Treatment	Gall index	Yield in kg/plant
Muskmelon grafted on *C. pustulatus*	1	2.276
Muskmelon non-grafted	5	1.612

In muskmelon, *C. metuliferus* rootstock reduced galling in nematode (*M. incognita*) infested soil (Kokalis-Burelle and Rosskopf, 2011).

17.1.4.5. Integrated methods

i) Cultural and botanicals: Combination of crop rotation (sorghum in 2005 followed by melon in 2006 and 2007), biocidal intercropping (*Eruca sativa* cv. Nemat) and biofumigant treatments (pellet based on formulated defatted seed meals of *Brassica carinata*), was effective for the management of root-knot nematodes. The results showed that sorghum cultivation as a non-host crop halved the nematode population in the soil. The biofumigant treatment in spring 2006, caused a strong decrease in galling index (G.I. based on a scale from 0 to 5) on the roots of the following melon crop in rotation (G.I. = 0.3) compared with the untreated control (G.I. = 4.2). These positive results were further confirmed in 2007, when an important increase in quality and quantity was recorded in the yield of melon fruits. The autumn cultivation of *E. sativa* cv. Nemat and its incorporation into the soil brought a good amount of organic matter, with positive effects on melon yield. The intrinsic characteristics of the pellet based on *B. carinata* defatted seed-meals in rotation with *E. sativa* highlighted an excellent biofumigant effect, fully comparable to chemical nematicides, and a noticeable contribution in both organic matter and nitrogen that played an important fertilizing effect. The full effectiveness of pellet in decreasing substantially the nematode population, suggest its application in alternate years with Eruca green manure, in the presence of low larval infestations in the soil (Curto *et al.*, 2008).

ii) Physical and chemicals: The yield of melon was significantly increased by soil solarization with or without sulphur and by sulphur as single treatment. The thickness of the polyethylene did not affect the yields, which were significantly increased by EVA 0.150 mm, compared to EVA 0.035 mm. The addition of either 750 or 1,000 kg/ha sulphur previously to soil solarization was beneficial compared with the polyethylene tarping only. The sulphur application under EVA tarping did not statistically increase the yield with respect to 500 kg/ha sulphur alone (Table 17.3). The nematode population was significantly suppressed by either solarization or sulphur treatments. No difference was found between the two films or among the sulfur dosages and no further suppression derived by the combined use of solarization and sulfur.

Table 17.3. Effect of soil solarization and sulfur on *Meloidogyne incognita* on melon.

Sulfur (kg/ha)	Soil solarization	Crop yield (tons/ha)	Final nema popn. (eggs & J_2/ml soil)
-	-	12.6 ab	11.7 a
500	-	16.4 b	3.2 c
500	Polyethylene 0.050 mm	16.2 b	3.1 c
500	EVA 0.035 mm	19.0 bc	2.9 c
750	-	17.8 bc	4.3 c
750	Polyethylene 0.050 mm	20.4 c	2.6 c
750	EVA 0.035 mm	20.2 c	4.3 c
1000	-	17.0 bc	3.7 c
1000	Polyethylene 0.050 mm	20.4 c	3.4 c
1000	EVA 0.035 mm	20.4 c	1.6 c
-	Polyethylene 0.050 mm	17.0 bc	2.1 c
-	EVA 0.035 mm	16.4 b	4.9 bc

* Means followed by the same letters in the same column are not significantly different according to Duncan's Multiple Range Test (P = 0.01).

Chapter - 18

PUMPKIN, *Cucurbita moschata*

18.1. INTRODUCTION

Pumpkin occupies a prominent place among vegetables owing to its high productivity, nutritive value, good storability, long period of availability, better transport qualities and extensive cultivation in sub-tropical and tropical parts of the world. Fruits are rich in vitamin A. It is also consumed as processed and stock feed. The fruits can be used in preparing sweets, candy, fermented into beverages. Yellow or orange-fleshed pumpkins are rich in carotene. In India, it is grown mainly in Assam, West Bengal, Tamil Nadu, Karnataka, Madhya Pradesh, Uttar Pradesh, Orissa and Bihar.

The root-knot nematode is the most important pest on pumpkin.

18.2. ROOT-KNOT NEMATODES, *Meloidogyne* spp.

18.2.1. Crop losses

The root-knot nematode, *M. incognita* is responsible for 27% yield loss in pumpkin (Singh and Kumar, 2013).

18.2.2. Symptoms

The very broad-leaved plant like pumpkin which is infected with root-knot nematodes may show day time wilting and develop much larger galls. In cucurbits, the roots react to the presence of *Meloidogyne* spp. by the formation of large, fleshy galls. Above-ground, plants affected by root-knot nematode appear yellowed, stunted or generally unthrifty. Affected roots are disfigured, swollen and stubby in appearance. The galls formed on roots range in size from 0.75 cm. to more than 1.25 cm. in diameter, depending on the species present and the degree of infestation.

18.2.3. Survival

The root-knot nematodes survive as eggs or juvenile larvae in the soil. Eggs remain dormant in soil in the absence of plant roots.

18.2.4. Management

18.2.4.1. Physical methods: Pulverizing the planting pit soil and exposing to sunlight by repeated raking and heaping 10 to 15 cm thick dry trash and burning it before application of compost helps in reducing the nematode population. Frequent cultivation may be given in May-June to expose the soil to sun.

18.2.4.2. Cultural methods: Adding organic matter or using soil management practices that favor organic matter accumulation in soil may help reduce nematode levels. Application of 250g of fresh neem or karanj cake at the time of sowing reduces the nematode population in soil. Crop rotation with rice, oats, wheat and taramira is also effective.

18.2.4.3. Chemical methods: Seed treatment with carbosulfan (Marshall 25 ST) at 6% w/w is effective against root-knot nematodes. Application of carbofuran at 0.2g a.i./pit at sowing is also effective against root-knot and reniform nematodes.

Carbofuran @ 1 kg a.i. /ha gave the best result in respect of yield (15.512 t/ha) as well as reducing the nematode infestation (59.67 percent) (Lalhruaitluanga and Hazarika, 2020).

18.2.4.4. Biological methods: *Pseudomonas fluorescens* @ 2.5 kg/ha was found effective in increasing the yield of pumpkin to an extent of 61.5% over control. It was followed by *Trichoderma viride* which was significantly different from *P. fluorescens* but was at par with *T. viride,* whereas *Purpureocellium lilacinum* and *Bacillus subtilis* were at par with each other. Application of carbofuran was found effective in reducing the nematode population but the effect was at par with *P. fluorescens* which recorded 53.1%, 49.8% reduction in soil and root population in pumpkin (Shanthi, 2016).

18.2.4.5. Host resistance: Pumpkin cvs. Jaipuri and Dasna were found resistant to root-knot nematodes.

18.2.4.6. Integrated methods

i) Bioagents and botanicals: Addition of 250 g fresh neem cake or karanj cake enriched with *Trichoderma harzianum/Paecilomyces lilacinus* per planting pit effectively controlled the nematodes.

Combined application of Mustard cake at 375 kg/ha applied 15 days prior to sowing, and *Trichoderma harzianum* at 10 kg/ha at time of sowing brought about a significant reduction in number of galls (63.333), egg masses (56.42) and final nematode population (235.05 per 200 cc of soil) in soil with increased yield (14.007 t/ha) compared to control (Table 18.1) (Lalhruaitluanga and Hazarika, 2020).

Table 18.1. Management of *Meloidogyne incognita* on Pumpkin cv Karbi-I

Treatment	No. of galls/ plant	No. of egg masses/plant	Yield (t/ha)
Trichoderma harzianum @ 20 kg/ha at sowing	77.12	66.50	9.379
Mustard cake @ 750kg/ha 15 days prior to sowing	86.25	73.83	11.904
T. harzianum @ 10 kg/ha at sowing + Mustard cake @ 375kg/ha 15 days prior to sowing	63.33	56.42	14.007
Carbofuran @ 1 kg ai/ ha	59.67	46.71	15.512
Control	115.62	135.17	6.937
CD (P = 0.05)	3.51	9.44	—

ii) Solarization and chemicals: The yield of cantaloupe was significantly increased by soil solarization with or without sulfur and by sulfur as single treatment. The thickness of the polyethylene did not affect the yields, which were significantly increased by EVA 0.150 mm, compared to EVA 0.035 mm. The addition of either 750 or 1,000 kg/ha sulfur previously to soil solarization was beneficial compared with the polyethylene tarping only. The sulfur application under EVA tarping did not statistically increase the yield with respect to 500 kg/ha sulfur alone (Table 18.2). The nematode population was significantly suppressed by either solarization or sulfur treatments: No difference was found between the two films or among the sulfur dosages and no further suppression derived by the combined use of solarization and sulfur.

Table 18.2. Effect of soil solarization and sulfur on *Meloidogyne incognita* on cantaloupe.

Sulfur (kg/ha)	Soil solarization	Crop yield (MT/ha)	Final nematode population (eggs & J_2/ml soil)
-	-	12.6 ab	11.7 a
500	-	16.4 b	3.2 c
500	Polyethylene 0.050 mm	16.2 b	3.1 c
500	EVA 0.035 mm	19.0 bc	2.9 c
750	-	17.8 bc	4.3 c
750	Polyethylene 0.050 mm	20.4 c	2.6 c
750	EVA 0.035 mm	20.2 c	4.3 c
1000	-	17.0 bc	3.7 c
1000	Polyethylene 0.050 mm	20.4 c	3.4 c
1000	EVA 0.035 mm	20.4 c	1.6 c
-	Polyethylene 0.050 mm	17.0 bc	2.1 c
-	EVA 0.035 mm	16.4 b	4.9 bc

* Means followed by the same letters in the same column are not significantly different according to Duncan's Multiple Range Test (P = 0.01).

18.3. INTERACTION OF ROOT-KNOT NEMATODE WITH ROOT ROT/WILT

18.3.1. Interaction

The association of the root-knot nematode, *Meloidogyne javanica* with root rot infecting fungi like *Macrophomina phaseolina, Fusarium oxysporum,* and *F. solani* causes a disease complex in pumpkin. The disease complex produces greater losses than either pathogen alone.

18.3.2. Management

18.3.2.1. Biological methods: Chahal and Chahal (1998) reported beneficial effects of *T. harzianum* in reducing the number of galls and disease complexes formed by root-knot nematodes and fungi on cucurbits.

18.3.2.2. Integrated methods: *Pseudomonas aeruginosa* and *Paecilomyces lilacinus* when used together significantly reduced infection of the disease complex on pumpkin (Perveen *et al.*, 1998). Use of *P.*

aeruginosa and *P. lilacinus* significantly (P<0.05) increased plant height and fresh shoot weight of pumpkin (Table 18.3). *P. aeruginosa* and *P. lilacinus* used alone or together significantly (P<0.05) reduced infection of root knot nematode *M. javanica* and root infecting fungi viz., *M. phaseolina, R. solani, F. solani* and *F. oxysporum* on pumpkin. *P. aeruginosa* was more effective than *P. lilacinus* in reducing the *M. javanica* infection (Table 18.4). Combined use of *P. lilacinus* and *P. aeruginosa* was more effective in reducing the infection of *M. phaseolina* and *F. oxysporum* on pumpkin than either used alone (Perveen *et al.*, 1998).

Table 18.3. Effect of *P. aeruginosa* and *P. lilacinus* on growth of pumpkin.

Treatment	Germination %	Fresh weight of shoot (g)	Plant height (cm)
Control	40	4.2	26.5
P. lilacinus (PL)	50	6.7	32.2
P. aeruginosa (PA)	55	7.2	30.0
PL + PA	50	5.9	28.2
CD (P = 0.05)	11.1	0.55	2.2

Table 18.4. Effect of *P. aeruginosa* and *P. lilacinus* for the management of root-knot and root disease complex in pumpkin

Treatment	Root-knot index	Infection %			
		M. phaseolina	***R. solani***	***F. solani***	***F. oxysporum***
Control	3.1	87	25	56	75
P. lilacinus (*PL*)	1.2	94	0	44	62
P. aeruginosa (*PA*)	0.6	50	6	19	44
PL + PA	0.5	19	0	19	6
CD (P = 0.05)	0.34	6.1	6.1	6.1	6.1

Chapter - 19

BOTTLE GOURD, *Lagenaria siceraria*

The root-knot nematode is an important limiting factor in successful cultivation of bottle gourd.

19.1. ROOT-KNOT NEMATODES, *Meloidogyne incognita, M. javanica*

19.1.1. Crop losses

Root-knot nematodes are responsible for annual yield loss of 22% in bottle gourd amounting to Rs. 403.78 million in India (Walia and Chakraborty, 2018).

19.1.2. Symptoms

The common symptoms of root-knot nematode on plants include general stunting, low vigor, chlorosis, necrosis, defoliation, and twig dieback. Infected plants show fewer, small rootlets, reduction in aerial growth and highly deformed and longer size galls are formed which can be easily attacked by secondary pathogens to form disease complex.

19.1.3. Management

19.1.3.1. Cultural methods: Application of Neem cake at 10 g/pit 10 days before sowing in bottle gourd proved effective.

Neem, Castor and Jatropa cakes were evaluated @ 30 g/m^2 in bottle gourd; yield was higher in all the treatments (1.186 to 1.224 t/ha) compared to 1.149 t/ha in untreated plots. The benefit: cost ratio was favorable in all the treatments ranging from 2.2 to 2.5 (Anon, 2011).

Effective management of root-knot nematode and enhanced yield of bottle gourd was obtained by soil incorporation of Neem cake at 4.286 t/ha or Mustard cake at 3.333 t/ha (ICBR 1: 16.10) or Press mud at 10 t/ha (ICBR 1: 18.95) one week before sowing (Dahiya *et al.*, 1998).

Spot application of mustard cake at 15 g/plant one week before sowing was effective in controlling root-knot nematodes and significantly improved the yield (8.30 tons/ha) followed by neem cake (8.23 tons/ha) (Paruthi *et al.*, 2010).

Soil application of aak (*Calotropis procera*) and neem (*Azadirachta indica)* leaves at 1.5% w/w proved to be more effective in improving plant growth characters and reducing root-knot index and final nematode population (Singh and Patel, 2015).

19.1.3.2. Chemical methods: Seed treatment with Carbosulfan 25DS or Carbofuran 25ST @ 3% a.i. w/w (120 g/kg seed) was found effective for the control of root-knot nematodes.

Seed dressing with Carbosulfan or Benfurocarb each @ 1.5 and 3% increased yield and reduced root galling (*M. javanica*) in bottle gourd.

Fenamiphos, Quinalphos, Carbosulfan and Monocrotophos, each @ 500 and 1000 ppm as seed dip for 12 hours lowered the root-knot index in bottle gourd.

Seed treatment with carbofuran at 3 per cent was effective against root-knot nematodes infecting bottle gourd (Kandasamy and Sivakumar, 1981). Carbosulfan and benfuracarb at 3% w/w as seed treatment increased yield and reduced root galling in bottle gourd.

19.1.3.3. Integrated methods: Seed treatment of bottle gourd with *Gluconoacetobacter diazotrophicus* strain 35-47 @ 50 ml/2 kg seed and soil incorporation combined with soil incorporation of Neem cake @ 30 g/spot was found effective for the management of root-knot nematodes, *Meloidogyne* spp.

Under field conditions, the bottle gourd seeds that were treated either with *Bacillus amyloliquefaciens*, *B. megaterium*, *Pseudomonas fluorescens*, or *P. putida* at 10 g/kg seed along with the application of neem cake (1 ton/ha) significantly resulted in reduced root gall index and the number of nematode juveniles in soil and roots than the non-treated control (Table 19.1) (Rani *et al.*, 2022).

Table 19.1. Efficacy of different microbial formulations against root-knot nematode in bottle gourd (Pooled for 2018–2019).

Treatment	Root gall index	Yield (t/ha)
Neem cake @ 1 t/ha	2.83	14.39
P. putida @ 10 g/kg of seed	2.41	15.23
B. megaterium @ 10 g/kg of seed	2.31	16.12
B. amyloliquefaciens @ 10 g/kg of seed	2.53	15.50
P. fuorescens @ 10 g/kg of seed	1.98	16.05
Neem cake @ 1 t/ha + *P. putida* @ 10 g/kg of seed	1.83	17.33
Neem cake @ 1 t/ha + *B. megaterium* @ 10 g/kg of seed	2.03	17.56
Neem cake @ 1 t/ha + *B. amyloliquefaciens* @ 10 g/kg of seed	2.24	16.51
Neem cake @ 1 t/ha + *P. fuorescens* @ 10 g/kg of seed	1.68	17.66
Carbofuran @ 1.0 kg a.i. /ha	2.30	17.08
Control (untreated check)	3.42	13.87
CD (P = 0.05)	**0.76**	**2.61**

Solarization of nematode infested field + seed dressing of bottle gourd with Carbosulfan 25 DS @ 3% a.i. (w/w) was found effective for management of root-knot nematodes.

The integration of soil amendment with Tricho-composts [(Tricho-compost-1 containing Trichoderma inoculums mixed with vermi-compost @ 1:20 (w/w) and kept for 7 days for multiplication of *Trichoderma*; and Tricho-compost-2 containing *Trichoderma harzianum* (TM7), mixture of rice bran, wheat bran and mustard oilcake mixed with vermi-compost @ 1:20 (w/w) and kept for 7 days for multiplication of *Trichoderma*)] with Furadan 5G, poultry refuse with Furadan 5G and neem oil cake with Furadan 5G drastically suppressed gall index values caused by root-knot nematode *M. incognita* and improving plant growth parameters such as shoot length, shoot weight, root length and root weight as well as fruit yield of bottle gourd compared to control (Table 19.2) (Faruk, 2021).

Table 19.2. Effect of integration of bioagents, neem cake and marigold as intercrop for the management of *Meloidogyne incognita* infecting pointed gourd

Treatment/Dose	No. of galls /5 g root		Yield (tons/ha)	
	2017-18	**2018-19**	**2017-18**	**2018-19**
Tricho-compost-1 @ 2 kg/pit	2.33	1.50	65.00	59.38
Tricho-compost-2 @ 2 kg/pit	2.08	1.21	68.13	64.38
Tricho-compost-1 @ 2 kg/pit + Furadan 5G @ 20 g/pit	1.70	1.00	68.75	66.25
Tricho-compost-2 @ 2 kg/pit + Furadan 5G @ 20 g/pit	1.68	0.86	72.50	66.88
Poultry refuse @ 5-6 kg/pit + Furadan 5G @ 20 g/pit	1.58	0.73	75.00	75.50
Neem oil cake @ 500 g/pit + Furadan 5G @ 20 g/pit	1.63	1.15	72.50	68.75
Saw dust burning +Furadan 5G @ 20 g/pit	2.31	1.19	67.25	63.75
Control	5.13	3.50	43.13	51.25
CD (P = 0.05)	0.496	0.637	9.101	9.48

Chapter - 20

POINTED GOURD, *Trichosanthes dioica*

The root-knot and reniform nematodes are the major limiting factors in successful production of pointed gourd.

20.1. ROOT-KNOT NEMATODE, *Meloidogyne incognita*

20.1.1. Crop losses

M. incognita is responsible for 30-40% loss in yield of pointed gourd (Verma, 2001). The quantitative loss in fruit yield due to root-knot nematode had been 43.8 per cent in pointed gourd (Anwar *et al.*, 2021).

20.1.2. Symptoms

The common symptoms of root-knot nematode on pointed gourd plants include general stunting, low vigor, chlorosis, necrosis, defoliation, and twig dieback. Infected plants show fewer, small rootlets, reduction in aerial growth and highly deformed and longer size galls are formed which are easily attacked by secondary pathogens to form disease complex.

In pointed gourd, very large galls which are commonly seen outside the soil level, is a major characteristic symptom of this nematode.

20.1.3. Management

20.1.3.1. Cultural methods: Removing the roots of each crop as soon as harvest is being completed, followed by tilling or summer deep plowing of the soil two to three times is very effective in reducing nematode population (Anwar and Varma, 1994).

Nematode free planting material should be used for propagation. The root-knot nematodes on pointed gourd were controlled by planting of marigold, *Tagetes patula* at 3 to 5 plants/pit.

Application of Neem cake @ 100 g/pit during mid-February to mid-March when rootlets start to initiate from the nodes of the vine cuttings in pointed gourd prevent the attack of *M. incognita* and increase the yield by 5000 kg/ha. In other studies, application of Neem cake @ 30 g/pit in pointed gourd as spot application at 10 days prior to planting achieved maximum average yield along with simultaneous reduction (52%) of J2 population *M. incognita* in soil.

Crop rotation with mustard helped to impede root-knot nematode invasion in pointed gourd. Cruciferous crops are moderately resistant to root-knot nematodes which can be used in rotation with pointed gourd.

20.1.3.2. Chemical methods: Dipping vines in Carbosulfan 25 EC at 500 ppm for 6 hours + Carbofuran 3G at 2.5 g/ pit 45 days after planting was found effective to enhance fruit yield and the greatest reduction of root-knot nematode population in soil (Mukhopadhyay *et al.*, 2006).

Rooted cuttings should be given a 6-hr. dip in 0.1% carbosulfan 25 EC or triazophos 40 EC before planting. In the ratoon crop, carbofuran/phorate should be applied at 0.2g a.i. /plant.

20.1.3.3, Host resistance: Pointed gourd variety BP-2 has some tolerance against root-knot nematode, *M. incognita* (Varma and Anwar, 1992).

20.1.3.4. Integrated methods

i) Bioagents and botanicals: *Purpureocillium lilacinum* (PL) formulation (talc based) enriched with FYM at 15 kg in 375 kg/ha (formulation of PL mixed with FYM and incubated for two weeks in shade and the mixture gets ready for pit application during planting of cuttings/roots of pointed gourd) reduced nematode infestation and increased yield to the extent of 16%.

ii) Bioagents, botanicals and cultural: Increase in pointed gourd plant growth and yield (6.6 kg/ plant as compared to 1.4 kg in control), and reduced final nematode population in soil (66.6/100g as compared to 86660/100g in control), root galling (14.3/5g as compared to 75.0/5g in control), number of egg masses (40/5g roots as compared to 150/5g in control) by intercropping with marigold (3 plants/pit) combined with soil incorporation of Neem cake (250g) + *Trichoderma harzianum* (100g) + *Purpureocillium lilacinum* (50 g) (Table 20.1) (Verma *et al.*, 2005).

Table 20.1. Effect of integration of bioagents, neem cake and marigold as intercrop for the management of *Meloidogyne incognita* infecting pointed gourd

Treatment	No. of galls /5 g root	No. of egg masses/5 g root	Nema. popn./100 g soil	Yield/ plant (kg)
Neem cake + *P. lilacinum* (50 g) + *T. harzianum* (100g) + (250 g) + marigold (3 plants)/pit	10.4	30	57	8.5
P. lilacinum (50 g/pit)	15.3	53	70	6.4
Carbofuran (3 g/pit)	22.5	72	95	5.8
Neem cake (250 g/pit)	22.4	75	120	5.0
T. harzianum (100 g/pit)	25.6	45	130	4.2
Marigold (3 plants/pit)	35.3	113	135	4.0
Control	80.4	186	901	1.3
CD (P = 0.05)	3.5	**9.6**	**25.3**	**0.5**

iii) Bioagents, botanicals, physical, and cultural: Integrated approaches, i.e. cultural control (deep plowing followed by soil solarization for 15 days and stubble burning) with or without minimum synthetic chemical (Carbofuran at 2.0 or 3.0 kg a.i. /ha) and with the inclusion of Neem components like leaf, cake and Azadirachtin, were highly effective in checking the nematode population build-up and infection. The eco-friendly integrated treatments recorded very low final nematode population and good yield (Chakraborti, 2000).

iv) Botanicals and chemicals: Neem cake @ 50 g/pit + Carbofuran 3G @ 0.3 g a.i./pit or drenching of rhizospheric soil with Carbosulfan 25EC @ 500 ppm in the infested field was found to be effective in reducing root-knot nematode population as well as to increase yield of pointed gourd.

Dipping vines in carbosulfan 25 EC at 500 ppm for 6 hours + carbofuran 3G at 2.5 g/pit 45 days after planting was the best treatment as it produced the highest yield of 20.216 t/ha and the greatest reduction of root-knot nematode gall index in both growing seasons. The largest benefit/cost ratio was obtained by dipping vines in monocrotophos (11.9) (Table 20.2) (Mukhopadhyay *et al.*, 2006).

Table 20.2. Effect of treatments on the yield of pointed gourd in a field infested with *M. incognita* and *M. javanica* in 2002-03 and 2003-04 (Average of 4 replications).

Treatment	Yield (tons/ha)	Gall index	Benefit: cost ratio
Vines dipped in carbosulfan 25EC at 500 ppm for 6 hrs	16.66	3.75	8.4
Vines dipped in monocrotophos 36EC at 500 ppm for 6 hrs	17.35	4.20	11.9
Carbofuran 3G, 2 kg a.i. /ha at planting (broadcasting)	18.19	4.40	4.6
Carbofuran 3G, 2.5 g/pit at planting	17.07	4.00	3.0
Neem cake, 1500 kg/ha at planting (broadcasting)	17.41	4.05	1.0
Neem cake, 50 g/pit at planting	17.60	3.30	1.2
Neem cake, 50 g/pit at planting + carbofuran 3G 2.5 g/pit 45 DAP	18.18	3.35	1.2
Vines dipped in carbosulfan 25EC at 500 ppm for 6 hrs + carbofuran 3G 2.5 g/pit 45 DAP	20.22	2.75	6.5
Untreated control	14.78	4.05	—
CD (P = 0.05)	—	—	0.885

v) Bioagents, botanicals and chemicals: Adoption of vine dipping in monocrotophos 36SL at 1000 ppm followed by soil inoculation of *Trichoderma viride* at 10g / pit once at planting and second dose at 40 days after planting reduced root galling caused by *M. incognita* and gave fruit yield almost double of the untreated plots (Table 20.3) (Khan *et al.*, 2009).

Table 20.3. Management of root-knot nematode (*Meloidogyne incognita*) infecting pointed gourd in West Bengal

Treatment	Yield (kg/plant)	Gall index
Vine dipping (VD) with monocrotophos 36SL at 1000 ppm for 6 hrs + decomposed organic matter at 20 t/ha	1.93	22.10
VD + Vermicompost at 2 t/ha	1.78	25.53

VD + *Purpureocellium lilacinum* at 10 g/pit (in two split doses)	2.54	11.93
VD + *Trichoderma viride at* 10 g/pit	1.75	33.80
VD + carbofuran at 1 kg a.i/ha	3.60	16.90
VD + neem cake at 500kg/ha	2.16	17.43
VD alone	3.27	12.60
Untreated control	3.67	16.83
CD (P=0.05)	0.93	6.02

20.2. RENIFORM NEMATODE, *Rotylenchulus reniformis*

20.2.1. Symptoms

Pointed gourd (*parwal*) infested by *R. reniformis* shows slower growth, reduced plant size, thinner veins, reduced fruit size resulting in substantial loss of yield. The symptoms produced on different hosts are not easily discernible. Root discoloration, leaf shedding and malformation of fruits and seed can also occur.

20.2.2. Bio-ecology

Reniform nematodes can survive in air-dried soil for a long period. Young females, males and fourth stage juveniles can survive in soil in a coiled anhydrobiotic state formed within the encrusted cuticles retained from previous molting. Survival of reniform nematode is positively correlated with decreasing soil moisture. Alternate drying and wetting of soil result in a sharp decline of nematode population density. It takes 3 to 4 weeks to complete life cycle depending upon temperature and host suitability. The pathogenic potential of *R. reniformis* in pointed gourd was 2 nematodes per ml of soil (Roy *et al.*, 2008).

20.2.3. Interaction with other pathogens

R. reniformis can interact with other pathogens like *Fusarium* spp., *Verticillium* spp., *Sclerotium rolfsii*, and *Rhizoctonia solani* to generate disease complexes.

20.2.4. Management

20.2.4.1. Cultural methods: Crop rotation with non-host crops like chilli, mustard, maize, sugarcane, marigold and restricting susceptible crops in winter seasons in the multiple cropping systems proved beneficial.

20.2.4.2. Chemical methods: Seed dressing with Carbosulfan 25DS @ 3% a.i. w/w has been found quite effective.

20.2.4.3. Integrated methods

i) Botanicals and cultural: Application of oilcakes like Neem, Karanj, Mustard, FYM, etc. and two to three deep summer plowings during hot months (May-June) helps to reduce soil nematode populations.

ii) Botanicals and chemicals: Use of organic amendments including Neem and Castor cakes @ 1 t/ha 10 days before sowing in infested fields and their combinations with seed treatment with Carbosulfan @ 3.0% a.i. (w/w) has been found to reduce nematode damage.

iii) Two bioagents: Among various bioagents, seed treatment with *P. fluorescens* + *Trichoderma viride* each @ 5 g/kg seed was found most effective for the management of reniform nematode infecting pointed gourd.

iv) Bioagents and botanicals: Applying biocontrol formulations like *Purpureocillium lilacinum* either as seed treatment (10 g/kg seed) or FYM-enriched formulation (1 kg in 25 kg FYM) in nursery bed and field treatment was found effective.

20.3. CONCLUSION

Reduction in crop yield due to nematodes can be greatly managed by using available management practices. Crop rotation is one of the oldest and most economic methods of controlling nematodes. However, these management practices must be taken before planting or propagating the crops through its vine nature. Once the nematodes persist inside the roots/twigs, effective and potential treatments are not available. Therefore, control strategies need to be preventive rather than curative in nature and aimed from the onset at preventing the build-up of high population densities. Many techniques used for managing root–knot invasion on pointed gourd simultaneously control other phytoparasitic nematode affecting the crop. Combining and effective rotational scheme and selected cultural practices and use of chemical give excellent control with little added cost. In severely affected field, chemical/ nematicides may be very useful in lowering down the nematode population with its threshold level.

Chapter - 21

RIDGE GOURD, *Luffa acutangula*

21.1. INTRODUCTION

Ridge gourd (*Luffa acutangula*) or ribbed gourd is a monoecious viny vegetable. It has some medicinal value and has all the nutrients. The fibre from well grown fruits has commercial value, as it can be used for making brushes, for cleaning. It is mainly cultivated in Andhra Pradesh, Tamil Nadu, Karnataka, Gujarat, Assam, West Bengal and Konkan region of Maharashtra.

The root-knot nematode is the major limiting factor (causing major yield loss) in successful production of ridge gourd.

21.2. ROOT-KNOT NEMATODES, *Meloidogyne* spp.

21.2.1. Crop losses

The avoidable loss in the yield of ridge gourd ranged from 16.32 to 28.64 per cent (Waykule *et al.*, 2023). The root-knot nematodes cause severe damage thus leading to dramatic yield losses up to 24 per cent (Sikora and Fernandez, 2005). The preventable yield losses due to root-knot nematode in ridge gourd was up to 74.52 percent under field conditions with application of Furadon 3G at 2 kg a.i. /ha (Pandey and Nayak, 2018a).

21.2.2. Symptoms

The root-knot nematode is sedentary endoparasite and produces disease symptoms both on above and below ground plant parts. Foliar symptoms include stunting, premature wilting, and leaf chlorosis and below ground symptoms include root galls or knots on the roots, which affect nutrient uptake by plant.

21.2.3. Management

21.2.3.1. Chemical methods: Carbofuran 3G was found to be best compared to all other treatments and recorded the increased growth parameters viz., plant height (14.66 and 16.26 cm) and number of leaves (6.33 and 7.83), the root length (13.38 cm), fresh and dry root weight (0.60 and 0.22 g) and reduced nematode parameters, minimum number of galls per root system (21.00), nematode population (712.00) and gall index (3.03) at 30 and 60 DAS, respectively (Shobha *et al.*, 2016).

21.2.3.2. Host resistance: Ridge gourd cvs. Panipati and Meerut Special were found resistant to root-knot nematodes.

Out of fifty two varieties/cultivars screened against root-knot nematode (*Meloidogyne incognita*), only 6 (Priya, Challenger, Jaipur long, BSS-1009, Tauri, and Pallishree) showed resistant reaction with 7-10 number of galls per plant (RKI - 1.1 to 2.0), and 7 showed (Narayana, Aneeta, NHRG-1001, Ridge gourd 12 pata, 12 Pata Jhinga, Dhenkanal local, Nayagarh local, Yeshvi-38, and Indo-US-216) moderately resistant reaction with 23-31 number of galls per plant (RKI - 2.1 to 3.0) (Pandey and Nayak, 2018b).

21.2.3.3. Integrated methods: Neem cake + carbofuran 3G proved best and recorded the increased growth parameters viz., plant height (15.80 cm and 17.42 cm) and number of leaves (6.66 and 8.00), the root length (14.30 cm), fresh and dry root weight (0.55 and 0.20 g) and reduced nematode parameters minimum number of galls per root system (17.33), nematode population and gall index (696.33 and 3.0) at 60 DAS, respectively (Shobha *et al.*, 2016).

Chapter - 22

BITTER GOURD, *Momordica charantia*

The root-knot nematode is an important limiting factor in successful cultivation of bitter gourd.

22.1. ROOT-KNOT NEMATODES, *Meloidogyne incognita, M. javanica*

22.1.1. Crop losses

Root-knot nematodes are responsible for annual yield loss of 13.5% in bitter gourd amounting to Rs. 252.82 million in India (Walia and Chakraborty, 2018).

22.1.2. Symptoms

The root-knot nematode infested bitter melon plants showed above-ground symptoms of wilting and dying. The root systems were found to be heavily infected with root-knot nematodes (100 % of root system) forming large coalescing root galls.

The common symptoms of root-knot nematode on plants include general stunting, low vigor, chlorosis, necrosis, defoliation and twig dieback. Infected plants show fewer, small rootlets, reduction in aerial growth and highly deformed and longer size galls are formed which can be easily attacked by secondary pathogens to form disease complex.

22.1.3. Management

22.1.3.1. Cultural methods: The use of green manure (Sudax sorghum) and soil amendments (saw dust and chicken manure) to manage root-knot nematodes has had some success.

Organic amendment viz, acacia compost at 1 kg/m^2 gave the least root - knot indices in cumber indicating efficacy in managing the root-knot nematodes (Trimbake and Mail, 2021).

Rotation of bitter gourd with marigold cv. Single Gold gave effective management of root-knot nematode (Wang, 2010).

22.1.3.2. Chemical methods: The green biomass was highest with carbofuran treatment (6326 kg/ha), which was at par with carbosulfan (5884 kg/ha), followed by poultry manure (6175 kg/ha). The highest grain yield was also obtained from carbofuran (3148 kg/ha) and poultry manure (3124 kg/ha). Significantly the lowest soil nematode population of *M. incognita* was recorded with carbofuran (313/100 g soil), which was at par with mustard cake (318/100 g soil), followed by poultry manure (319/100 g soil) and jatropha cake (330/100 g soil) (Venkatesan and Patel, 2013).

22.1.3.3. Biological methods: *Purpureocillium lilacinum* at 12 g/kg seed was found most effective followed by *Pochonia chlamydosporia* at 12 g/kg seed and *Glomus fasiculatum* at 12 g/kg seed for enhancing plant growth of bitter gourd and to reduce the infection of *M. incognita*. Seed treatment with *Trichoderma viride* at 12 g/kg seed was found most effective for enhanced plant growth of bitter gourd and management of *M. incognita* (Bhati *et al.*, 2019).

22.1.3.4. Host resistance: The five *M. balsamina* accessions (MB-3, MB-5, MB-6, MB-7 and Local-1) showed resistant reaction to root-knot nematode, *M. incognita* in terms of low root galls, soil nematode population and reproduction factor. On the other hand, most of *M. charantia* accessions were found moderately susceptible or susceptible to root knot nematode except two lines (PAUBG-1 and PAUBG-13) which gave moderate resistant reaction (Sharma *et al.*, 2019).

Screening of bitter gourd germplasms against root-knot nematode, *Meloiodogyne incognita* revealed that four germplasms viz., 'IC 33275', 'IC 44426', 'IC 45346' and 'Pusa Hybrid' were found moderately resistant (Machal *et al.*, 2022).

Three rootstocks viz., Kumatikai (*Citrulus colocynthis*), African horned cucumber (*Cucumis metuliferus*) and pumpkin (*Cucurbita moschata*) showed resistant reaction followed by two rootstocks viz., sponge gourd (*Luffa cylindrica*) and mithipakal (*Momordica charantia* var. *muricata*) which are moderately resistant (Tamilselvi *et al.*, 2017).

Bitter gourd variety DVBTG-1 was moderately resistant to anthracnose and resistant to root-knot nematode (Jadhav *et al.*, 2010).

22.1.3.5. Integrated methods

i) Bioagents and botanicals: Soil application of vermicompost @ 2.5 tons/ha enriched with *P. lilacinum* @ 2.5 kg and vermicompost @ 2.5 tons/ha enriched with *P. fluorescens* @ 2.5 kg/ha proved effective to manage root-knot nematode, *Meloidogyne* spp. and hence increased bitter gourd fruit yield (Table 22.1) (Singh *et al.*, 2019).

Table 22.1. Efficacy of bio-agents in the management of *Meloidogyne* species in bitter gourd.

Treatment	Root-knot index	Yield (tons/ha)
T1 - *Purpureocillium lilacinum* (cfu 2x 10^6) @ 2.5 kg/ha along with 2.5 tons of Vermicompost/ha	1.77ef (3.12)	10.433a
T2 - *Pochonia chlamydosporia* (cfu 2x 10^6) @ 2.5 kg/ha along with 2.5 tons of Vermicompost/ha	1.98bc (3.90)	8.923ab
T3 - *Pseudomonas fluorescens* (cfu 2x 10^6) @ 2.5 kg/ha along with 2.5 tons of Vermicompost/ha	1.93bcd (3.73)	7.493bc
T4 - T1 + T3 (Pl + Vermi + Pf+ Vermi)	1.74f (3.03)	10.483a
T5 - T2 + T3 (Pc + Vermi + Pf+ Vermi)	1.84de (3.37)	9.428ab
T6 - 2.5 tons of Vermicompost/ha	1.90cd (3.62)	8.800ab
T7 - Mulching with green leaves of Sunhemp @ 3 kg/m^2	2.16a (4.68)	8.476abc
T8 - *Bacillus pumilus* @ 2.5 kg + 2.5 t FYM/ha	1.84de (3.37)	8.967ab
T9 - Carbofuran 10 g/m^2	2.01b (4.02)	7.223bc
T10 - Untreated Control	2.20a (4.02)	6.436c

Figures in parentheses are retransformed values of Log and $\sqrt{x}$ trans.

Figures indicating common letters do not differ significantly at 5% level of significance according to DNMRT.

The best protection of bitter gourd root-knot disease was achieved through integrated application of neem cake, *P. flourencens* and *T. harzianum*. Significantly better results (in terms of vine height and reduction in *M. incognita* population) were recorded under consequent second year trial (84 and 94% respectively) as compared to first year trial (47 and 73% respectively). It also enhanced marketable yield of bitter gourd significantly (P>0.5) (up to 108%) in the second season trial, as compared to farmers' practice. It is suggested that integrated approach using organic amendment with bio-agents to manage root-knot disease of bitter gourd under natural infestation is environmentally friendly and beneficial to growers/farmers' having small holdings of land (Singh, 2019).

Purpureocellium lilacinum at 250 g/m^2 (cfu 2 x 10^6) both individually and in combination with acacia compost (1 kg/m^2) performed very close to the recommended check carbofuran 3G at 3g a.i. /m^2 (Trimbake and Mail, 2021).

ii) Physical and chemical: Solarization of nematode infested field + seed dressing of direct sown crops with Carbosulfan 25 DS @ 3% a.i. (w/w) was found effective for management of root-knot nematodes.

22.2. INTERACTION OF ROOT-KNOT NEMATODE WITH ROOT ROT

22.2.1. Interaction

The major problem with bitter gourd production in India is Fusarium wilt (*Fusarium solani*) and root knot nematode (*Meloidogyne incognita*) disease complex. It is the most devastating soil-borne disease complex and one of the major yields limiting constraint which cause profound economic losses ranging from 30 to 50 per cent under dry warm conditions (Tamilselvi and Pugalendhi, 2015). The maximum Fusarium wilt disease incidence and the root-knot nematode severity was recorded in Manakkadavu village with 71.85 per cent of wilt disease incidence and mean gall index of 5.0 (Sreegayathri *et al.*, 2018).

22.2.2. Management

22.2.2.1. Host resistance: Three genotypes viz., Kumatikai (*Citrulus colocynthis*), African horned cucumber (*Cucumis metuliferus*) and pumpkin (*Cucurbita moschata*) encompass the lowest value for number of galls per gram root, egg masses per gram root and numbers of females per gram of root and showed resistant reaction followed by two rootstocks viz., Sponge gourd (*Luffa cylindrica*) and mithipakal (*Momordica charantia* var. *muricata*) with bare minimum values and are moderately resistant (Tamilselvi *et al.*, 2016).

Memordica balsamina, due to its resistance to *M. incognita* race 2, could serve as an alternative crop for use as vegetable, ethnomedicinal and/or neutraceutical plant in rural communities with high populations of *M. incognita* race 2 (Fopu *et al.*, 2016).

22.2.2.2. Integrated methods: The field trial revealed that the soil application of neem cake @ 250 kg ha^{-1} + *Streptomyces reticuli* (BG 01) @ 10 ml/ lit + *Pseudomonas fluorescens* (Pf 1) @ 2.5 kg ha^{-1} + *Trichoderma viride* (Tv 1) @ 2.5 kg ha^{-1} recorded a significantly lower wilt incidence of 14.4 per cent as against 42.5 per cent in the control, which was found to be 66.1 per cent reduction over control and gall index was also significantly reduced to 1.6 as against 5.0 in control. The treatment recorded a higher fruit yield of 15.40 tons ha^{-1} as against 10.75 tons ha^{-1} in control (Table 22.2) (Sreegayathri *et al.*, 2018).

Table 22.2. Effect of biocontrol agents and neem cake on the wilt incidence and severity of root-knot nematode in bitter gourd under field conditions.

Treatment	Fusarium wilt incidence	Root gall index
Soil application of neem cake alone @ 250 kg ha^{-1}	21.0c (27.27)	1.8
Soil application of neem cake @ 250 kg ha^{-1}+ *Streptomyces reticuli* (BG 01) @ 10ml/ lit	17.2c (24.50)	2.1
Soil application of neem cake @ 250 kg ha^{-1}+ *Pseudomonas fluorescens* (Pf 1) @ 2.5 kg ha^{-1}	20.4e (26.85)	2.8
Soil application of neem cake @ 250 kg ha^{-1}+ *Trichoderma viride* (Tv 1) @ 2.5 kg ha^{-1}	19.1d (25.91)	3.2
Soil application of neem cake @ 250 kg ha^{-1}+ *Streptomyces reticuli* @ 10 ml / lit + *Pseudomonas fluorescens* (Pf 1) 2.5 kg ha^{-1} + *Trichoderma viride* (Tv 1) @ 2.5 kg ha^{-1}	14.4b (22.30)	1.6
Carbendazim 0.1% soil drenching + carbofuran 3G soil application @ 1kg a.i/ha.	13.1a (21.21)	2.2
Control	42.5f (40.68)	5.0

Values are mean of three replications.

Values in parentheses are arcsine transformed values.

In a column, means followed by a common letter are not significantly different at the 5% level by DMRT.

Chapter - 23

SPONGE GOURD, *Luffa cylidrica*

23.1. ROOT-KNOT NEMATODE, *Meloidogyne incognita*

23.1.1. Crop losses

The root-knot nematode, *M. incognita* is responsible for 15% yield loss in sponge gourd (Singh and Kumar, 2013).

23.1.2. Symptoms

Chlorotic patches were seen on the leaves in sponge gourd field infected with root-knot nematodes.. The leaf size was reduced, the newer leaves were bunched, flowering was reduced and fruits were deformed. Fruit set was completely abnormal. Deformed fruits were present and growth of the new branches was retarded. The roots of these plants had galls. Entire roots were deformed due to gall formation. Certain gall just near the surface of soil was corky and larger in shape (Fatma and Singh, 2020).

23.1.3. Management

23.1.3.1. Cultural methods: Eco-friendly, non-synthetic chemical, means of control measure like field sanitation, plowing during summer, crop rotation with non-hosts, organic amendment, co-cultivation of plants having antagonistic effect on nematodes, trap crops *etc.* have been recommended.

23.1.3.2. Chemical methods: Pots treated with oxamyl gave effective control of root-knot nematodes (Abd El-Aal *et al.*, 2021).

23.1.3.3. Biological methods: Supernatant of *Streptomyces* sp. from the uninfected soils exhibited approximately 67.59% nematocidal activity against *M. incognita* pre-J2s. *Bacillus amyloliquefaciens*, *Bacillus* sp., and *Melia azadirachtae* repelled the nematodes. *M. incognita* invasion significantly influenced the microbiota composition in rhizosphere soils and plant roots. These results provided insights into exploring novel microbiological agents for safe control of these harmful plant nematodes (Qu *et al.*, 2023).

Khan *et al.* (1992) investigated the importance of seed dressing of *Luffa cylindrica* with *Purpureocellium lilacinum* for the control of root-knot nematodes during seed germination. Spores of *P. lilacinum* in polyvinyl alcohol and gelatin with vegetable charcoal mixtures as vehicles were used separately for seed dressing. The medium and high doses of *P. lilacinum* (50 and 75 mg spores, respectively) significantly reduced root galling.

In the pot experiment, *Luffa cylidriaca* treated with *Serratia* spp. and *Pseudomonas* spp. displayed significantly higher ($P \leq 0.05$) levels of growth (as indicated by root length, fresh roots weight and fresh shoot weight) compared to control plants and significantly ($P \leq 0.05$) suppressed galling (number of galls) and reproduction (as indicated by number of egg-masses on roots and number of eggs and juveniles in pot soil). Meanwhile, among the treated plants, *Serratia* spp. and *Pseudomonas* spp. gave the best results in shoot weight of pots infected by eggs of *M. incognita* than those infected with IJs as compared with positive control (Table) (Abd El-Aal *et al.*, 2021).

Table 5. Effect of *Serratia* spp. and *Pseudomonas* spp. in comparison with oxamyl on galling and reproduction of *M. incognita* in relation to growth of *Luffa aegyptiaca* under greenhouse conditions.

Treatments	Shoot weight (% increase)	Root weight (g) (% increase)	Root gall index (GI)	Egg mass index (EI)
Uninoculated control	6.98a	5.22a	0.0	0.0
Inoculated control (infective juveniles)	3.21d	3.89d	4.0	5.0
Inoculated control (eggs)	4.30c	3.75d	4.0	4.3
Serratia spp. (infective juveniles)	5.39b (67.91)	4.55c (16.96)	4.0	4.0
Serratia spp. (eggs)	5.41b (25.18)	4.76bc (22,36)	4.0	4.0
Pseudomonas spp. (infective juveniles)	5.99ab (86.60)	4.57c (17.48)	4.0	4.6
Pseudomonas spp. (eggs)	5.16b (20.0)	4.78bc (22.87)	4.0	4.0
Oxamyl (infective juveniles)	6.07ab (89.09)	4.95b (27.24)	3.3	4.0
Oxamyl (eggs)	6.48a (50.69)	4.87b (29.86)	3.3	4.0

*Reported numbers represent means of 5 replicates.

**Different bacterial isolates in the same column with the same letters are not significantly different from each other (Tukey's test).

Pre-, post- and simultaneous inoculation with VAM fungi showed reduced incidence of root-knot disease (*M. incognita*) in sponge gourd (lower nematode reproduction and plant growth reduction) (Hazra *et al.*, 2015).

23.1.3.4. Integrated methods: The lowest galling (gall index), number of eggs and juveniles in soil was observed in the treatment with mixture of *Serratia* spp. and *Pseudomonas* spp. as well as, enhanced growth of sponge gourd more than their individual application (Abd El-Aal *et al.*, 2021).

Table 6. Effect of mixture of *Serratia* spp. and *Pseudomonas* spp. in comparison with oxamyl on galling and reproduction of *M. incognita* in relation to growth parameters of *Luffa aegyptiaca* under greenhouse conditions.

Treatments	**Shoot weight (% increase)**	**Root weight (g) (% increase)**	**Root gall index (GI)**	**Egg mass index (EI)**
Uninoculated control	6.98a	5.22a	0.0	0.0
Inoculated control (infective juveniles)	3.21d	3.89c	4.0	5.0
Inoculated control (eggs)	4.30c	3.75c	4.0	4.3
Serratia spp. + *Pseudomonas* spp. (infective juveniles)	6.19ab (91.90)	4.70b (20.82)	3.00b	3.66b
Serratia spp. + *Pseudomonas* spp. (eggs)	6.23ab (44.88)	4.89b (30.40)	3.00b	3.66b
Vydate (infective juveniles)	6.07ab (89. 09)	4.95b (27.24)	3.3	4.0
Vydate (eggs)	6.48ab (50.69)	4.87b (29.86)	3.3	4.0

*Each value is a mean of 5 replicates.

**Treatments in the same column with the same letters are not significantly different from each other (Tukey's test).

The field data revealed that maximum plant growth parameters viz., fresh weight of shoot, dry weight of shoot, fresh weight of root and dry weight of root, number of fruits and yield per plot whereas minimum root-knot index and nematode population over control were recorded in the treatment with *G. fasciculatum* @ 300 spores/m^2 + Org-Trichojal @ 1 l/ha + vermicompost @ 1.25 t/ha (Table 7) (Sonowal *et al.*, 2020).

Table 7. Effect of *Glomus fasciculatum*, Org-Trichojal, vermicompost and carbofuran 3G on root-knot index and yield

Treatments	**Root-knot index**	**Yield/ 2.25 m^2**
T1= *Glomus fasciculatum* @ 600 spores/m^2	3.67b	814.43cd
T2 = Org-Trichojal (5ml/1kg of vermicompost) @ 2 l/ha	3.33bc	798.14d
T3 = Vermicompost @ 2.5 t/ha	3.33bc	896.93bc
T4 = *Glomus fasciculatum* @ 300 spores/m^2 + vermicompost @ 1.25 t/ha	3.00bc	776.10de
T5 = Org-Trichojal 1 l/ha + vermicompost @ 1.25 t/ha	3.33bc	908.27b
T6 = Org-Trichojal @ 1 l/ha + *Glomus fasciculatum* @ 300 spores/m^2	3.00bc	809.98d
T7 = *Glomus fasciculatum* @ 300 spores/m^2 + Org-Trichojal @ 1 l/ha + vermicompost @ 1.25 t/ha	2.67cd	994.29a
T8 = Carbofuran 3G @ 1 kg a.i. /ha prior to planting of cuttings	2.00d	697.80e
T9 = Control	4.67a	587.49f
CD (P = 0.05)	0.84	83.98

Mean followed by the same letter in the superscript(s) are statistically at par.

Chapter - 24

OKRA, *Abelmoschus esculentus*

Root-knot, *Meloidogyne* spp. and reniform, *Rotylenchulus reniformis* nematodes are the major nematode problems on okra which interact with *Fusarium* wilt and *Rhizoctonia* root rot in causing disease complexes.

24.1. ROOT-KNOT NEMATODES, *Meloidogyne incognita, M. javanica*

24.1.1. Crop losses

M. incognita was responsible for 28.08 to 90.90% loss in fruit yield of okra (Bhatti and Jain, 1977; Parvatha Reddy and Singh, 1981), while *M. javanica* caused 20.20 to 41.20% loss in yield (Jain *et al.*, 1986).

Root-knot nematodes are responsible for 19.5% annual yield loss in okra amounting to Rs. 2480.86 million in India (Walia and Chakraborty, 2018).

24.1.2. Symptoms

Meloidogyne species cause similar upper ground symptoms (stunting, leaf yellowing and wilting) as in other vegetable crops. On roots prominent galls are formed which are large in size and easily attacked by secondary pathogens to form disease complexes.

24.1.3. Biology and life cycle

Root-knot nematodes are soil-borne and sedentary endoparasites. After penetration of the second-stage larvae in the root, they feed on the vascular tissues and develop endoparasitically. Adult females are sac-like and lay eggs in masses on the surface of galls, or in advanced stages inside the compound galls, while the vermiform males leave the roots and come out in soil. During summer months, the nematode takes about 25-30 days to complete its life cycle.

24.1.4. Interaction with other pathogens

Root-knot nematodes cause wilt disease complex in association with *Fusarium oxysporum* f. sp. *vasinfectum* and root rot disease complex in association with *Rhizoctonia solani*.

24.1.5. Spread

Nematodes disseminate through infected transplanting material, irrigation water, and agricultural implements.

24.1.6. Management

24.1.6.1. Physical methods: There was 71.8% reduction in nematode population in okra by soil solarization alone.

24.1.6.2. Cultural methods: Decrease in *M. incognita* population in okra occurs when the field was left fallow (Khan *et al.*, 1975). There was 56.2% reduction in nematode population in okra by fallowing alone

i) Organic amendments: Application of Neem/Eupatorium leaves @ 15 t/ha two weeks before sowing proved beneficial for controlling root-knot and reniform nematodes.

Shoot length (23.14 cm), root length (12.08 cm) and shoot weight (5.81 g) were significantly increased in soil amended with chopped leaves of *Ricinus communis* at 20 g/kg soil while minimum number of galls was recorded in treatment where leaves of *Azadiachta indica* at 20 g/kg soil was applied (Sujata *et al.*, 2022).

Singh and Sitaramaiah (1966) applied finely divided oil cakes to root-knot infested soil and noted a reduction in disease incidence in okra. Soil application of neem cake at 2 t/ha recorded lowest root-knot index (1.6), followed by fresh neem leaves at 5 t/ha (1.6) and silkworm litter at 2.5 t/ha (1.7). Significant highest yield was also obtained with neem oil cake (6.2 t/ha) followed by neem leaves (6.1 t/ha) (Sahoo *et al.*, 2004b).

Best results in respect of root-knot reduction due to *M. javanica* and increase in yield of okra were obtained by amending the soil with saw dust at 2.5 tons per ha 3 weeks before planting and then applying N through urea at 120 kg per ha (Singh and Sitaramaiah, 1971b).

Application of saw dust/paddy husk @ 500 g per plant/Neem leaves/Eupatorium leaves @ 250 g/plant in the basins 3 weeks prior to planting and watering daily proved beneficial for managing root-knot nematode.

Spot application of groundnut cake and sunflower cake each at 0.25 t/ha, saw dust and coal ash each at 5 t/ha significantly reduced the root-knot nematode population (Bhosle *et al.*, 2006).

Furrow application of neem cake at 100g/2m furrow, spot application of neem cake at 15g/spot and spot and furrow application of mustard and castor cakes gave 58.88, 48.36, 35.86, 35.53, 34.87 and 31.91% increase in yield, respectively and reduced the nematode infestation (Talukdar, 1993). Application of Neem cake @ 1 t/ha at sowing in the main field was found effective.

Poultry manure at 100 kg/ha can be used in the management of *Meloidogyne incognita*-infected okra in view of the outstanding performances recorded in growth, yield, nematode management and for eco-friendly reasons. Therefore, poultry manure can be used in place of carbofuran to manage *M. incognita* on okra in view of its little or no cost, no known hazard to plant, man and his environment (Tanimola and Akarekor, 2014).

Application of press mud at 15 t/ha, a week prior to sowing is recommended for effective management of root-knot nematodes in kharif season with benefit: cost ratio of 3.1. Application of fresh neem or karanj cake at 1 t/ha at the time of final land preparation is recommended.

Among organic amendments, the application of biosolids produced the highest okra yield and biomass, and greatly suppressed root-knot nematodes, *Meloidogyne incognita*, in the soil (Wang *et al.*, 2007).

ii) Green manuring: Use of *toria* crop as green manure after okra and before tomato was observed to keep the nematode infestation low in the okra-*toria*-tomato rotation.

iii) Crop rotation: Rotation with non-host crops like cereals was found effective.

A 44 and 21% decline in nematode numbers was recorded in plots under cluster bean (guar) or sweet corn and babycorn, respectively. The galling index on a crop of okra that followed sweet corn was 3.3 compared to 8.6 in the control, which was continuously under okra, resulting in an increase in yield within a range of 60-92%. This underscores the potential of rotating highly susceptible crops with poor hosts in the management of root-knot nematodes (Table 24.1) (Mweke *et al.*, 2008).

Table 24.1. Response of okra plots previously planted with selected nematode suppressive crops.

Treatment (Previous crop)	Root-knot index		Yield (tons/ha)	
	Season I	Season II	Season I	Season II
Baby corn	4.20	4.00	4.40	4.80
Cluster bean (Guar)	3.80	3.60	4.80	5.30
Sweet corn	3.50	3.00	4.00	5.00
Pioneer	4.50	4.20	4.50	5.20
Control (Okra)	8.60	8.50	2.50	3.00
CD (P = 0.05)	0.68	0.64	1.10	1.20

Decrease in root-knot nematode population in okra occurs following marigold, spinach and bottle gourd (Khan *et al.*,1975). Out of 15 different cropping sequences studied, tomato-onion-resistant tomato cv. Hisar Lalit-okra was found to be the best in terms of root-knot nematode reduction and economic returns (Kanwar, 1990). Crop rotations having okra-garlic-cluster bean and okra-coriander-resistant tomato cv. Hisar Lalit effectively managed the root-knot nematode population (Anon, 1993b).

Rotation of okra with sweet potato (cv. Sree Bhadra) reduced the root-knot nematode population by 21% and increased the fruit yield by 20.57% (Sheela *et al.*, 2002). Crop rotation with cabbage, marigold and wheat significantly reduced the root-knot nematode population. Crop rotation with cereals and marigold helps in reducing the nematode population in soil. In cropping sequence involving cabbage-marigold-wheat-kochia, the population of root-knot nematode larvae reduced to such an insignificant level that growing okra following the above sequence did not bring about increase in number of larvae (Alam *et al.*, 1977).

The cropping sequence of okra-wheat-cluster bean decreased soil population of root-knot nematode by 52-56.8% in comparison to 57.3% increase in susceptible check (okra-brinjal-okra). There was a 72.3% decrease in final nematode population by growing a combination of onion/cluster bean/resistant tomato variety, Hisar Lalit for three years.

Okra-rice-fallow or okra-cucumber-mustard cropping sequence suppressed soil population of root-knot nematode (*M. incognita*) in okra.

Cauliflower-onion-okra has been recommended in root-knot nematode infested fields and is economically beneficial.

The following rotations are beneficial for the management of root-knot nematodes on okra:

» First year: Planting of okra in kharif and tomato in rabi season.
» Second year: Root-knot resistant cowpea variety 'Anand Vegetable Cowpea 1' in *kharif* and root-knot resistant tomato variety 'Hisar Lalit' in *rabi*.
» Third year: okra in kharif and tomato in rabi.

iv) Intercropping: Atwal and Mangar (1969) showed that root exudates from sesame (*Sesamum orientale*) have nematicidal properties against *M. incognita*. When okra was grown in *M. incognita* infested soil it was only slightly attacked and there were fewer nematodes compared with when okra was grown in the absence of sesame.

Intercropping of okra with sesame decreased root-knot nematode penetration in okra and also delayed the life cycle of *M. incognita* (Tanda and Atwal, 1988). Similarly, intercropping of okra with marigold, margosa and Persian lilac have been recorded to reduce nematode population and root galling.

v) Biofumigation: Biofumigation using crop residues of cabbage and cauliflower @ 2.5 kg/m^2 reduced the nematode population and increased fruit yield in okra.

The highest decrease in nematode population, root gall index and increase in yield was observed when cabbage leaves @ 50 t/ha were incorporated in soil (Table 24.2) (Patil *et al.*, 2020).

Table 24.2. Management of root-knot nematode in okra by bio-fumigation (2018-2019)

Treatment	Root-knot index at harvest	Fruit yield (tons/ha)
Cabbage leaves @ 25 t/ha	2.8a, b	65.5c
Cabbage leaves @ 50 t/ha	2.5a	78.2a
Cauliflower leaves @ 25 t/ha	3.3a,b	63.2c,d
Cauliflower leaves @ 50 t/ha	3.0a,b	71.8b
Carbofuran @ 1.0 kg a.i. /ha	3.5b	63.4c,d
Untreated check	4.8c	57.1e
CD (P = 0.05)	**0.89**	**5.1**

Incorporation of cabbage leaves @ 5 t/ha 15 days before sowing is effective against root-knot nematode in okra (Anon, 2017).

vi) Cover cropping: Between two cover crops, sun hemp was superior to sorghum Sudangrass in improving okra production and in suppressing root-knot nematodes. The result indicates that growing sun hemp as a cover crop and applying certain organic amendments can improve okra production and suppress root-knot nematodes, which are very damaging to okra plants. Such combined practices show a significant potential for application in organic farming and sustainable agriculture systems in a tropical or subtropical region (Wang *et al.*, 2007).

24.1.6.3. Chemical methods: Seed treatment with Carbofuran at 6 and 12 per cent was effective against the root-knot and reniform nematodes infecting okra (Sivakumar *et al.*, 1976). Jain and Bhatti (1981) reported effective control of *M. javanica* on okra by seed treatment with Fenamiphos at 2, 4 and 6 per cent.

Alam *et al.* (1973) reported that oxamyl was effective for the control of root-knot nematodes infecting okra. Carbofuran, Ethoprophos and Fenamiphos each at 1 to 2 kg a.i. per ha were found effective in reducing the root-knot nematode population and in increasing fruit yields of okra (56 to 87 per cent)

Seed dressing of okra with Carbosulfan, Benfuracarb (Oncol 50 WP) and Triazophos @ 1.5 and 3% inhibited the gall index and increased the yield.

Seed treatment with Abamectin 40% FS @ 0.4 g a.i. /kg seed achieved highest yield of okra (6.87 t/ha) and proved better that Carbosulfan 25DS @ 3% a.i. w/w with reference to the gall index and population of *M. incognita*.

Fenamiphos, Quinalphos, Carbosulfan and Monocrotophos, each @ 500 and 1000 ppm as seed dip for 12 hours lowered the root-knot index in okra.

Soaking of seeds of okra cv. MDU 1 for 12 hr. in 0.1% solutions of Monocrotophos and Phosphamidon caused higher shoot weight of 30.71 and 30.69 g/plant and lower root-knot index of 2.33 and 2.40, respectively when compared to control which recorded a shoot weight of 19.12 g/plant and root-knot index of 4.83.

Soil application of Carbofuran 3G @ 2 kg a.i. /ha was found to be most effective in reducing root-knot nematode population (59.49 %), number of root galls (69.57 %) and gall index (39.33 %) and increasing the length of root (64.28 %) and shoot (71.37 %), fresh weight of root (93.41) and shoot (83.67 %), dry weight of root (103.13 %) and shoot (87.58 %) and the fruit yield of okra (32.94 %) in micro plots (Shendge *et al.*, 2010).

Root-knot nematodes in okra can be effectively managed by soil application of Carbofuran 3G @ 1 kg a.i. /ha (Choudhury *et al.*, 2011).

Application of Carbofuran 3G in two split doses of 1.5 kg +1.5 kg a.i. /ha at 30-day interval in okra was observed to be more effective in reducing root-knot infestation as compared to single dose application of Carbofuran 3G at the time of sowing.

24.1.6.4. Biological methods

i) Antagonistic bacteria: Seed treatment with *Bacillus macerans* at 3% w/w of seed was found effective for the management of root-knot nematodes. Root-knot nematodes in okra can be managed by the application of *Bacillus macerans* at 1.2 x 10^8 cells/m^2 before sowing. *Azotobacter* sp. significantly reduced root-knot nematode infestation in okra (Amalendu Chatterjee, 1995).

Seed treatment with *B. macerans* @ 3% w/w of seeds + drenching pits with 2% bioagent solution may be used as sticking agents for seed treatment in the endemic areas of root-knot infestation.

ii) Antagonistic fungi: *Purpureocellium lilacinum* reduced root-knot nematode egg masses by 91% after 60 days and by 96% after 90 days after inoculation, respectively in okra. In general, *P. lilacinum* treatment greatly suppressed the numbers of nematode galls and egg masses on okra plants (Ibrahim and Rezk, 1988).

The soil incorporation of *P. lilacinum* (in rice hulls and bran soil mix) was effective in reducing the gall index (2.7 as against 5.0 in control), nematode population in soil (77.3% over control) and in increasing the number of okra fruits (121.7 as against 86.0 in control) and fruit weight (3.26 kg as against 2.11 kg in control) (Davide and Zorilla, 1986).

Shahzad and Ghaffar (1987) found that carbofuran at 1 kg a.i. /ha was less effective than *P. lilacinus* against *M. incognita*. This fungus also effectively controlled nematodes on okra in subsequent seasons but the nematicide was no longer effective. Similarly, in the fungus treated plots, higher yields were recorded than carbofuran at 2.5 kg a.i. /ha.

A statistically significant difference (P = 0.05) was obtained regarding the plant growth parameters of okra with treatments *Streptomcyes avermentilis* and *Purpureocellium* sp. Treatments *Streptomcyes avermentilis* and *Paecillomyces* sp. reduced gall formation on the roots, number of juveniles in the soil, number of eggs, number of eggs mass, number of adults compared to control treatment (Hnoosh and Aljuaifari, 2020).

Soil application of *Trichoderma viride* @ 2.5 kg/ha at the time of sowing successfully managed root-knot nematode *Meloidogyne* spp. and increased the okra yield by 16.4-18.7%.

iii) Arbuscular mycorrhizal fungi: The early establishment of *Glomus fasciculatum* in okra roots could reduce the number of galls and egg masses per plant (Sharma and Trivedi, 1994). The number of galls and egg masses per plant and final soil population of *M. incognita* reduced significantly in AMF treated plants.

24.1.6.5. Host resistance: The reductions in yields were the minimum and ranged from 2.9 to 6.5% in moderately resistant (MR) cultivars of okra (Sanam, Dikshah, Arka Anamika, Ikra-1 and Ikra-2) to *M. incognita*. They are recommended for cultivation in fields heavily infested with *M. incognita* (Mukhtar *et al.*, 2017).

24.1.6.6. Integrated methods

i) Bio-agents and botanicals: Increase in okra fruit yield and reduction in root-knot nematode population (*M. incognita*) is achieved by soil incorporation of *P. lilacinum* @ 2.5 kg/ha in 2.5 t FYM.

Application of 5 per cent inoculum of *Arthrobotrys conoides* to the pot soil amended with FYM, effectively reduced the larval penetration of *M. incognita* and root galling was reduced by 30-40 per cent in okra (Srivastava and Swarup, 1986).

Soaking of okra seeds in 10% Castor cake suspension mixed with spores of *P. lilacinum* (1.5 x 10^6 spores/ml) for 30 min. and sowing in soil drenched with 10% Castor cake suspension at 20 L/6 m^2 was effective in reducing root galling (4.7 as compared to 9.2 in control), final nematode population of *M. incognita* and increasing the fruit yield (6.8 kg/6 m^2 as compared to 4.2 kg/6 m^2 in control), root colonization, propagule density in soil and parasitization of eggs by *P. lilacinum* (Table 24.3) (Rao *et al.*, 1997).

Table 24.3. Effect of integration of bioagents and botanicals for the management of *Meloidogyne incognita* infecting okra

Treatment	Root-knot index	Yield (kg)/6 m^2	Propagule density on root (cfu/g)	Propagule density in soil (cfu/g)	% egg parasiti-zation
Seed treat. with *P. lilacinum*	7.6	5.2	1865	1676	43
Seed treat. with castor cake suspension	7.2	5.0	---	---	---
Soil drenching with castor cake suspension	7.5	5.4	---	---	---
Seed treat. with *P. lilacinum* + castor cake suspension	5.5	5.7	2676	2349	46
Seed treat. with *P. lilacinum* & castor cake suspension + soil drenching with castor cake suspension	4.7	6.8	3445	3576	53
Seed treat. with castor cake suspension + soil drenching with castor cake suspension	5.9	6.3	---	---	---
Castor cake – 1 ton/ha	5.6	6.5	---	---	---

Castor cake – 2 ton/ha	4.6	7.0	---	---	---
Carbofuran – 1 kg a.i. /ha	5.2	6.4	---	---	---
Carbofuran – 2 kg a.i. /ha	4.5	6.7	---	---	---
Control	9.2	4.7	---	---	---
CD (P = 0.05)	0.87	0.72	284.57	340.85	2.24

Seed treatment with *Purpureocillium lilacinum* @ 5 ml/kg + soil application of vermicompost @ 2.5 ton/ha enriched with *P. lilacinum* @ 10 ml/kg recorded highest increase of 51.28 %,87.0%, 55.53%, 67.62% in plant height, root length, shoot dry weight, root dry weight over untreated check, recpectively with reducing final nematode population in soil (171.0 J2/200cc soil) and in roots of okra (41.25/ 5g. root) with the lowest root-knot index (2.0) followed by seed treatment with *Pochonia chlamydosporia* @ 5 ml/ kg+ soil application of vermicompost @ 2.5 tons/ha enriched with *P. chlamydosporia* @ 10 ml/kg. Moreover, seed treatment with *P. lilacinum* @ 5 ml/kg + soil application of vermicompost @ 2.5 ton/ha enriched with *P. lilacinum* @ 10 ml/kg gave the highest fruit yield (7.19 ton/ha) which was 36.6% higher than untreated plot (5.26 ton/ha) and performed as the most economical treatment than other treatments except chemicals for root-knot nematode management in okra with highest incremental benefit: cost ratio of 2.75 (Table 24.4) (Mahalik and Sahu, 2018).

Table 24.4. Effect of bio-agents on *M. incognita* and fruit yield in okra (cv.Utkal Gaurav)

Treatment	Root-knot index	Yield (tons/ha)	Benefit: cost ratio
Seed treat. with *P. lilacinum* @ 5 ml/kg followed by soil application of vermicompost @ 2.5 ton/ha enriched with *P. lilacinum* @ 10 ml/kg	2.00	7.19	2.75
Seed treatment with *P. chlamydosporia* @ 5 ml/ kg followed by soil application of vermicompost @ 2.5 ton/ha enriched with *P. chlamydosporia* @ 10 ml/kg	2.75	7.10	2.62
Seed treatment with *P. lilacinum* @ 2.5 ml/kg + *P. chlamydosporia* @ 2.5 ml/kg followed by soil application of vermicompost @ 2.5 ton/ ha enriched with *P. lilacinum* and *P. chlamydosporia* each @ 5 ml/kg	3.25	6.97	2.44
Seed soaking with carbosulfan 25 EC @ 0.2% for 12 h before sowing followed by soil application of carbofuran @ 1 kg a.i. /ha	2.50	6.02	3.04
Untreated control	4.50	5.26	—
CD (P = 0.05)	0.83	0.31	—

Soil application of jatropha oil cake @ 1.0 t/ha 15 days before sowing + seed treatment with *T. viride* + *P. fluorescens*, each @ 5g/ kg seed + soil application of *P. lilacinum* @ 2.5 kg/ha 15days after sowing was the most effective integrated combination which recorded the maximum increase in plant height (68.1%), root length (95.3%), fresh shoot weight (57.0%), fresh root weight (92.8%), shoot dry weight (90.9%), root dry weight (77.5%), biomass yield (87.6%) and fruit yield (65.5%) with maximum reduction in number of galls per plant (84.7%), number of egg masses per plant (88.3%), number of eggs per egg mass (48.3%), nematode population (81.1%) and root knot index (34.0%) over untreated check (Table 24.5) (Mahalik *et al.*, 2020).

Table 24.5. Effect of oil cakes and bioagents on root-knot index and yield of okra.

Treatment	Biomass yield (g)	Fruit yield (tons/ha)	No. of gallls/ plant	Root-knot index
T1 - Soil appln. of Jatropha cake at 1 t/ha	36.5	7.63	64.0	4.0
T2 - Soil appln. of Neem cake at 1 t/ha	35.4	7.40	70.1	4.1
T3 - Seed treat. with *T. viride* + *P. fluorescens* + soil appln. of *P. lilacinum* at 2.5 kg/ha	40.4	8.00	49.1	3.8
T4 - Seed treat. with *T. viride* + *P. fluorescens* + soil appln. of *T. viride* at 2.5 kg/ha	31.9	7.90	52.8	3.9
T5 - T1 + T3	48.8	9.77	31.0	3.3
T6 - T1 + T4	45.6	9.23	38.3	3.5
T7 - T2 +T3	47.8	9.67	34.3	3.4
T8 - T2 + T4	42.9	8.90	42.0	3.6
T9 - Soil appln. of Carbofuran at 1 kg a.i. /ha (Standard check)	41.7	8.97	44.1	3.7
T10 - Untreated check	26.0	5.90	203.4	5.0
CD (P = 0.05)	1.6	0.15	4.17	0.12

Seed treatment with *T. harzianum* @ 20 g/kg of seed + main field application of *T. harzianum* @ 2.5 kg along with 2.5 tons of FYM/ha showed 18.8% increase in okra yield.

Seed treatment with *P. lilacinum* @ 20 g/kg (2x10^6 cfu/g) seed in combination with its application @ 2.5 kg along with 2.5 tons of FYM/ha prior to sowing gave effective control of root-knot nematodes.

Reduction in nematode population in okra was obtained with seed treatment of *P. lilacinum* @ 5 ml/kg followed by soil application of Vermicompost @ 2.5 t/ha enriched with *P. lilacinum* @ 10 ml/kg.

The highest yield with greater suppression of soil nematode population was obtained in the treatment with *P. lilacinum* at 2.5 kg in 2.5 tons FYM and followed by soil application with *Pseudomonas fluorescens* at 2.5 kg/ha, *Pochonia chlamydosporia* at 2.5 kg/ha and *Trichoderma harzianum* at 2.5 kg/ha.

The combined application of *P. chlamydosporia* with neem cake at reduced dose gave better recovery of okra fruit yield and significantly suppressed *M. incognita* in terms of galling, egg production and soil population of nematodes (Dhawan and Singh, 2009).

Walia *et al.* (1991) reported that application of *P. lilacinum* through wheat bran was superior than carbofuran at 1 kg a.i. /ha in gall reduction on okra plants infected with *M. javanica.*

Increase in fruit yield and reduction in nematode population is obtained by seed treatment with the biopesticide formulation of *P. lilacinum/Trichoderma harzianum* (2 x 10^6 cfu/g) @ 20 g/kg seed + soil application of *P. lilacinum/T. harzianum* (2 x 10^6 cfu/g) @ 2.5 kg along with 2.5 tons of FYM/ ha prior to sowing of okra.

ii) Botanicals and chemicals: An integrated management of *M. incognita* infecting okra using neem or karanj oil cake at 0.5 t/ha along with carbofuran at 1 kg a.i. /ha gave maximum reduction in root galling with consequent increase in okra fruit yield (Parvatha Reddy and Khan, 1991).

Application of subabool (*Leucaena leucophila)* leaves at 40 g/kg soil which are allowed to decompose for four weeks before sowing + carbofuran at 1 kg a.i. /ha while sowing of okra seeds resulted in minimum galling (35.4 galls/plant) in comparison to control (64 galls/plant) and better plant growth parameters in okra (Paruthi *et al.*, 1987).

iii) Bio-agents and chemicals: Shahzad and Ghaffar (1987) have used *P. lilacinus* and carbofuran for the effective management of root-knot nematodes under field conditions on okra.

Integration of seed treatment with *T. viride* (1% w/w with 2 x 10^8 cfu/g) with 50 ppm zinc and potash significantly reduced number of galls, while *Aspergillus niger* (1% w/w with 2 x 10^8 cfu/g) with 50 ppm zinc and calcium and 100 ppm phosphorus was effective in reducing root galling on okra infected with *M. incognita* (Sharma *et al.*, 2007).

The combined application of *P. chlamydosporia* with carbofuran at reduced dose gave better recovery of okra fruit yield and significantly suppressed *M. incognita* in terms of galling, egg production and soil population of nematodes (Dhawan and Singh, 2009).

iv) Bio-agents, Chemicals and Botanicals: The combined treatment with *P. lilacinus* at 4 g/kg soil + carbosulfan 25 EC at 0.2% + poultry manure at 2.5 t/ha + FYM at 2.5 t/ha gave maximum increase in plant growth parameters and yield (9.2 t/ha compared to 2.0 t/ha in control), and decrease in number of galls (83.5 compared to 183.6 in control), egg masses per root system (23.6 compared to 70.5 in control), and final nematode population in soil (200 compared to 585 in control) (Das and Sinha, 2005).

The trio combination of *P. chlamydosporia*, carbofuran and neem cake yielded significantly not only gave higher fresh fruit weight with a recovery of 53% but also suppressed root-knot disease severity in terms of galling, egg mass production and nematode population (89%, 90% and 81%, respectively) (Dhawan and Singh, 2009).

The combined application of a 100% recommended dose of nitrogen fertilizer (100 kg N/ha) along with *Azospirillum* and neem seed cake is recommended for better growth and yield of okra with better control of nematodal population (Ansari *et al.*, 2021).

v) Cultural and chemical: Summer ploughing + mulching transparent polythene sheet + seed treatment with carbosulfon at 3% a.i. w/w led to 32.5% higher yield. Summer solarization + treated seeds + use of neem cake at 400 kg/ha was most effective treatment and gave 50% higher okra yield.

vi) Botanicals and physical: Integration of soil solarization for 15 days in summer and application of neem cake at 200 kg/ha is effective in the management of root-knot nematodes and in getting higher okra fruit yields.

vii) Physical and chemical: Soil solarization with single layer of polyethylene mulch for 20 days during June and application of carbofuran at 0.5 kg a.i. /ha gave least root galling (42/plant compared to 245/plant in control) (Sharma *et al.*, 2005).

In okra, summer plowing + seed treatment with Carbosulfan @ 3% a.i. + Carbofuran application @ 1 kg a.i. /ha led to 76-79% decrease in nematode population and 35.1% higher yield over untreated check.

Solarization of nematode infested field + seed dressing of direct sown crops with Carbosulfan 25 DS @ 3% a.i. (w/w) was found effective for management of root-knot (*Meloidogyne* spp.), reniform (*Rotylenchulus reniformis*) and lesion nematodes (*Pratylenchus* spp.) in okra.

In okra cv. Pusa Sawani, the integration of seed treatment with Carbosulfan 3% with either mulching with polythene sheet after plowing and exposure for 20 days or plowing and exposure for 20 days reduced the nematode population by 57.8 and 48.6% and also increased the fruit yield by 46% and 34%, respectively.

viii) Two bioagents: *Bacillus circulans* and *B. macerans* and their combinations were effective for the management of root-knot nematodes and increased the fruit yield ranging from 27 to 31%.

P. penetrans has been used in combined applications with *P. lilacinum* and other soil inoculants such as *B. subtilis* and *Talaromyces flavus* for the control *M. javanica* on okra (Zaki and Maqbool, 1991).

24.2. INTERACTION OF ROOT-KNOT NEMATODE WITH FUSARIUM WILT

24.2.1. Interaction

The interaction of the wilt fungus, *Fusarium oxysporum* f. sp. *vasinfectum*, and the rot-knot nematode, *Meloidogyne incognita* in causing wilt disease complex has been considered important on okra leading to the reduction in its productivity.

The individual, simultaneous, and sequential inoculation of second stage juveniles (at 1000 J2 /kg soil) of *Meloidogyne incognita* and *Fusarium oxysporum* (1.1×10^6 cells/kg soil) resulted in significant reduction in plant growth parameters (plant height, plant girth, chlorophyll content, fresh and dry shoot weights, and fresh and dry root weights). Maximum suppression in plant growth parameters was observed in plants that received NF21 (*Fusarium oxysporum* inoculated 21 days after *Meloidogyne incognita* on okra) and F21 (fungus inoculated 21 days after seedling emergence) treatments for pot and field experiments, respectively (Agbaglo *et al.*, 2020).

24.2.2. Management

24.2.2.1. Biological methods: Among chemicals, botanicals and biocontrol agents evaluated, carbofuran was the most effective treatment in improving plant growth followed by *Trichoderma harzianum*, neem seed powder, neem cake, *Aspergillus niger* and bavistin respectively. Highest suppression of root colonization of the fungus (*F. o.* f. sp. *vasinfectum*) was achieved in plants treated with bavistin (5.0) followed by neem seed powder (7.5), neem cake (12.5), *A. niger* (15.0), *T. harzianum* (17.5) and carbofuran (20.0). The greatest suppression of root-knot development was achieved by carbofuran (0.25) followed by *T. harzianum* (0.75), neem seed powder (0.90), neem cake (1.90), *A. niger* (2.20) and bavistin (2.75). In general, carbofuran and *T. harzianum* were highly effective against the nematodes, bavistin against the fungus and neem seed powder and neem cake against both the pathogens (Table 24.6) (Abuzar and Haseeb, 2006).

Table 24.6. Effect of chemicals, botanicals and bioagents on plant growth and disease complex in okra

Treatment	Plant height (cm)	Fresh weight of plant (g)	Root-knot index	Disease index (%)
Uninoculated control	71.0	112.7	---	---
Inoculated control (N + F)	48.7	62.4	3.75	42.5
Carbofuran	67.7	107.2	0.25	20.0
Bavistin	55.9	93.5	2.75	5.0
Neem seed powder	62.9	101.9	0.90	7.5
Neem cake	60.6	98.0	1.90	12.5
T. harzianum	65.1	104.0	0.75	17.5
A. niger	57.5	95.6	2.10	15.0
CD (P = 0.05)	0.61	1.38	0.36	1.45

24.2.2.2. Integrated methods

i) Bioagents and botanicals: Soil application of 25 g/m^2 of *Pseudomonas fluorescens* (2 x 10^6 cfu/g) or *Pochonia chlamydosporia* (2 x 10^6 cfu/g) enriched deoiled neem cake has proved to be an effective treatment in combating the damage caused by *M. incognita* and *F. oxysporum* f. sp. *vasinfectum* to the tune of 68 and 57%, respectively. This treatment also increased the yield of okra fruits by 24% under field conditions (Chaya *et al.*, 2010).

24.3. INTERACTION OF ROOT-KNOT NEMATODE WITH ROOT ROT

24.3.1. Interaction

Plants in untreated field soil or in sterilized soil inoculated with both organisms, developed a root rot in about 42 days. If the nematode preceded the fungus by three weeks, the root rot was more severe and appeared within 14-21 days. The fungus penetrated either directly or through ruptures in the root created by the mature female nematode. *R. solani* colonized nematode giant cells and root xylem cells. Vascular discoloration occurred both in roots and stem, however no fungus was isolated from stems.

M. incognita predisposed roots to *R. solani* which resulted in severe root rot and subsequent plant death. Okra plants inoculated with either *R. solani* or *M. incognita* alone were free of root decay for the entire period of six-week study. Three weeks after nematode and fungus inoculation, black sclerotia of *R. solani* were visible on nematode- induced galls, while on non-galled portions on the same root system were free of sclerotia. Prior to root rot development, *R. solani* demonstrated marked preference for root galls on nematode infected roots. It is hypothesized that the leakage of nutrients from the root was responsible for attracting the fungus to the galls and for initiating sclerotial formation.

Five weeks after inoculation, distinct brown lesions were observed only on the galls of plants inoculated with both *M. incognita* and *R. solani*. Lower leaves of plants were chlorotic and suffered premature leaf drop.

M. incognita and *R. bataticola* when inoculated simultaneously in soil, reduce the germination of seeds in okra (Chhabra and Sharma, 1981). Combined attack of both these pathogens causes significantly greater damage to the crop than that of the damage caused by either pathogen alone (Bhagawati *et al.*, 2007). If the nematode preceded the fungus by three weeks, the root rot was more severe and appeared within 14-21 days. *M. incognita* predisposed roots to *R. solani* which resulted in severe root rot and subsequent plant death.

24.3.2. Management

24.3.2.1. Chemical methods: Seed treatment of okra with Carbofuran at 3% a.i. (w/w) or Carbendazim 0.2% a.i. (w/w) + Difolaton 0.25% a.i. (w/w) was effective in the control of reniform nematode/root-knot nematode - Rhizoctonia wilt complex.

Soil fumigation usually decreased disease severity due to *R. solani* and *M. incognita* disease complex in okra (Golden and Van Gundy, 1975).

24.3.2.2. Biological methods

i) Antagonistic fungi: *Purpureocellium lilacinum* can be effectively used against *R. solani* – *M. incognita* complex of okra (Shahzad and Ghaffar, 1989).

24.3.2.3. Integrated methods

i) Bioagents and botanicals: Chaitali *et al.* (2003) observed that *T. viride* combined with neem cake controlled the disease complex better than *T. viride* combined with groundnut cake in okra.

Integration of *P. lilacinum* and *T. viride* with mustard cake gave significant reduction in root galling, egg mass production, fecundity, reproduction factor and nematode population both in soil and roots coupled with increase in plant growth parameters (Tripathi and Singh, 2006).

ii) Bioagents and chemicals: The lowest pre-emergence (5.01%) and post-emergence (7.73%) damping-off were observed in the treatment where both carbosulfan 25 SD and carbendazim 50 WP were applied together as seed treatment which was at par with the treatment with dual application of *T. harzianum* (6.26%) and *P. fluorescens* (7.73%) as seed treatment. As evidenced from the results, the seed treatment was found to be significantly superior to soil application of bioagents (Bhagawati *et al.*, 2009) (Table 24.7).

Table 24.7. Effect of bioagents and chemicals for the management of disease complex caused by *Meloidogyne incognita* and *Rhizoctonia solani*.

Treatment	Pre- emergence damping- off (%)	% decrease over inoculated control	Post-emergence damping- off (%)	% decrease over inoculated control
Uninoculated & untreated control	32.98	---	34.34	---
T. harzianum (seed treatment)	11.05	66.49	13.99	59.26
P. fluorescens (seed treatment)	10.86	67.07	12.52	63.54
T. harzianum + *P. fluorescens* (seed treatment)	6.26	81.02	7.73	77.45
T. harzianum (soil application)	20.45	37.99	24.95	27.34
P. fluorescens (soil application)	18.69	43.33	25.14	26.79
T. harzianum + *P. fluorescens* (soil application)	18.88	42.75	23.38	31.92
Carbosulfan (seed treatment)	20.26	38.57	24.94	27.37
Carbendazim (seed treatment)	18.78	43.06	24.76	27.88
Carbosulfan + carbendazim 50 (seed treatment)	5.01	84.81	7.73	77.45
Uninoculated & untreated control	0.00	---	0.00	---
CD (P=0.05)	3.94	---	4.25	---

The maximum plant height, dry weight of shoot and root were recorded in the uninoculated and untreated control followed by the treatment receiving *T. harzianum* and *P. fluorescens* as seed treatment which were at par with the treatment receiving carbosulfan 25 SD and carbendazim 50 WP as seed treatment (Bhagawati *et al.*, 2009) (Table 24.8).

Table 24.8. Effect of bioagents and chemicals on plant growth parameters

Treatment	Plant height (cm)	% increase over control	Shoot weight (g)		Root weight (g)	
			Dry	% increase over control	Dry	% increase over control
Inoculated & untreated control	27.30	---	1.80	---	0.32	---
T. harzianum (seed treatment)	36.40	33.33	3.12	73.33	0.48	50.00
P. fluorescens (seed treatment)	36.82	34.87	2.96	64.44	0.47	46.88
T. harzianum + *P. fluorescens* (seed treatment)	44.26	62.12	3.98	121.11	0.72	125.00
T. harzianum (soil application)	31.85	16.67	2.24	24.44	0.53	65.63
P. fluorescens (soil application)	32.76	20.00	2.28	26.67	0.47	46.88
T. harzianum + *P. fluorescens* (soil application),	36.10	32.24	2.90	61.11	0.50	56.25
Carbosulfan (seed treatment)	32.58	19.34	2.34	30.00	0.46	43.75
Carbendazim (seed treatment)	31.40	15.01	2.38	32.22	0.51	59.34
Carbosulfan + carbendazim 50 (seed treatment)	43.60	59.70	4.10	127.78	0.68	112.50
Uninoculated & untreated control	45.94	---	4.25	---	0.74	---
CD (P=0.05)	**3.26**	---	**0.38**	---	**0.13**	---

The minimum number of galls and egg masses in roots and nematode population in soil were recorded in the treatment receiving carbosulfan 25 SD and carbendazim 50 WP as seed treatment which was on par with the treatment receiving *T. harzianum* and *P. fluorescens* as seed treatment. Further, the treatment with *T. harzianum* and *P. fluorescens* as seed treatment was found to be significantly better in reducing the host infection and nematode multiplication than soil application of both these bioagents (Bhagawati *et al.*, 2009) (Table 24.9).

Table 24.9. Effect of bioagents and chemicals on root galling, egg mass production and nematode population in soil

Treatment	No. of galls/ root system	% decrease over control	No. of egg masses/ root system	% decrease over control	Nema popn./ 200 ml soil	% decrease over control
Inoculated & untreated control	236.00	---	138.25	---	1846	---
T. harzianum (seed treatment)	70.50	70.13	44.68	67.68	595	67.77
P. fluorescens (seed treatment)	74.75	68.33	46.25	66.55	614	66.74
T. harzianum + *P. fluorescens* (seed treatment)	54.25	77.01	26.25	81.01	348	81.14
T. harzianum (soil application)	102.25	56.67	58.75	57.50	815	55.85
P. fluorescens (soil application)	107.50	54.45	60.25	56.42	792	57.10

T. harzianum + *P. fluorescens* (soil application),	72.50	69.28	42.50	69.26	831	54.98
Carbosulfan (seed treatment)	76.00	67.80	40.26	70.88	820	55.58
Carbendazim (seed treatment)	146.00	38.14	62.30	54.94	830	55.04
Carbosulfan + carbendazim 50 (seed treatment)	51.75	78.07	30.75	77.76	336	81.80
Uninoculated & untreated control	0.00	---	0.00	---	0.00	---
CD (P=0.05)	0.71	---	0.67	---	2.59	---

24.4. RENIFORM NEMATODE, *Rotylenchulus reniformis*

24.4.1. Symptoms

General symptoms include reduced root systems, leaf chlorosis, overall stunting of host plants, and reduced yields and plant longevity. Female nematodes and their eggs are often visible when plant roots are viewed under a dissecting microscope.

The reniform nematode causes root necrosis resulting in severe root pruning and subsequent dwarfing of plants. Fibrous or feeder roots are mostly attacked which may reduce the absorption ability and other physiological functions of the plant. This may lead to stunting, yellowing, and wilting.

24.4.2. Management

24.4.2.1. Cultural methods: Application of potash in combination with phosphorus or nitrogen or potash alone checks the reniform nematode multiplication on okra to a great extent (Sivakumar and Meerazainuddin, 1974).

Application of recommended doses of urea coated with Nimin and U-Coatm significantly reduced soil population of *R. reniformis* and increased plant growth parameters. In field trial, all the three treatments (Nimin, U-Coatm and Modified-neem-oil) reduced the soil populations of reniform nematodes (14–25%) compared to the check, and increased the yield/plot (Jumder *et al.*, 2004).

Neem cake @ 10 g^{-kg} of soil was found effective in reducing the reniform nematode population and increasing overall plant growth of okra (Jagadeeswaran and Singh, 2011).

24.4.2.2. Chemical methods: Treatment of okra seeds with carbofuran at 1.25g a.i. w/w economized the nematode management by reducing the quantity of nematicide required (Sivakumar *et al.*, 1973).

24.4.2.3. Biological methods: The plants treated with avermectin (100%) (secondary metabolite of *Streptomyces avermitilis*) as seed treatment recorded maximum shoot length, fresh shoot weight, dry shoot weight and fruit yield. It was on par with carbofuran 3G @ 1kg a.i. /ha and avermectin (75%) as seed treatment. Highest reduction in *R. reniformis* population was observed in plants treated with avermectin 10% as seed treatment. Avermectin treated plants also recorded significantly increased total phenol content and enzymatic activities such as peroxidase and indole acetic acid oxidase compared to control plant. (Jayakumar *et al.*, 2005).

24.4.2.4. Integrated methods

i) Boagents and botanicals: A combination of seed dressing treatment with culture filtrate of *Purpureocellium lilacinum* and soil amendment with neem cake gave synergistic effect with respect to nematode management and improvement in plant growth parameters. Soil application of neem oil cake and *P. lilacinum* showed significant suppression of the reniform nematodes, both singly as well as concomitantly, with the consequent improvement in different plant growth parameters (Anver, 2020).

ii) Botanicals and chemicals: Application of carbofuran at 1 kg a.i./ha + neem cake at 0.5 ton/ha proved effective in reducing the *R. reniformis* population and in increasing the growth of okra plants (Krishna Rao *et al.*, 1987).

iii) Cultural and chemicals: Deep plowing (20 cm) followed by fallowing for one month, fallowing for one month after weeding, integration of nematicide application at 1.0 kg a.i. per ha at sowing after either of the cultural practices or deep plowing (20 cm) together with carbofuran seed treatment resulted in the control of the reniform nematode and better yield of okra (Lakshmanan and Sivakumar, 1981).

24.5. INTERACTION OF RENIFORM NEMATODE WITH ROOT ROT

24.5.1. Interaction

Glasshouse studies proved that when the reniform nematode, *R. remformis* and the root rot fungus *R. solani* were present, okra plants succumbed to root rot at an early stage and when the nematode infection preceded the fungus, root rot appeared earlier than when the plants were infected first with the fungus. The fungus caused root rot 45–60 days after germination in the absence of the nematode, depending upon the level of inoculum and the age of the plant, while the initial injury by the nematode shortened the period to 34–43 days after germination. When both the pathogens were inoculated at high level 15 days after germination, root rot occurred 23 days after germination. A low initial/nematode inoculum preceding the fungal inoculum also caused early root rot irrespective of the density of the latter (Kumar and Sivakumar, 1981).

24.6. CONCLUSION

The production of okra throughout the globe is threatened by root-knot nematodes (*Meloidogyne incognita* and *M. javanica*). Nematodes interact with Fusarium wilt and Rhizoctonia root rot pathogens in causing disease complexes. Of all the pathogens, root-knot nematodes (*Meloidogyne* spp.) are among the most serious sedentary endoparasitic nematodes worldwide. Although, chemical usage is a common and popular practice to manage nematodes, search for better options has been emphasized. A viable approach for the management of plant parasitic nematodes is the adoption of sequential cropping system using suppressive crops that have been tested for host status under local nematode populations. There is a need for development of most effective and economical method i.e., the use of resistant cultivars. An economical practice to be adopted by the farmers for gradual decrease in nematode populations in the infested fields is by growing resistant cultivars of crops in crop rotations. Integrated approaches to root-knot nematodes management be further developed using resistant/moderately resistant crop varieties combined with sequential cropping. They are more likely to be adopted by small-holder farmers due to the multiple benefits associated with them.

Chapter - 25

ONION, *Allium cepa* and GARLIC, *Allium sativum*

Stem and bulb, root-knot, and stubby root nematodes have been recognized as the limiting factors in successful production of onion and garlic.

25.1. STEM AND BULB NEMATODE, *Ditylenchus dipsaci*

The stem and bulb nematode cause great damage to the onion and garlic crops in temperate areas of many countries. The nematode is primarily observed in heavy soils. The nematode problem becomes serious with abundant precipitation or high soil humidity. The nematode causes serious damage at inoculum level of 10 or more nematodes/500g of soil. *D. dipsaci* was responsible 'bloat' disease of onion and garlic.

25.1.1. Crop losses

In Canada, *D. dipsaci* can reduce yields of garlic by as much as 100% (Celetti, 2009).

25.1.2. Symptoms

Infected young seedlings are stunted, pale and have swollen areas along the cotyledons. Leaves can develop yellowish-brown spots, short and thickened, and stem swelling (bloating) can occur. As the disease progresses the foliage collapses and the stems and necks of bulbs soften. The bulb scales become soft and light gray. Infected bulbs are light in weight, may be malformed or produce sprouts and double bulbs. The common secondary infections due to bacteria and fungi are responsible for decay and bad smell.

In garlic, stunting, distortion, yellowing, wilting, and premature dieback of leaves may be among the first symptoms observed. Swelling and distortion of bulbs and stems are characteristic. Infected bulbs are light in weight and shrunken, dark brown, and soft. Tissue death predisposes bulbs to colonization by soft rot bacteria and saprophytic fungi. Roots may be partially or completely absented from part of the basal plate. Although they are microscopic, large numbers of nematodes can congregate and form a visible "wool" between the scales of infected bulbs.

The leaves of garlic plants turn yellow and die but no thickening or twisting is evident. The infected garlic bulbs become loose (the cloves separate from the base) and cannot be used for seed even if they tolerate winter storage well because the plants which form often die during the vegetative period.

25.1.3. Host range

The species has an extensive list of host plants. There are more than 20 races of stem/bulb nematodes, each with a different selection of preferred host plants. The main race infesting onions and other Alliums also affects a wide range of crops including shasta peas, parsley, celery, lettuce, hairy nightshade, and salsify. Mixtures of more than one race may occur in the same field.

25.1.4. Life cycle

The fourth stage juveniles penetrate the stem and leaf tissue through the stomata. Egg laying begins at temperatures of 1-5 °C with the optimum at 13-18 °C. *D. dipsaci* completes one generation in 19-23 days at 15 °C. Nematode activity stops at 36 °C. The nematode prefers the cool moist climatic conditions existing in the upland tropics and wet winter seasons in the subtropics. *D. dipsaci* can parasitize plants in both heavy and light soils, although a higher incidence of infestation seems to occur on heavy soils.

25.1.5. Interaction with other pathogens

In garlic fields in Ontario, agricultural advisors have observed increased damage caused by *Fusarium oxysporum* f. sp. *cepae* (Foc) when *D. dipsaci* is in the same fields. This has led to the speculation that a nematode-fungus interaction may increase disease severity and yield losses of garlic. Since Foc and *D. dipsaci* are widely distributed throughout the major garlic production areas of the world, the possibility exists that these pathogens may act synergistically when present together in garlic fields. More damage, and higher severity of basal plate rot, could result from either an additive or synergistic interaction, that would result from infection with either pathogen alone. When Foc is applied prior to *D. dipsaci*, a suppressive effect is observed, indicating that the fungus is acting antagonistically towards the nematode (Ives, 2019).

25.1.6. Spread and survival

The nematode can survive in onion bulbs/garlic cloves used for seed as well as in the soil without a host plant for more than one year and the fourth juvenile stage can survive in anabiosis for many years. The nematodes on plant residues can tide over drought for several years which becomes active with the availability of moisture. Desiccated fourth stage larvae can survive for several years in the absence of growing host plants either in the soil or attached to plant tissue such as seeds or bulbs.

Infected transplants are the most common way in which these nematodes spread. The means by which the nematode spreads include soil adhering to bulbs, crop residues, and movement of infected bulbs. Seed-borne infections are also responsible for long distance spread in onion. Infestations can also be maintained by volunteer host plants and weed hosts in between onion and garlic crops.

25.1.7. Management

25.1.7.1. Physical methods: The nematodes can be exterminated by soaking the plants in water containing 0.5% formalin at 24 °C, followed by a hot water treatment (with addition of 0.5% formalin at 49 °C for 20 min.).

Soaking of onion sets or garlic cloves in cold water for 3 to 4 days will also kill the nematodes.

In fields heavily infested with *D. dipsaci* in Israel, soil solarization protected garlic bulbs throughout the growing season resulting in a greater yield increase compared with EDB and methyl bromide applications (Siti *et al.*, 1982). In Italy, control of *D. dipsaci* in a sandy soil increased with solarization periods and only 10, 6 and 2 percent nematodes were still viable after four, six, and eight weeks of solarization, respectively (Greco *et al.*, 1985).

25.1.7.2. Cultural methods: The diseased plants and their remnants should be gathered and destroyed.

Make every effort to ensure that seed, sets from onion and garlic bulbs are clean to prevent contaminating the field. Sets and bulbs should be examined carefully and only plant if they appear perfectly healthy should be used for planting. The important symptoms on infected sets include lighter in weight, soft, shrunken and discolored (dark brown) near the neck of the plant.

A four-year rotation is recommended for garlic production. The rotation of onion and garlic should be avoided with susceptible crops like beans, bulbous plants, carrot and pea. Onion and garlic may be rotated with wheat, barley, maize, rape, white clover, red clover, lucerne, potatoes, carrots, beets, crucifers, spinach, or lettuce to overcome the stem and bulb nematode damage because they are poor hosts for stem and bulb nematode. Control weeds and volunteers during rotations.

Ditylenchus dipsaci lacks the ability to efficiently survive and/or reproduce in soils in which wheat and soybean are grown; therefore, these plants may be categorized as being either active or passive antagonist, depending on whether they produce nema-toxic compounds or are simply unsuitable host for the nematodes. Due to the antagonistic nature of these plants, they may potentially be used in rotation with garlic areas infested by the nematode (Ives, 2019).

Use of green manure has been shown to reduce nematode populations. Mustard, rapeseed, oilseed, radish, and sorghum-Sudan grass can be used as a green manure which release nematode toxic chemicals when tilled into the soil. Planting bio-fumigants such as brown mustard and Sudan grass as cover crops proved beneficial.

25.1.7.3. Chemical methods: Row application of Thionazin at 4.5 g/meter row gives effective control of the stem and bulb nematodes.

A drench application of fluopyram (Velum Prime) to infested planted stock in in open furrows significantly decreased nematode survival resulting in a greater marketable yield (Table 25.1).

Table 25.1. Use pattern for Velum Prime on onion and garlic at planting.

Nematode suppressed	Rate	PHI	Use information
Stem and bulb nematode (*Ditylenchus dipsaci*)	500 mL/ha or 3.75 mL / 100 m row (based on 75 cm row spacing)	30 days	One application at planting in the fall or in the spring. Spray specified dosage in a 10-15 cm band in-furrow at planting and cover with soil. For best results, direct the in-furrow spray to the seed and soil. Apply in 50-300 L of water per hectare.

Fluopyram applied as a seed soak prior to planting or a drench at planting effectively reduced the nematode population density in soil and plant tissue, and consequently suppressed plant infection throughout the growing season, and resulted in high yields (Ives, 2019). Previous experiments have shown that fluopyram can inhibit nematode motility and reproduction (Faske and Hurd, 2015; Jones *et al.*, 2017). There is evidence to suggest that fluopyram treatments can protect garlic plants from *D. dipsaci*. In addition, when using clean seeds in infested soil, the most effective time to apply fluopyram is as a drench at planting. Fluopyram also increased yield to the extent of 58 to 67%

25.1.7.4. Biological methods: In Brazil, cool dip applications of 0.036 and 0.072g/L of abamectin controlled D. dipsaci and increased yields of garlic (Bekker, 1999). Abamectin as a seed soak reduced symptoms, and increased yield.

25.2. ROOT-KNOT NEMATODES, *Meloidogyne* spp.

Vadivelu and Rajendran (1986) reported *M. arenaria* on onion from Palladam district of Tamil Nadu. The infected plants showed patchiness and unthrifty growth with 'neck twisting' symptoms.

25.2.1. Symptoms

Affected crop show stunted growth, yellowing of leaves, smaller bulbs, delayed maturity, and wilting of the plants. Heavy infection of older plants causes wilting of plants unexpectedly and die-off early. Small, swollen galls 1-2 mm in diameter can be found on the roots when infected plants are carefully lifted from the ground and freed from soil particles without damaging the roots. Depending on the species causing infection, the shape of the galls can be round or spindly, and with or without short root branches that arise from the upper part of galls. White to dark brown egg masses can be observed on the roots.

25.2.2. Management

25.2.2.1. Physical methods: Hot water treatment of onion bulbs at 43.5 ^{0}C for 120 min. eliminates the nematode infection.

Soil solarization resulted in reductions of nematode populations. Table 25.2 shows that all treatments gave RF = 0, except the treatments of June (four weeks) and July (four weeks), while RF of non-solarized control plots was 2.35. Data in Table 25.3 show that solarization gave good control of nematodes. The RF in solarized treatments ranged from 0 to 0.47, while in non-solarized plots, reductions ranged from 1.18 to 1.57.

Table 25.2. Effect of solarization and timing on the control of nematodes on onion

Treatment	**Average number of nematodes per 200 g soil[1]**		**Reproduction factor (RF) RF = Pf/Pi**
	(Pi)	**(Pf)**	
June - 4 weeks	280	10	0.035
June-July - 6 weeks	140	0	0
July - 4 weeks	100	60	0.60
July-August - 6 weeks	200	0	0
August - 4 weeks	180	0	0
Control	170	400	2.35

1 Pi = Initial nematode population; Pf = Final nematode population.

Table 25.3. Effect of solarization on the control of nematodes on onion

Treatment	Average number of nematodes per 200 g soil*		Reproduction factor (RF) RF = Pf/Pi
	(Pi)	(Pf)	
Onion nursery solarized	235	0	0
Onion field - solarized	65	0	0
Onion nursery - non - solarized	165	260	1.57
Onion field - non-solarized	125	147	1.18

* Pi = Initial nematode population; Pf = Final nematode population.

25.2.2.2. Cultural methods: Crop rotation to a non-host or a long fallow period helps to reduce populations of root-knot nematodes. The lowest populations of plant parasitic nematodes were recorded in onion-cowpea-fallow cropping sequence (Vetrivelkalai and Subramanian, 2006).

Soil application of organic amendments enhances biological suppression of nematodes.

Planting onions when the soil temperatures are below 64° F can minimize root-knot nematode damage.

25.2.2.3. Integrated methods: Root dip treatment of onion seedlings in *P. lilacinum* + Neem cake suspension for 1 hr. before transplanting proved effective and increased bulb yield.

Seed treatment with *P. fluorescens* (10^9 cfu/g) at 10 g/kg and subsequent soil application of 5 tons of FYM enriched with 5 kg each of *Pseudomonas fluorescens* (10^9 cfu/g) and *Pochonia chlamydosporia* (10^6 cfu/g) per ha significantly reduced *M. incognita* population in roots by 69% and increased bulb yield by 21% (Anon, 2012).

Integration of neem cake with *P. lilacinum* was found to be effective and increased bulb yield (Table 25.4).

Table 25.4. Effect of neem cake and bioagents on the bulb yield of onion

Treatments	Yield (tons/acre)
Neem cake (NC)	12.972
Purpureocellium lilacinum (PL)	14.022
Pseudomonas fluorescens (PF)	13.746
NC+PL	17.347
NC+PF	16.842
PL+PF	14.749
Control	12.476

25.3. STUBBY-ROOT NEMATODES, *Paratrichodorus minor, P. allius*

Two species of stubby root nematodes, *Paratrichodorus minor* and *P. allius* have been reported to be associated with onion yield losses. *Paratrichodorus minor* appears to be the most common species recovered from onion fields in Georgia. Their high population pressures have been associated with areas of severely stunted onion seedlings (Hajihassani *et al.*, 2020).

25.3.1. Symptoms

The most obvious above-ground symptoms of nematode infection are poor, stunted growth of onion seedlings. Depending on nematode population density, infected plants will produce roots of varying lengths due to nematode feeding specifically on the root tips using a solid, curved stylet or spear. Reduced root growth, excessive lateral root growth, branched and "stubby"-appearing root systems are common symptoms of nematode damage. Stubby root nematodes also feed on the whole root system and cause swelling. Stunted root systems are not able to take up water and nutrients from soil, so plants show signs of nutrient or water deficiencies. Later in the season when warm weather appears, onion leaves become yellow with necrotic leaf tips, and variable sizes of onion plants can be seen across the field depending on the degree of nematode damage (Hajihassani *et al.*, 2020).

25.3.2. Host range

Stubby-root nematodes have a wide host range, including sugar beets, potatoes, corn, cotton, peanuts, wheat, onions, grasses and some other vegetable crops. Some weeds can also serve as good hosts or provide an alternate survival host for stubby-root nematodes (Hajihassani *et al.*, 2020).

25.3.3. Biology

Stubby-root nematodes spend their entire life cycle in the soil. The life cycle begins with an egg. Like other plant-parasitic nematodes, the stubby-root nematode has four vermiform (worm-like) juvenile stages that feed externally on root systems and develop. Each generation of a life cycle is completed between 16 to 22 days depending on the soil temperature (Hajihassani *et al.*, 2020).

25.3.4. Favorable conditions

Optimal conditions for nematode growth include sandy and moist soils with temperatures ranging from 20° to 30° C. Stubby-root nematodes appear to be very sensitive to low soil moisture, and in the absence of host crops, their populations can decline quickly in the soil. Coarse-textured soils such as sandy and sandy loam are ideal for nematode development and reproduction. Newly emerged onion seedlings are very vulnerable to nematode infection (Hajihassani *et al.*, 2020).

25.3.5. Management

25.3.5.1. Physical methods: Table 25.5 clearly shows the effect of solarization on the control of nematodes on onion, with RFs of 0.03 for *Trichodorus* spp. The corresponding RF values in non-solarized soil were 1.44 on the same crop and the same nematodes.

Table 25.5. Effect of solarization on the control of nematodes on onion

Treatment	Average number of nematodes per 200 g soil[1]		Reproduction factor (RF) RF = Pf/Pi
	(Pi)	(Pf)	
Onion- solarized	580	16	0.03
Onion-non-solarized	685	985	1.44

Pi = Initial nematode population; Pf = Final nematode population.

25.3.5.2. Cultural methods: Transplanting nematode infested seedlings and seeding or transplanting onions into nematode-infested fields should be avoided. Use certified nematode-free seedlings for planting.

Leaving the infested fields fallow for 4 to 6 weeks appears to suppress the nematode population enough to provide good control.

Thus, planting onions when the soil temperature is 64 ^{0}F can minimize nematode damage.

Resistant cultivars of oat, pearl millet, velvet bean, cowpea, and sun hemp can be grown as cover or green manure crops in rotation with onion for suppressing *P. minor* in soil.

25.3.5.3. Chemical methods: The use of non-fumigant nematicides like oxamyl (Vydate) and fluopyram (Velum Prime) have good efficacy on *P. minor* control (Hajihassani *et al.*, 2020).

25.3.5.4. Biological methods: MeloCon WG (*Purpureocellium lilacinum*) at 2.25 to 4.50 kg/ha at 4- to 6-week intervals for nematode suppression.

25.4. CONCLUSION

The most common nematodes on onion are root-knot nematode (*Meloidogyne hapla*), stem nematode (*Ditylenchus dipsaci*), and stubby root nematodes (*Paratrichodorus minor*). Considering the costs and environmental concerns of application of nematicide, planting resistant cultivars is an economically sound and environmentally viable option for sustainable nematode management in onion. Integrated management of nematodes using neem cake, trap crops, crop rotation, soil solarization, turmeric (*Curcuma longa*) grown as non-host antagonistic crop, crop sanitation, and application of carbofuran can be explored for effective nematode management.

Chapter - 26

CARROT, *Daucus carota*

26.1. INTRODUCTION

Carrot (*Daucus carota*) is a root vegetable rich in vitamin A and some minerals. It is grown all over India. Main carrot growing states are Uttar Pradesh, Assam, Karnataka, Andhra Pradesh, Punjab, and Haryana. It is taken raw as well as in cooked form. It is made into pickles and sweet meat. Carrot juice is a rich source of carotene (a precursor of vitamin A) and contains appreciable quantities of thiamine and riboflavin. It is sometimes used for coloring butter and other foods. Green carrot leaves are highly nutritive, rich in protein, minerals and vitamins and used as fodder and for preparation of poultry feed.

The root-knot nematodes are the major limiting factors in successful cultivation of carrot crop.

26.2. ROOT-KNOT NEMATODES, *Meloidogyne incognita, M. javanica*

26.2.1. Crop losses

The crop loss caused by root-knot nematode, *M. incognita* to carrots is to the extent of 56.64% (Devi, 1993).

The avoidable yield loss in carrot due to *M. hapla* was 35.95 per cent (Anita and Selvaraj, 2011).

Root-knot nematodes are responsible for annual yield loss of 34% in carrot amounting to Rs. 754.12 million in India (Walia and Chakraborty, 2018).

26.2.2. Symptoms

Nematode infestation causes qualitative loss in carrot along with quantitative loss. The infested plant exhibits reduced growth, yellowing of the leaves, and wilting symptoms. Galls are seen on the secondary roots of main tuberous tap root. Under severe infestation, the tuberous roots exhibit branching/splitting/spiking. The infestation paves way for secondary degrading organisms like *Erwinia*, which are responsible for further rotting of carrots. Carrot is particularly sensitive to root-knot nematode because the harvested product is directly affected, resulting in severe losses in quality and marketable yield.

Root-knot nematodes caused significant reduction in the yield and reduction in consumer preference due to forking of carrots.

26.2.3. Biology and life cycle

Root-knot nematodes are soil-borne, sedentary endoparasites. After penetration, the juveniles feed on conducting tissues endoparasitically and develop into adults. Adult females are sac-like and lay eggs within the galls or on the surface of root galls, while the males are vermiform, leave the roots and come out in soil. Under north Indian conditions, the nematode takes about 60 to 90 days to complete the life cycle.

26.2.4. Spread

Root-knot nematodes are disseminated through infected transplanting material, irrigation water, and agricultural implements.

26.2.5. Management

26.2.5.1. Physical methods: Soil solarization during summer months with 400 gauze polythene sheet for 3 to 4 weeks reduces nematode population in soil.

26.2.5.2. Cultural methods

i) Crop rotation: Rotation with non-host crops like wheat/barley/oats/sorghum/bajra/maize/ rice/ mustard/sesame was found beneficial.

The population of *M. incognita* larvae decreases in combination of spinach-marigold-carrot and spinach-radish-carrot (Hasan and Jain, 1998).

Significant carrot yield increase was observed in the plots previously rotated with *E. sativa* cv. Nemat, *B. juncea* and *R. sativus*. The beneficial effect of the rotation with mustard species was greater in biocidal crops sown in late spring rather than in late summer (Curto *et al.*, 2006).

ii) Organic amendments: Application of Neem cake or Mustard cake @ 300-400 kg/ha 15 days before sowing is effective for the management of root-knot nematode and in increasing yields of carrot (Choudhury *et al.*, 2011).

The oil cakes (neem, groundnut, mustard and castor) were found effective in reducing parasitic nematode population (*Hoplolaimus*, *Tylenchorhynchus*, *Meloidogyne* and *Helicotylenchus* species) and in increasing yields of carrot.

In a field trial, saw dust, Neem cake, and Poultry manure at 1.0 and 1.5 t/ha and Water hyacinth at 1 and 2 t/ha increased yield of carrot (28-66%) by reducing the infestation of *M. incognita*. Devi (1993) reported that soil application of Neem cake gave maximum reduction of *M. incognita* population in soil (38.5%) followed by Saw dust (23%), Poultry manure (22%), and Water hyacinth (14%) over control.

The number of galls, egg masses and eggs/egg mass gradually suppressed at the 30% concentration level of fly ash and promoted the growth of carrot plants as compared to untreated inoculated control (0% fly ash with nematode) (Haris *et al.*, 2019).

iii) Time of planting: Late sown carrots suffer less damage due to *M. incognita* in California, USA.

Nematodes are not very active in cool weather, so sometimes winter carrots will be successful even in soils where some nematodes are present.

iv) Biofumigation: The effectiveness of soil biofumigation treatments with *Brassica juncea* and *Eruca sativa* green manures, *B. juncea* hay, and a *Brassica* seed meal commercial pellet were investigated over two years in field experiments on a carrot - tomato rotation for the control of the root-knot nematode *Meloidogyne incognita*. Biofumigation treatments were compared with the standard nematicide fenamiphos, a quillay-derived nonchemical formulation and an untreated control. At the end of the first cycle, soil population density of *M. incognita* was significantly lower in the plots treated with green manure or hay of *B. juncea* or with green manure of *E. sativa*, either alone or combined with seed meal pellet, compared with untreated control. Nematicidal effects of biofumigation treatments did not differ from that of fenamiphos. Moreover, commercial pellet and *B. juncea* green manure and hay resulted in carrot yield almost double the control, whereas lower increases were provided by *E. sativa* green manure, either alone or combined with the pellet.

In the following tomato crop, the lowest *M. incognita* densities were recorded in soil treated with *B. juncea* hay or with *E. sativa* green manure plus pellet. Hay and green manure of *B. juncea* and integrated treatment with *E. sativa* green manure and pellet resulted in the highest tomato yields, always significantly higher than fenamiphos.

The *Brassica* crop arugula (*Eruca sativa* var. Nemat) also received interest as a potentially useful biofumigant crop to manage root-knot nematodes, as it appears to act as a trap crop for *M. hapla* (Melakeberhan *et al.*, 2006). Positive effects on yields of nematode-susceptible crops grown after arugula cultivation and biofumigation were reported for carrot.

26.2.5.3. Chemical methods: Soil application with Carbofuran 3G @ 16.6 g/plant proved effective.

26.2.5.4. Biological methods: Enhancement of yield (147.6%), reduction in forking of carrot roots (92.4%), reduction in nematode population in soil (45.23%), and least root-knot index (1.6 as against 5.0 in controls) was obtained by soil incorporation of *P. lilacinum* (containing 1 x 10^6 spores/g) at 10 g/ 2.5 m^2 (Sivakumar, 1998).

26.2.5.5. Host resistance: Carrot cv. Arka Suraj is reported to be tolerant to the root-knot nematode (Swamy and Sadashiva, 2004). Ten carrot genotypes namely New Kuroda, Selendro, Perfecta Elita, Zeno, Nantes Strong Top, Nantes, F-524, Danvers Half Long, Avenger and HYDC-4 were found resistant to *M. incognita* (Veere Gowda *et al.*, 2001).

26.2.5.6. Integrated methods

i) Bioagents and botanicals: Carrot seed treatment with *P. lilacinum* (2x10^6 cfu/g) @ 20 g/kg seed + application of *P. lilacinum* (2x10^6 cfu/g) @ 2.5 kg along with 2.5 tons of FYM/ha prior to sowing managed *M. incognita*.

Soil application of *P. lilacinum* (2 x 10^6 cfu/g) at 2.5 kg + FYM at 2.5 t/ha gave minimum number of galls and final nematode population (120 and 12.5, respectively) compared to control (440 and 235, respectively). The highest yield was also recorded in the above treatment (9.34 t/ha) over control (6.54 t/ha) (Somasekhara and Ravichandra, 2010).

Integration of Neem cake with *P. lilacinum* and *Trichoderma harzianum* was found effective and increased carrot yield (26.742 t/acre as compared to 18.245 t/acre in control) (Table 26.1).

Table 26.1. Effect of neem cake and bioagents on yield of carrot

Treatment	Yield (t/ha)
Neem cake	49.717
Neem cake + *Trichoderma harzianum*	56.414
Neem cake + *Purpureocellium lilacinum*	57.619

Neem cake + *P. lilacinum* + *T. harzianum*	66.081
Control	45.084

Seed treatment with *Pseudomonas putida* (10^9 cfu/g) at 10 g/kg and subsequent application of 5 tons of FYM enriched with 5 kg each of *P. putida* (10^9 cfu/g) and *T. harzianum* (10^6 cfu/g) per ha significantly reduced root-knot nematode population in roots by 77%. Significant increase in yield (24%) was also observed. Benefit: cost ratio calculated for marginal cost of biopesticides and returns accrued by application of biopesticides was 5.3 (Anon, 2012).

Seed treatment with *T. harzianum* at 10 g (with 1 x 10^6 cfu/g)/kg and *P. fluorescens* at 10 g (with 1 x 10^9 cfu/g)/kg and subsequent field application of 5 tons of enriched FYM with *T. harzianum* (with 1 x 10^6 cfu/g) and *P. fluorescens* (with 1 x 10^9 cfu/g) per ha significantly reduced root-knot nematodes in carrot roots by 79%, increased yield to the tune of 29.8% with a benefit: cost ratio of 13.6.

ii) Bioagents, botanicals and physical: Significantly lowest number of second stage juveniles (J2's) was recovered from the soil incorporated with broccoli left-over materials and *Trichoderma* inoculant. Galls and egg masses in secondary roots were highest in unamended-inoculated soil which was significantly different from broccoli-amended soil with solarization and *Trichoderma* inoculant. The yield was significantly highest in broccoli-amended soil with solarization and *Trichoderma* inoculant. In general, the treatments with broccoli residues and *Trichoderma* inoculant were able to decrease root-knot nematode population and significantly increase the yield relative to untreated soil (Pedroche *et al.*, 2009).

iii) Cultural and chemical: A soil fumigant at the beginning of two consecutive carrot crops, followed by two-year onion production and at last a cover crop sudax in the course of 5 year rotation can control *M. hapla* problem (Bird, 1981).

iv) Bioagents and chemicals: Combined application of *P. fluorescens* @ 1 kg/ha along with *P. lilacinum* @ 2.5 kg/ha and carbofuran 3G @ 0.25 kg a.i./ha at the time of sowing enhanced the yield of carrot to the tune of 43.58 per cent compared to untreated control and reduced the soil nematode population by 66.19 per cent (Anita and Selvaraj, 2011).

v) Bioagents and cultural: Integration of nematode resistant cover crops (Sudan grass, sorghum - Sudan grass, and cowpea), and coating carrot seed with a biologically based nematicide (abamectin) was found effective to minimize root-knot nematode damage in carrot without compromising air quality (Ploeg, 2012). Specific varieties of Sudan grass (var. Trudan 8), sorghum - Suda grass (var. SX-17), and cowpea (var. UCR CC36) have been reported resistant to root-knot nematodes. Sudan grass and sorghum - Sudan grass can be harvested for hay or ethanol production or incorporated into soil to increase organic matter and act as a biofumigant through the release of nematicidal metabolites. Cowpea can fix large amounts of nitrogen and can be harvested for forage or incorporated to increase the soil organic matter.

26.2.5.7. Success story: Management of root-knot nematode, *Meloidogyne incognita* in carrot was demonstrated in farmer Mr. Chaannappa's field, Uppinahalli village, Hosakote taluk, Bangalore North District, Karnataka. Soil solarization during summer months with 400 gauze polythene sheet, soil application with *Purpureocillium lilacinum* @ 25 g/m^2 along with Neem cake at 100 g/m^2, and soil application with Carbofuran 3G @ 16.6 g/ m^2 gave 77.7% benefit amounting to Rs. 1,60,000/acre to the farmer.

26.3. INTERACTION OF ROOT-KNOT NEMATODE WITH BACTERIAL WILT

26.3.1. Interaction

Carrot crop suffers severely (> 50% loss) in India from soft rot, *Pectobacterium carotovorum* s. sp. *carotovorum*, one of the most serious of bacterial diseases with world-wide occurrence. The bacterium causes soft rot disease in carrot by entering the aerial plant part or tap root through natural openings or wounds. The root-knot nematode, *Meloidogyne incognita* is one of the major pathogens of carrot throughout the world, affecting both quality and quantity of the marketable carrot yield.

The association of soft rot bacterium, *Pectobacterium carotovorum* s. sp. *carotovorum* with the root-knot nematode, *Meloidogyne incognita* was responsible for wilt disease complex (Sowmya and Rao, 2011). The infection by the nematode creates wounds that facilitate entry of the bacteria and cause disease complex in carrot.

26.3.2. Management

26.3.2.1. Integrated methods

i) Bioagents and botanicals: The neem cake enriched with *Pseudomonas fluorescens* and *Purpureocellium lilacinum* applied at 10 g/m^2 increased the root colonization of both the bioagents and reduced the incidence of *M. incognita* and *E. carotovora* s. sp. *carotovora* by 68 and 56%, respectively. There was also a significant increase in the yield of carrot to the tune of 23% (Sowmya *et al.*, 2010).

Table 26.2. Effects of single and combined applications of *Pseudomonas putida* and *Purpureocellium lilacinum* against disease complex caused by *Meloidogyne incognita* and *Erwinia carotovora* s. sp. *carotovora* in carrot.

Treatment	M. incognita J2 in 5 g root	% disease incidence by E. carotovora	Yield (kg/4m^2)
P. putida (cfu 2×10^8/g) seed treatment at 20 g/kg	12.24 e	31.00 g	25.18 a
P. putida substrate treatment at 20 g/m^2	10.44 d	23.25 d	26.99 b
P. putida seed treatment + substrate treatment	8.46 b	19.60 b	28.41 d
P. lilacinum (cfu 2×10^6/g) seed treatment at 20 g/kg	12.6 f	33.28 g	24.35 a
P. lilacinum substrate treatment-20 g/m^2	9.18 c	25.08 e	26.02 a
P. lilacinum seed treatment + substrate treatment	7.56 b	21.43 c	27.52 c
P. putida + *P. lilacinum* seed treatment each at 10 g/kg	11.34 e	28.27 f	25.73 a
P. putida + *P. lilacinum* substrate treatment each at 10 g/m^2	6.84 a	17.32 a	28.84 e
P. putida + *P. lilacinum* seed treatment + substrate treatment	5.58 a	15.50 a	29.65 e
Neem cake at 20 g/m^2	12.96 f	35.11 h	23.67 a

Treatment	M. incognita J2 in 5 g root	% disease incidence by E. carotovora	Yield (kg/4m²)
P. putida (cfu 2×10⁸/g) seed treatment at 20 g/kg	12.24 e	31.00 g	25.18 a
P. putida substrate treatment at 20 g/m²	10.44 d	23.25 d	26.99 b
P. putida seed treatment + substrate treatment	8.46 b	19.60 b	28.41 d
P. lilacinum (cfu 2×10⁶/g) seed treatment at 20 g/kg	12.6 f	33.28 g	24.35 a
Control	18.00 g	45.60 i	23.20 a

Means in the same column followed by a common letter are not significantly different ($P \leq 0.05$) according to Duncan's multiple range test.

Seed treatment with both bio-agents (*Pseudomonas putida* and *Purpureocellium lilacinum* at 20 g/kg seed and application of neem cake enriched with *P. putida* and *P. lilacinum* at 20 g/m² as a substrate treatment before sowing proved to be the best of all the treatments, leading to a reduction in the *M. incognita* (J2) population (in roots by 69% and soil by 47.6%) and *E. carotovora* by 66%, with a significant increase in the yield (27.8%) of carrot. *Pseudomonas putida* and *P. lilacinum* co-existed without affecting root colonization by either (Table 26.2) (Sowmya *et al.*, 2012).

Seed treatment together with soil application of *Bacillus subtilis* IIHR BS-2 (5L ha^{-1}) enriched with vermicompost (2 tons ha^{-1}) recorded the maximum increase in carrot yield (28.8%) and decrease in nematode population (69.3%) and disease incidence (70.2%) followed by treatment with chemicals (carbofuran 1 kg a.i. ha^{-1} + streptocycline- 0.02%) which recorded 62.2% and 68.15% decrease in disease incidence and nematode population, respectively (Rao *et al.*, 2017).

26.4. RENIFORM NEMATODE, *Rotylenchulus reniformis*

26.4.1. Symptoms

General symptoms include reduced root systems, leaf chlorosis, overall stunting of host plants, and reduced yields and plant longevity. Female nematodes and their eggs are often visible when plant roots are viewed under a dissecting microscope.

26.4.2. Management

26.4.2.1. Biological methods

i) Antagonistic bacteria: Seed treatment of carrot with *P. fluorescens* at 20 g/kg significantly reduced the reniform nematode population both in soil and roots and increased root colonization by the bioagent and yield (Rao and Shylaja, 2004) (Table 26.3).

Table 26.3. Effect of seed treatment with *Pseudomonas fluorescens* for the management of *Rotylenchulus* *reniformis* infecting carrot

Treatment/Dose/ kg seed	No. of nemas/ 10 g roots	No. of nemas/ 100 ml soil	Yield (kg)/4m^2	Root colonization (cfu/g)
P. fluorescens – 5 g	69	88	6.7	21745
P. fluorescens – 10 g	38	56	7.8	24574
P. fluorescens – 20 g	35	48	8.0	23789
Control (Untreated)	92	123	5.2	---
CD (P = 0.05)	13.67	17.34	1.47	124374

ii) Integrated methods

a) Bioagents and botanicals: Seed treatment with *P. fluorescens* at 10 g/kg (with 1 x 10^9 cfu/g) and subsequent field application of 5 tons of enriched FYM with *P. fluorescens* (with 1 x 10^9 cfu/g) per ha significantly reduced reniform nematodes in carrot roots by 75%. The yield increase was to the tune of 29.8% with a benefit: cost ratio of 13.6.

Seed treatment with *Pseudomonas putida* (10^9 cfu/g) at 10 g/kg and subsequent application of 5 tons of FYM enriched with 5 kg each of *P. putida* (10^9 cfu/g) and *T. harzianum* (10^6 cfu/g) per ha significantly reduced reniform nematode population in roots by 70 77%. Significant increase in yield (24%) was also observed. Benefit: cost ratio calculated for marginal cost of biopesticides and returns accrued by application of biopesticides was 5.3 (Anon, 2012).

26.5. CYST NEMATODE, *Heterodera carotae*

Carrot cyst nematode is a key pathogen of carrots in many temperate regions of the world. Carrot cyst nematode is a mono-specific and sedentary endoparasite with high degree of aggregation.

26.5.1. Crop losses

H. carotae caused an average 70.8% yield loss in carrots (Escobar-Ávila, 2021). Carrot cyst nematode crop losses can range from 20% in fertile soils to 80% in poor, sandy soils (Greco, 1987).

26.5.2. Symptoms

Typically, the disease symptoms appear as signs of nutrient deficiency because the roots are unable to absorb sufficient nutrients. The carrot cyst nematode infested crops appear stunted with reddish tops. Affected plants often appear in small, circular patches. The tap roots are small and unmarketable, with a "bearded look" due to abnormal rootlet growth. Typical symptoms are characterised by stunted growth with yellowish foliage of the plants. In the field, symptoms first appear in a small circular area but may extend further and destroy the entire crop in case of a major infestation. Tap root turns small with abnormal proliferation of rootlets. In infected plants, death of distal part of the root induces to produce more new rootlets and enhance the infection sites for the nematode. White females and cyst attached to secondary roots sometimes giving the roots a "bearded" appearance.

26.5.3. Host range

Compared to other cysts nematode, carrot cyst nematode has very limited host range. The nematode is mainly restricted to cultivated and wild carrots and other *Daucus* spp. However, other plant such as spreading hedge parsley (*Torilis arvensis*), spreading hedge parsley, bristle fruit hedge parsley and erect hedge parsley are found in Europe, Africa, Asia and North America (GRIN, 2008). Bristle fruit hedge parsley (*Torilis leptophylla*) and erect hedge parsley (*T. japonica*) may act as reservoir for the cysts survival and the nematode may partially complete its development in these plants (Vallotton, 1980).

26.5.4. Spread

This cyst nematode is dispersed with soil debris, plant material and machinery contaminated by the cysts.

26.5.5. Management

26.5.5.1. Physical methods: The efficacy of scil solarization increased with the length of mulching period. Only 17.5% of *H. carotae* eggs survived in the eight-week solarized plots, resulting in suppression of carrot feeder root invasion by the nematode and greater yield increase than with aldicarb and SIP 5561 (Table 26.4) (Greco and Grandoisio, 1990).

Table 26.4. Effect of soil solarization on the survival of eggs in the top 30 cm soil, infestation of carrot feeder roots by *Heterodera carotae* and on the yield of marketable carrots.

Treatment	% surviving eggs	Nematodes/3g feeder roots	Yield (kg/m²)
Polyethylene mulching for 4 weeks	67.1b	808b	3.8abcd
Polyethylene mulching for 6 weeks	40.7c	406bc	4.4ab
Polyethylene mulching for 8 weeks	17.5cd	209c	4.6a
SIP 5561 10 kg a.i. /ha	—	396c	4.0abc
SIP 5561 5 + 5 kg a.i. /ha	—	122c	3.5abcd
Aldicarb 10 kg a.i./ha	—	1215a	3.0de
Aldicarb 5 +5 kg a.i. /ha	—	1484a	3.1cde
Untreated control	100.0a	1278a	2.6e

Figures in each column followed by the same letter are not significantly different for $P \leq 0.05$ according to Duncan multiple range test.

Better control was achieved in Italy, where only 24-38% of eggs survived in cysts of *H. carotae* from the top 30 cm soil solarized for four to eight weeks (Greco *et al.*, 1985). In another trial, the yield of carrots from plots solarized for eight weeks was 4.6 kg/m^2, compared with 5.5 kg in plots treated with DD at 400 1/ha and only 2.3 kg in the control.

26.5.5.2. Cultural methods: In a field infested by the carrot cyst nematode *H. carotae*, biofumigation with *B. juncea* hay or *Brassica* seed meal pellet was compared with commercial liquid nematicidal formulations of quillay, tagetes or neem, and with the chemical fenamiphos and an untreated control. At the end of the first year, soil populations of *H. carotae* were significantly suppressed by all the treatments, though neem and fenamiphos provided the lowest nematode populations and the highest carrot yield. At the end of carrot cycle in second year, *H. carotae* population did not increase in plots treated with hay or fenamiphos, whereas pellet and quillay resulted in significantly higher, though lower than untreated control, densities. Pellet and fenamiphos provided the highest total yield and percentage of commercial tap roots (De Mastro *et al.*, 2008d).

26.6. CONCLUSION

Root knot nematodes are generally considered the most important nematode parasites of carrot worldwide, being geographically widespread and capable of complete crop loss. Management of nematodes is heavily reliant upon the use of nematicides. With potential loss in availability of nematicides there is a need to move back to a cropping system approach to nematode management. Alternative strategies will be required for nematode control in the future. An integrated strategy for control of nematodes is advocated involving pre-plant nematode counts to determine the need for nematicides, manipulation of planting date, and suitable crop rotations or break crops to reduce nematode populations in soil prior to carrot.

Chapter - 27

BEET ROOT, *Beta vulgaris* sub sp. *vulgaris*

The root-knot and cyst nematodes are the major limiting factors in the production of beet root crop.

27.1. ROOT-KNOT NEMATODES, *Meloidogyne* spp.

27.1.1. Symptoms

Beets with root-knot nematode can exhibit a variety of symptoms like stunting of plants, yellowing of foliage and withering despite the presence of adequate water. Root-knot nematodes cause galls and swellings on the main root and secondary roots that are present, which may have a warty appearance. Heavy infestation by root-knot nematodes in sandy soils may cause plants to wilt and collapse.

27.1.2. Interaction with other pathogens

The root-knot nematode, *Meloidogyne incognita* in association with *Fusarium solani* is reported to cause greater losses than either pathogen alone (Francl and Wheeler, 1993). Infection of roots by the root-knot nematode predisposes crop plants to infection by soil-borne root-infecting fungi, resulting in the development of root-rot disease.

27.1.3. Management

27.1.3.1. Cultural methods: Since beets will mature in temperatures around 13 ^{0}C, early planting means they can escape most of the root-knot nematode's life cycle. Young beet root seedlings are more susceptible than healthy, maturing beets which better tolerate root-knot nematode attack.

Maintain a 2- to 3-year crop rotation with non-hosts like fescue, marigold or small grains which can slowly decrease the number of nematodes present in the soil.

Incorporating organic matter into the soil can help to decrease nematode activity by enhancing the activity of nematode-destroying bacteria.

27.1.3.2. Biological methods: MeloCon WG (*Purpureocellium lilacinum*) at 6.725 to 10 kg/ha as an in-furrow application at planting is recommended for root-knot nematode suppression.

Soil treatments with Bio-Nematon (*P. lilacinum*) and Priority (*P. fumosoroseus*), when applied in the field, were promising in suppressing the incidence of root-knot (*M. incognita*) and root-rot (*Fusarium solani*) disease complex on beet, increasing soil microflora, and improving growth, yield (El-Nagdi *et al.*, 2011).

27.1.3.3. Host resistance: Beet root cv. 'Detroit Dark Red' was tolerant to *M. incognita*, whereas cv. 'Crimson Globe' was resistant to *M. javanica*.

Three beet cvs. CG, DDR and Red Ace were moderately resistant, with the minimum number of galls (4.0, 8.0 and 7.0, respectively). The moderately resistant cultivars can be recommended for cultivation as an eco-friendly approach for the disease management programs in nematode infested fields (Khan *et al.*, 2022).

27.2. CYST NEMATODE, *Heterodera schachtii*

27.2.1. Symptoms

Localized infestations may result in well-defined circular or oval areas where plant stands and growth are poor. Seedlings infested by sugar beet cyst nematode may have longer petioles than normal, with green or yellow leaves depending on the severity of infestation. Plants are likely to be stunted and wilted. Typically, storage roots will not be well developed, and will have excessive fibrous roots. Mature female nematodes can be seen on the root surface as tiny, pinhead size, lemon-shaped bodies that are white in the earlier stages and turn into brown, egg-filled cysts on aging.

27.2.2. Host range

The main host crops of *H. schachtii* are in the families Chenopodiaceae (e.g. garden beets, Swiss chard, spinach, mangold) and Brassicaceae (e.g. broccoli, radish, Brussels sprouts, rapini, cauliflower, kale, bok choy, mustard, canola, kohlrabi, and others). Several common weeds are also hosts such as common lambs' quarters, shepherd's purse, pigweed, chickweed, dock, and others.

27.2.3. Life cycle

Eggs and juveniles remain dormant and survive inside the cyst, which is the body of the dead female. When the root of a host plant contacts or grows near the cyst, and soil moisture is sufficient, root exudates stimulate juveniles to hatch and emerge from the cyst. Juveniles are attracted and migrate to the fibrous roots, infecting near the root tips. After entering the root, they migrate a short distance within the root cortex and begin developing into adults that become sedentary. After three additional molts, adult males emerge from the root and enter the soil. After a fourth molt. the females become lemon-shaped and can be seen as small white-yellow dots attached to fibrous roots. Males are attracted to the females where fertilization occurs. An average of 200 eggs are produced by one female, a few of which are laid outside the body in the soil. However, most of the eggs remain inside the female. At maturity, the female dies, and her body wall hardens and is transformed into a light brown to reddish brown cyst, completing the cycle. The cycle requires four to six weeks, depending on soil temperature.

27.2.4. Survival and spread

Cysts, containing eggs and/or hatched juveniles, may remain viable in irrigated fields for several years. The annual rate of decline of cysts in soil during rotation is estimated at 40-60 percent. Cysts may survive even longer in fallowed soil.

Long-distance spread has most likely been from cysts in soil. Other means of spreading soil such as on machinery or animal hooves, can also result in the spread of *H. schachtii.* Short distance spread occurs through irrigation water. Other means of spreading include wind-blown soil and cysts.

27.2.5. Management

27.2.5.1. Cultural methods

i) Sanitation: Thoroughly clean machinery and equipment with water between fields to mitigate the risk of spreading plant-parasitic nematodes to non-infested fields. This is especially important for *H. schachtii* as eggs in cysts are well protected against adverse environmental conditions.

ii) Crop rotation: Non-host crops that can be included in rotation include wheat, barley, corn, beans, and alfalfa. Non-infested fields cannot be cropped to sugar beets more than two years in a row and not more than four out of ten years. In infested fields, sugar beets can be grown only once every four years.

iii) Planting dates: Planting when soil temperatures are below 50 °F is recommended for the management of *H. schachtii.*

iv) Trap cropping: Oil seed radish is currently very effective as a beet cyst nematode trap crop. By planting this cover crop, the nematodes are tricked into hatching and then are unable to reproduce on the roots of the radish. This has been effective in lowering populations of beet cyst nematode by 80 to 90 percent and improving yields.

27.2.5.2. Biological methods

i) Antagonistic fungi: *Dactylella oviparasitica* suppressed beet cyst nematode. This fungal hyperparasite attacks developing juveniles, females, and eggs of the nematodes, thereby diminishing their population development. Introduction of this fungus into other beet cyst nematode-infested soils caused similar population declines.

27.2.5.3. Integrated methods: A combination of disposal or proper handling of tare soil, rotation with non-host crops, good weed control, and the possible fall planting of a trap crop will all reduce the soil population of *H. schachtii.*

27.3. DOCKING DISORDER, *Longidorus* spp., *Trichodorus* spp.

Docking disorder of beet was named after a village in the north-west of Norfolk. Although recognized and documented for many years it was not until the early 1960`s that free-living nematodes were identified as a prime causal agent of this condition.

The main free living nematodes species responsible for causing damage to beet are the stubby root nematodes (*Trichodorus* spp. *Paratrichodorus* spp.) and the needle nematodes (*Longidorus* spp.).

Longidorus leptocephalus, L. attenuatus, L. elongatus, Trichodorus spp. and viruses are involved in Docking disorder of beet (named for Docking region in south of England). Symptoms include stunted growth in spring due to nematodes and virus symptoms on foliage. Effects are most pronounced in spring, heavy rainfall in May seems to increase the problem. In July, the affected plants start to grow again and may achieve almost normal foliage, but a much-reduced tap root. The severity of Docking disorder varies from year to year with the climate.

27.3.1. Crop losses

It has been estimated that yield losses of up to 17.5 tons/ha can occur.

27.3.2. Symptoms

Hatched larvae are attracted to the emerging crop plants to feed on the roots tips by inserting their piercing mouth parts (hollow stylets) into cells of the root and the root hairs. Enzymes are injected into these cells and the resulting substrate is withdrawn into the gut of the nematode to provide nourishment.

Feeding on the tap root, at an early stage of plant development, can result in its growing tip being killed which encourages lateral roots to take over. The net result is short stubby stunted plants with numerous laterally growing side roots which gives rise to the condition described as `fanging` of roots and ultimately yield loss.

Affected plants can be stunted resulting in so called `hen and chick symptoms` where a large healthy plant has a small plant as its neighbor.

27.3.3. Life Cycle

The nematodes lay eggs near host plants and after a few days the larvae wriggle out in search of food. Several generations occur each season and populations comprise a mixture of both adults and juveniles. *Longidorus* spp. can live in the soil for over a year.

27.3.4. Ecology

Crops grown on light sandy soils in wet, cold conditions are most at risk.

27.3.5. Threshold

The current threshold is 200 nematodes per litre of soil but with late drilled crops when soils are warm the threshold for *Trichodorus* spp. can probably be safely doubled to 400 nematodes per litre of soil. However, *Longidorus* nematodes are larger than most species and are consequently damaging at relatively low populations (50 per liter of soil).

27.3.6. Management

Management by rotation is difficult as several nematode species are involved, each with a differing host range. Also host ranges of these nematodes are not completely known.

Beet growers can now use a garlic-based product to manage free-living nematodes, thereby reducing yield losses resulting from the soil pest.

NEMguard DE worked as well as Vydate for the management of beet docking disorder. NEMguard DE is a granule formulation containing 450 g/kg garlic extract which is applied at drilling via a granule applicator. NEMguard DE at 10kg/ha gave mean root fanging score 0.6 as compared to 3 in check.

27.4. CONCLUSION

Root-knot nematodes are generally considered the most important nematode parasites of beet root worldwide, being geographically widespread and capable of complete crop loss. Management of nematodes is heavily reliant upon the use of nematicides. With potential loss in availability of nematicides there is a need to move back to a cropping system approach to nematode management. Alternative strategies will be required for nematode control in the future. An integrated strategy for control of nematodes is advocated involving pre-plant nematode counts to determine the need for nematicides, manipulation of planting date, and suitable crop rotations or break crops to reduce nematode populations in soil prior to beet root.

Chapter - 28

PARSNIPS, *Pastinaca sativa*

28.1. INTRODUCTION

The most encountered nematode species include the root-knot (*Meloidogyne arenaria, M. hapla, M. incognita*, and *M. javanica*), stubby-root (*Paratrichodorus* and *Trichodorus* spp.), needle (*Longidorus* spp.), root-lesion (*Pratylenchus* spp.), and stunt (*Tylenchorhynchus dubius*) nematodes that are most likely to damage parsnips.

The most prevalent species in the UK recorded as damaging carrots and parsnips are needle nematodes (*Longidorus* spp.), and stubby-root nematodes (*Paratrichodorus* and *Trichodorus* spp.). The most frequently occurring species were the needle nematode *Longidorus elongatus*, spiral nematode *Rotylenchus goodeyi* and stunt nematode *Tylenchorhynchus dubius* (Boag, 1979).

28.2. STUBBY-ROOT NEMATODES, *Trichodorus viruliferus*

Trichodorus viruliferus, for example, is known to cause root browning, stunting, and occasional swelling of the root tips. Stubby-root nematodes have short stylets which can only pierce and extract sap from the outer layer of root cells. They are attracted by actively growing roots (as are most plant parasitic nematodes) and congregate behind the growing tips where their feeding causes browning and cracking. Feeding induces a proliferation of short, stunted, stubby, lateral, and coarse rootlets and excessive development of upper roots. All stubby-root nematodes are capable of damaging roots by feeding on the root tips and by acting as virus vectors (Decraemer, 1995). Severely attacked seedling root systems cease to grow, and the rootlets become stubby-ended, turning grey-brown and later black as they die and decay. New roots are put out near the soil surface and may escape heavy attack, possibly because the soil is drier there and less favorable for nematode movement. If the main rootlet is damaged, other rootlets replace its function and grow horizontally or diagonally but rarely straight down, thickening and so producing a "fanged" root.

Gratwick (1992), in a discussion of Docking disorder, states that parsnips are also damaged by needle and stubby-root nematodes, and that the symptoms are like those described above.

Cool, wet weather encourages the pathogenicity of stubby-root nematodes (Whitehead *et al.*, 1971).

28.3. NEEDLE NEMATODES, *Longidorus* spp.

Needle nematodes individually are larger and more injurious, and fewer are needed per seedling to retard growth. Feeding by these nematodes results in normal colored, if abnormally long tap roots, and many short lateral roots with darkened and swollen tips, often grouped in tufts. Occasionally a lateral root that has grown normally may thicken and extend abnormally in a horizontal direction just below the surface.

Needle nematodes have long, slender stylets which they insert into the growing points of root tips. Few of the lateral rootlets of attacked plants develop properly, and most have dead or dying tips with swellings immediately behind the tip.

Needle nematodes are numerous in sandy peats which have a high organic matter content. Sandy soil provides the most suitable environment for the nematodes.

Needle nematodes are the largest nematodes that are likely to attack parsnips. They are ectoparasites that feed principally on root tips, which are usually galled as a result. Several species have life cycles that last more than one year, and the adults may live for more than one year.

Plowing and cultivation may affect the abundance of needle nematodes in the top soil.

28.4. ROOT-KNOT NEMATODES, *Meloidogyne* spp.

Detection of damage by root-knot nematodes recognized by the formation of characteristic root galls. The symptoms include deformed growth with galls on the main tap root, and root proliferation from infested areas, with these roots often with galls themselves. The smaller discrete galls with adventitious root proliferation formed by the northern root-knot nematode are distinctive.

28.5. ROOT-LESION NEMATODES, *Pratylenchus* spp.

Identification of root-lesion nematode damage requires careful inspection of root symptoms under magnification. Some species do not cause the typical scratch-like lesions on the roots, because cell damage does not spread or result in color changes associated with polyphenol deposition. They are migratory endoparasites that feed in the root cortex, migrating through root tissue, and killing cells as they feed and deposit eggs.

28.6. STUNT NEMATODES, *Tylenchorhynchus dubius*

Tylenchorhynchus dubius usually feeds on epidermal cells in the vicinity of root hairs and the cell elongation region. These species are migratory ecto-, semi-ecto- or even endoparasites on young root tips.

28.7. Management

Much more is now known about non-chemical and cultural control, including use of rotation, leaving land fallow and delayed drilling. Although current guidelines recommend nematicide use where numbers of stubby-root nematodes exceed 200/L soil, it is suggested that 1,000/L soil may be more appropriate. *Tagetes patula* appears to decrease the number of root-lesion nematodes but increases numbers of stubby-root nematodes.

28.7.1. Cultural methods

28.7.1.1. Precision farming: Precision farming is now being investigated as a means of reducing the quantity of nematicide applied. Technology exists to enable the mapping of soil types (e.g., the Magna scan) which might be used to indicate the most likely sites of nematode activity (areas of lighter soils). Control measures could then be targeted on specific areas of the field.

28.7.1.2. Fallowing: Fallowing improves soil fertility and may encourage biodiversity and thus may encourage the multiplication of nematode parasites and predators. This strategy is being adopted in Norway, where heavy infestations of the northern root-knot nematode in carrot crops have led to recommendations to leave one year of fallow in the rotation as the only viable option. Fallowing is also recommended in The Netherlands where root-knot nematodes are a significant problem. For it to be effective, the land must be kept scrupulously free of weeds (Hockland, 2000).

28.7.1.3. Crop rotation: Rotation with grass or cereals can reduce populations of root-knot and root-lesion nematodes (Diamond *et al.*, 1991). Field trials in Oregon, USA which investigated the increasing problem of stubby-root nematodes in potatoes (Ingham et al., 2000), concluded that the crop was not a good host. Nematode populations declined during each season in which potatoes were planted. Rotation with potato can reduce stubby nematode population (Ingham *et al.*, 2000). The numbers of root lesion nematodes increased less rapidly under crops such as sugar beet, spring barley, ryegrass, chicory, and broccoli than under winter rye, white clover, lucerne, and maize.

A number of plants belonging to the Gramineae and Compositae are non-hosts of root-knot nematode, and can dramatically lessen soil infestations (Whitehead, 1998). Plants in the Compositae have been shown to be suppressive to root-lesion nematodes and are being considered by many researchers (Potter and Olthof, 1993). In continental Europe, an infestation of the root-lesion nematode *Pratylenchus penetrans* was reduced 80-90% by growing *T. patula* for three to four months (Oostenbrink *et al.*, 1957), and later work with this same nematode species repeated this success (Caubel *et al.*, 1978, 1981).

28.7.1.4. Plowing: Kuiper (1977) reported that over 80% of populations of *Paratrichodorus teres* may be killed after two rotavations. Plowing soil late in the autumn to expose nematodes to freezing has been effective in reducing populations of root-knot nematodes in Ontario, Canada (Potter and Olthof, 1993).

28.7.1.5. Organic amendments: Additives such as sewage sludge have produced a reduction of *M. incognita* (CastagnoneSereno and Kermarrec, 1991) probably because of stimulating resident soil antagonists. Organic matter may also improve the soil structure and condition, and lead to the introduction of beneficial organisms (such material contains hundreds of thousands of microbivorous nematodes which in turn attract parasites and predators).

Neem cake has also been used recently by organic growers in Israel to control nematodes.

28.7.1.6. Green manuring: Green manures of *Tagetes minuta* cv. Nemanon, *Hesperis matronalis*, rye or *Lolium multiflorum* sown in early May in "set-aside" land decreased northern root-knot nematode by 80-100% in trials in The Netherlands (Lamers and Roosjen, 1993). Green manures of Sudangrass were found effective in suppressing the populations of the northern root-knot nematode and its damage in New York State (Abawi *et al.*, 1999).

28.7.2. Chemical methods

Few carrot growers have experience of using carbosulfan. It is generally considered that aldicarb is more effective against nematodes than carbosulfan.

28.7.3. Biological methods

There are several projects investigating control of root-knot nematodes by fungi such as *Pochonia chlamydosporia* (Rothamsted Research; Agricultural Research Centre, Belgium). The new formulation, BioSafe (*Bacillus firmus*), is recommended for the control of root-knot nematodes in organic crops.

Chapter - 29

TURNIP, *Brassica rapa*

29.1. INTRODUCTION

Turnip (*Brassica rapa*) is grown in temperate, sub-tropical and tropical regions of India. It is extensively cultivated in Bihar, Haryana, Himachal Pradesh, Punjab, and Tamil Nadu. Turnip is rich in vitamins A, B, C and Calcium and iron. It can be eaten raw.

Root-knot and reniform nematodes are recognized as the major limiting factors in successful production of turnip crop.

29.2. ROOT-KNOT NEMATODES, *Meloidogyne arenaria, M. incognita, M. javanica*

29.2.1. Symptoms

The root-knot nematodes build up in the soil to the point where the soil becomes "sick." Plants growing in such soil show signs of nutrient deficiency because of serious root damage by way of extensive root galling.

29.2.2. Management

29.2.2.1. Cultural methods: The oil cakes (neem, groundnut, mustard and castor) were effective in reducing parasitic nematode population (*Hoplolaimus, Tylenchorhynchus, Meloidogyne* and *Helicotylenchus* species) and in increasing yields of turnip.

29.2.2.2. Chemical methods: Preplant applications of phenamiphos protected roots of turnips from damage by root-knot nematodes, *Meloidogyne incognita*. Concentrations of phenamiphos at application in the 0-15-cm soil layer were near 6 microg/g on turnips. After 30 days, concentrations were approximately 1 microg phenamiphos/g of soil (Johnson *et al.*, 1980).

29.2.2.3. Biological methods: *Trichoderma harzianum* induced significant resistance against pathogen, root knot nematode (RKN) and growth rate were comparatively high for turnip (*Brassica rapa*). Findings illustrated that *Brassica rapa* grown under the treatment of *Trichoderma harzianum* as compared to *Azadirachta indica*, produce best result in case of resistance level and in development also (Table 29.1) (Ibrahim *et al.*, 2012).

Table 29.1. Effect of *Trichoderma harzianum* and *Azadirachta indica* on the growth of turnip and resistance against Meloidogyne incognita

Treatment/Conc.	Plant height (cm)		Shoot dry weight (g)		Root dry weight (g)	
	T. harzianum	*A. indica*	*T. harzianum*	*A. indica*	*T. harzianum*	*A. indica*
Control	7.76± 0.02c	5.77±0.02d	1.02±0.06g	0.92±0.09f	0.19±0.4g	0.03±0.02g
T. harzianum - 15 ml *A. indica* - 60%	16.67±1.61a	17.07±1.31a	2.19±0.12f	1.05±1.23f	0.162±0.8g	0.07±1.32g
T. harzianum - 10 ml *A. indica* - 40%	14.83±1.35b	17.7±0.96a	1.62±0.79f	1.34±0.96f	0.18±0.4g	0.061±1.21g
T. harzianum - 5 ml *A. indica* - 20%	15.13±1.06b	13.93±1.12b	1.03±0.36f	0.87±1.08f	0.34±1.03g	0.04±1.22g

29.3. RENIFORM NEMATODE, *Rotylenchulus reniformis*

29.3.1. Symptoms

Above ground symptoms on host plants include dwarfing, shedding of leaves, formation of malformed fruit and seeds, and general symptoms of an impaired root system. Below ground symptoms include discolored and necrotic roots (dead) with areas of decay. Plant mortality is possible in heavy infestations.

29.3.2. Management

29.3.1.1. Host resistance: Turnip cvs. Purple Top and White Globe are reported to be resistant to reniform nematode.

Chapter - 30

LETTUCE, *Lactuca sativa*

Root-knot nematodes are recognized as the major limiting factors in successful cultivation of lettuce crop.

30.1. ROOT-KNOT NEMATODES, *Meloidogyne* spp.

The root-knot nematode cause large decreases in yield of lettuce.

30.1.1. Symptoms

Root knot nematodes feed within the roots and cause characteristic galls on roots. The northern root-knot nematode, *M. hapla* generally occurs in cooler regions than the other three *Meloidogyne* species that prefer hot summer climates. Galls formed by *M. hapla* are spherical, distinct, and generally smaller than those caused by the three "warm climate" species. Other symptoms include patchy appearance of infected fields during middle of the crop season, stunted growth of seedlings, and drastic reduction in lettuce yields.

30.1.2. Management

30.1.2.1. Physical methods: Steaming of greenhouse soil is used for the management of root-knot nematodes in lettuce crop.

30.1.2.2. Cultural methods: Plow under infested plants after harvest to prevent further reproduction of nematodes.

Provide adequate fertilizer and irrigation for crop development to reduce the impact of nematode feeding. Sprinkler irrigation tends to be more effective than furrow irrigation for supplying even water to nematode-damaged plants.

Thoroughly clean soil from all equipment with water to prevent the spread of infestation. Plant transplants free from nematodes. Do not allow irrigation water to flow from an infested field to other fields without impounding. Prevent animal grazing and movement from infested to non-infested fields.

i) Crop rotation: Crop rotation is not very effective against root knot nematodes due to their wide host range. Strawberries may be a suitable rotation crop in fields in which the species of root-knot nematode has been identified. They are non-hosts to *M. incognita* and most populations of *M. javanica* but allow reproduction of *M. hapla*. Some varieties of cereals, grasses, maize, and sorghum are non-hosts or poor hosts for RKN and can be grown prior to the crop to reduce nematode numbers.

ii) Intercropping: Planting of 4 and 5 seedlings of *Tagetes erecta* in a pot around a lettuce plant reduced root galling by 56 and 72% and nematode population by 73 and 94%, respectively (Pervez *et al.*, 1988).

iii) Cover cropping: Using cover crops, like oats, wheat, or rye grass, which do not host nematodes may also help. Incorporation of green manure crops such as Sudan grass, sesame, rapeseed, white mustard, or perennial ryegrass the season before a lettuce is planted can reduce nematode populations.

iv) Biofumigation: The influence of organic amendments applied as green manures and composts were tested against root-knot nematode *M. incognita* in resistant and susceptible lettuce cultivars. Chopped leaves, stems and roots of six plants namely, broccoli and cauliflower (*Brassica oleracea* var. *botrytis*), cabbage (*B. oleracea* var. *capitata*), mustard (*B. juncea*), radish (*Raphanus sativus*), and *Tithonia diversifolia*, including processed chicken manure were incorporated at 20g per kg of soil previously inoculated with 10,000 second stage juveniles of root-knot nematode. A treatment using fallow for 3 weeks was included as control. Significant increase ($P \leq 0.05$) in plant height and top weight were recorded in the two lettuce cultivars incorporated with organic amendments. Highest top weight and plant height were noted in broccoli- and cabbage-amended soil. Root weight was highest in broccoli- and cauliflower- amended soil. This was significantly different from the other treatments. A significantly lower number of galls was recorded in resistant cultivar Great Lakes as compared to that of susceptible cultivar Tyrol. The highest number of galls was observed in fallow and processed chicken manure which was significantly different from the rest of the treatments. No egg masses were produced in cultivar Great lakes while in cultivar Tyrol, fallow gave the highest number of egg masses significantly different from the amended soil. Significant reductions in second stage juveniles in the soil were found in mustard and radish amendment but comparable with the other treatments (Pedroche *et al.*, 2008).

v) Soil biodisinfestation: Soil biodisinfestation is the process generated after the incorporation of organic amendments followed by a plastic cover to control root-knot nematodes in lettuce. In the first crop after biodisinfestation, lettuce yield increased in the treated plots and root galling indices were significantly lower. One year after treatment, differences between treatments could be observed in the incidence of the damage caused by *M. incognita* that remained lower in the treated plots. The addition of beer bagasse and rapeseed cake along with fresh manure in biodisinfestation treatment demonstrated nematicidal effects against *M. incognita*. Moreover, the compounds released during the degradation of these by-products and the sub-lethal temperatures achieved in this trial during biodisinfestation (< 42 °C) were the key to develop suppressive soils in the long-term (Gandariasbeitia *et al.*, 2021).

30.1.2.3. Chemical methods: The soil application of Carbofuran at 1 kg a.i. /ha at 2-4 lbs. was found effective.

30.1.2.4. Biological methods

i) Antagonistic bacteria: The number of root galls was significantly decreased in *Pasteuria penetrans* treated pots with 75 (77.2%) and 150/root (54.4%) in 200,000 and 100,000 *P. penetrans* endospores/1 g medium, respectively, compared to the 330/root in inoculated control when harvested at 10 weeks after planting (Table 30.1).

Table 30.1. Suppression effect of *Pasteuria penetrans* KW1 (PP) on reproduction of *Meloidogyne incognita* in lettuce

Treatments	No. of root galls		Wt. of leaves harvested (g)	No. of total leaves harvested	Root wt. (g/plant)
	No. of galls/ root	Control effect (%)			
MI J2 5,000/plant (N)	330 c	---	54.6 a	40.9 a	11.9 ab
N + PP 100,000 endospores/g medium	150 b	54.4	55.5 a	38.4 a	11.6 b
N + PP 200,000 endospores/g medium	75 a	77.2	65.2 a	40.6 a	11.6 b
Uninoculated control	0 a	---	57.8 a	39.5 a	14.2 a

Data are means of 6 replicates.

Root galls, leaves weight, root weight, and number of total leaves were examined 10 weeks after planting.

Means within a column followed by the same letter are not different according to Duncan's multiple-range test ($P<0.05$).

The bioagents *Bacillus thuringiensis* and *Streptomyces costaricus* reduced egg numbers of *M. hapla* on lettuce as compared to control (Chen *et al.*, 2000).

ii) The fungi *Hirsutella rhossiliensis* and *Pochonia chlamydosporia* applied against *M. hapla* on lettuce were effective in reducing second stage juveniles in seedlings when applied alone or in combination (Viaene and Abawi, 2000).

Effective reduction of number of eggs of *M. hapla* on lettuce was obtained by soil incorporation of biological control agents like *Streptomyces costaricus* and *Bacillus thuringiensis* (Chen *et al.*, 2000). Soil application of MeloCon (*Purpureocellium lilacinum* strain 251) at 2-4 lbs. was found effective.

30.1.2.5. Host resistance

The cultivars Veronica, Grand Rapids and Crespa para Verão are resistant to the three nematode species. The cultivars Thaís, SRV 2005 and Marisa are resistant to *M. incognita* race 3 and *M. javanica*. The cultivar Black Seed Simpson is resistant to *M. enterolobii*. The cultivars Vanda and Mônica SF 31 are resistant to *M. incognita* race 3. The cultivars Crespa, Rubia, Cinderela and Veneranda are resistant to *M. javanica*.

Lettuce cv. 'Salinas 88', a USDA crisp-head lettuce cultivar, presents resistance to *M. incognita* as well as to *M. javanica* (Maluf *et al.*, 2003; Wilcken *et al.*, 2005).

Based on the host response, lettuce cultivars Great Lakes and Gilaben were rated highly resistant and resistant to *M. incognita*, respectively (Pedroche *et al.*, 2007).

30.1.2.6. Integrated methods

Biofumigation with mustard under black polythene for 30 days followed by soil application of *Pseudomonas fluorescens* at 2.5 kg/ha at the time of planting was effective in enhancing the lettuce leaf yield by 23.29% and decreasing the population of root-knot nematodes by 76.39% which was on par with Carbofuran at 1 kg a.i./ha.

In areas with hot summers, integration of soil solarization along with chicken manure was found most effective against root-knot nematode, while solarization alone gave partial control. Solarization can also reduce populations of soil-borne fungi pathogens and weed seed.

30.2. CONCLUSION

The root knot nematodes, *Meloidogyne* spp. are the most important pests of lettuce. In view of the short life span and the systemic nature of the available nematicides, field sanitation and other cultural control methods are given preference over chemicals. Identification of resistant indigenous lettuce varieties would contribute greatly to the management of the pests in the cropping systems. The identified resistant varieties can be used as rotation crops in agricultural production systems as a component of root-knot nematode suppression in the soil. Amaranth, sun hemp, and *Cleome* spp. (spider plant) were poor hosts with some level of resistance resulting in reduced incidence and prevalence of root-knot nematodes and could be cultivated in areas where the pathogenic nematodes are endemic. The farmers are advised to grow *Crotalaria* spp. in rotational and mixed cropping systems to reduce nematode populations for subsequent and companion susceptible crops.

Chapter - 31

CELERY, *Apium graveolens*

Root-knot nematodes are recognized as the major limiting factors in successful cultivation of celery crop.

31.1. ROOT-KNOT NEMATODES, *Meloidogyne hapla, M. incognita, M. javanica*

31.1.1. Symptoms

Celery fields infected by the nematode showed severe decline and heavily damaged and distorted roots. Celery infected with root-knot nematodes cause stunting, uneven stand of plants, and characteristic galls (numerous knobby growths) on roots. Roots showed numerous large galls that usually contained one or more *Meloidogyne* females, males and eggs. Small feeder roots will have been destroyed, with irregular galls growing in their place. Even in the presence of enough moisture in soil, the nematode - infected plants are likely to wilt earlier as compared to healthy plants. Plants begin to wilt, when nematodes interfere with moisture uptake. New growth comes to a standstill. Since the deranged roots are not in a position to absorb enough water and nutrients from the soil, the plant leaves become chlorotic. Under severe nematode infestation during extended warm growing season the plant stand is reduced.

31.1.2. Host range

Apart from celery, common vegetable hosts are beans, capsicum, carrot, cucumber, eggplant, ginger, lettuce, potato, tomato, and yam; common fruit crop hosts are melon, papaya, and pineapple, and many ornamentals are also hosts.

31.1.3. Life cycle

After hatching from eggs, the young - called juveniles - invade roots near the root tip. Most are females as males are rare. Once inside the roots, the tissues enlarge, galls form, eggs are laid on the root surface, and they hatch to start the life cycle once more.

31.1.4. Histopathology

Comparative histological observation of healthy and *M. incognita*-infected celery roots showed cellular alterations in tissues of the cortex, endodermis and pericycle, but the greater damage was caused in the vascular system by the nematode-induced cellular expansion of nurse cells and by the expanding bodies of the nematode female. In the permanent feeding sites, nematode-induced formation of large multinucleate giant cells (three to six giant cells), adjacent to the vascular tissues, was observed in all infected celery galls. This formation led to distortion and crushing of tissues. Giant cells showed dense cytoplasm and variable numbers of hypertrophied nuclei and nucleoli. Hyperplasia of tissues adjacent to giant cells contributed to the formation of root galls (Vovlas *et al.*, 2008).

31.1.5. Spread

Spread over short distance occurs as the juveniles move in soil water in search of new roots to infect; over larger distances, spread occurs in the roots of seedlings, young plants or vegetative propagating materials. It is also possible for nematodes to spread in rain water, and soil on tools, machinery or foot wear.

31.1.6. Management

31.1.6.1. Physical methods: Soil solarization shall kill root-knot nematodes (plus other pathogens and weeds) in the top 30 cm of (especially sandy) soil by using transparent polyethylene covers. Moisten soil and cover for 4-6 weeks.

31.1.6.2. Cultural methods: The most important cultural control methods employed for root-knot nematode are crop rotation with a non-preferred host plant or natural soil fumigant (i.e., cole crops).

In areas where cover crops can be utilized between growth periods, hairy indigo may reduce populations as well (Univ Florida, 1999).

Keep soil bare for 4-6 months, but with frequent cultivation; make sure that weeds are controlled. The method exposes the nematodes to drying by the sun, and to starvation.

Grow a solid planting of French marigolds for several months.

Add manure or compost to the soil, especially to sandy soils, to increase ability of the soil to hold water and nutrients.

Aim for optimum plant growth by frequent, deep watering, and adding mulches and organic matter.

Rotating with non-host crops can reduce nematode population levels. Some grasses have been recommended e.g., sorghum x Sudan grass hybrids (cv. Jumbo), and Green panic (a type of Guinea grass)]. Maize and peanuts are poor hosts of *Meloidogyne incognita*, and onions, cabbages and cauliflowers have tolerance.

Farm equipment and machinery operated in infested fields to be thoroughly cleaned before moving those to non-infested fields.Celery grown in flooded soil has lower nematode damage.

31.1.6.3. Chemical methods: All Nemitz applications to celery must be incorporated either physically or via drip or overhead irrigation. Make preplant applications at a rate of 4 to 8 L/ha, a minimum of seven days before planting.

31.1.6.4. Biological methods: The treatment of endophytic bacteria reduce number gall per g celery roots was 19.67–31.00 and control (46.67); juvenile population per g root was 9.33-25.67 and control 50.67; and juvenile population per 10 g of soil was 11.33-17.00 and control 29.33 (Althaf *et al.*, 2022).

Aplication of endophyte bacteria on Celery seedling detected that the treatment of bacteria (AB12) had the lowest number of galls, the lowest number of nematode populations in gall compared to other isolates. This result indicates the AB12 isolate similar characterize from genus *Pseudomonas* spp. have good potential as a biological control agent for *Meloidogyne* spp. in celery plants (Afriyani *et al.*, 2019).

31.1.6.5. Integrated methods: Biofumigation with sulfur containing cruciferous vegetable waste at the rate of 1kg/5 kg soil in pots covered with 50-micron thick black LLDPE sheets was found to reduce significantly the root knot nematode, *M. hapla* infecting celery and enhance plant growth and yield. Among the various sources evaluated radish leaf residue was the most effective resulting in 60.6 % reduction in nematode population in soil and 41.9% increase in celery green leaf and stalk yield compared to untreated control (Table 31.1) (Anita, 2012).

Table 31.1. Effect of crucifer vegetable leaf residue on plant growth, root-knot disease and yield of celery

Treatment	Plant height (cm)	Green leaf yield/plant (g)	Number of females/g root	Root gall index
Cabbage leaf waste @1 kg/5kg soil	52.5b	339.8b	16.5b	3
Cauliflower leaf waste @1 kg/5kg soil	43.6d	321.8c	18.9c	3
Radish leaf waste @1 kg/5kg soil	54.9a	369.6a	12.6a	2
Chinese cabbage leaf waste @ 1 kg/5kg soil	45.5c	298.5d	29.3d	3
Control (Untreated)	39.6e	214.5e	39.5e	5
CD (P = 0.05)	1.5	3.52	1.4	—

31.2. CONCLUSION

The root-knot nematodes, *Meloidogyne* spp. are the most important pests of celery. In view of the short life span and the systemic nature of the available nematicides, field sanitation and other cultural control methods are given preference over chemicals. Identification of resistant indigenous celery varieties would contribute greatly to the management of the pests in the cropping systems. The identified resistant varieties can be used as rotation crops in agricultural production systems as a component of root-knot nematode suppression in the soil. Amaranth, sun hemp, and *Cleome* spp. (spider plant) were poor hosts with some level of resistance resulting in reduced incidence and prevalence of root-knot nematodes and could be cultivated in areas where the pathogenic nematodes are endemic. The farmers are advised to grow *Crotalaria* spp. in rotational and mixed cropping systems to reduce nematode populations for subsequent and companion susceptible crops.

Chapter - 32

CORIANDER, *Coriandrum sativum*

Root-knot nematodes are the major limiting factors in the production of coriander crop.

32.1. ROOT-KNOT NEMATODS, *Meloidogyne* spp.

32.1.1. Symptoms

» Infested plants appear in patches in the field.

» Symptoms such as chlorosis and stunting appear after sufficient damage is inflicted.

» Formation of galls on host root system is the primary symptom. Galls on roots which can be up to 3.3 cm in diameter but are usually smaller

» Roots branch profusely starting from the gall tissue causing a 'beard root' symptom.

» Infected roots become knobby and knotty.

» In severely infested plants, the root system is reduced, and the rootlets are almost completely absent hampering their function of uptake and transport of water and nutrients.

» Reduction in plant vigor and yellowing of plants.

» Plants wilt during the hot part of day, especially under dry conditions and are often stunted.

» Nematode infestation predisposes plants to fungal and bacterial root pathogens.

32.1.2. Survival and spread

32.1.2.1. Primary: Egg masses in infested plant debris and soil or collateral and other hosts like Solanaceous, Malvaceous and Leguminous plants act as sources of inoculum.

32.1.2.2. Secondary: Autonomous second stage juveniles that may also be water dispersed.

32.1.3. Favorable conditions

» Nematodes prefer sandy loam light soils and damage in areas of field or garden with this type of soil is most likely.

» Soil temperatures of 77 to 82°F are the optimal range for the development of root-knot nematode species, they complete their life cycle within 3 to 4 weeks.

32.1.4. Management

- Soil application of neem cake.
- Solarizing soil can reduce nematode populations in the soil and levels of inoculum of many other soil-borne pathogens.
- Plant when soil temperatures are below 60 to 64 ^{0}F to minimize the early attack on seedlings.
- Plant resistant varieties if nematodes are known to be present in the soil.

Chapter - 33

AMARANTHUS, *Amaranthus* spp.

Root-knot nematodes, *Meloidogyne* spp. are the major limiting factors to produce amaranthus leafy vegetable crop.

33.1. ROOT-KNOT NEMATODES, *Meloidogyne* spp.

Kaspary *et al.* (2022) reported that root-knot nematodes, *M. incognita* parasitizing *Amaranthus hybridus* and *A. tuberculatus*; and *M. javanica* parasitizing *A. palmeri, A. hybridus* and *A. tuberculatus.*

33.1.1. Symptoms

Root-knot nematode infested plants were found to be associated with poor and unhealthy growth and diminish yields. The affected plants produce fewer and smaller leaves. Infected plants revealed more root-knots or galls in the plants, with corresponding decrease in flowers along with stunted growth. The occurrence of root-knot nematodes caused serious yield reduction in *Amaranthus hybridus* (Bawa *et al.*, 2016).

Heffes *et al.* (1991) reported root-knot nematode infested plants of *A. viridis* demonstrated reductions in shoot height, leaf area and shoot and root dry weight in comparison to control plants.

33.1.2. Management

33.1.2.1. Cultural methods: Twenty seedlings each of eight-week-old African marigold (*Tagetes erecta*) or Siam weed (*Chromolaena odorata*) incorporated into soil prior to planting has potentials to reduce root-knot nematode (*M. incognita*) population densities, increase soil fertility and enhance leaf yield of associated amaranthus leafy vegetable crop (Table 33.1) (Ogundele *et al.*, 2016).

Table 33.1. Effects of African marigold and siam weed as amendments on percentage change in soil population densities of *M. incognita* infecting *Amaranthus cruentus* under field conditions.

Treatment	% change in population		Green leaf yield (t/ha)	
	Early rainy season 2014	Late rainy season 2014	Early rainy season 2014	Late rainy season 2014
African marigold at 10 seedlings/plot	-45.96	-55.75	8.10	3.21
Siam weed at 10 seedlings/plot	-44.93	-55.90	8.26	3.53
African marigold at 20 seedlings/plot	-72.27	-66.05	10.99	3.71
Siam weed at 20 seedlings/plot	-72.35	-57.33	11.22	3.38
Control	+25.31	+9.20	6.13	1.78
CD (P = 0.05)	8.48	10.36	0.96	1.45

Various organic amendments like tea, *Camellia sinensis* residues and rejects effectively suppressed RKN infection followed by *Tithonia diversifolia* manure comparable to that of Mocap nematicide on amaranthus. Hence, their potential for adoption into RKN management programs as non-chemical measure for controlling RKN (Nchore, 2012).

Amaranth failed to develop root galls by nematodes when cultivated after a fallow crop of African marigold.

33.1.2.2. Biological methods: Soil application of arbuscular mycorhizal fungus, *Glomus fasciculatum* resulted in significantly lower number of galls on amaranthus. The population of the nematodes in soil and in the root system was also significantly reduced.

33.1.2.3. Host resistance: Amaranthus resistant genotype Thohoyandou significantly reduced population densities of both *M. incognita* and *M. javanica* under controlled (glasshouse) and prevailing environmental conditions in micro plots as well as on-farm fields compared to its susceptible counterpart Arusha. For *M. javanica,* only egg-mass numbers per root system were significantly lower for genotype Thohoyandou compared to those of Arusha. Use of resistant genotype Thohoyandou will benefit producers and can contribute to food security, household income and sustainable crop production (Ntidi *et al.*, 2019).

One of the ten species screened, Local 33 had an Rf-value ≤1, when screened against *M. incognita* race 2, indicating possible resistance to this population of *M. incognita* race 2 (Steyn *et al.*, 2013).

Amaranthus cruentus and *A. hypochondriacus* were found moderately resistant to root-knot nematodes with 1.59 and 0.90 reproduction factor; and with 1.0 and 0.5 root gall index, respectively (Nchore, 2012).

Chapter - 34

MUSHROOMS, *Agaricus bisporus*, *Volvariella volvacea*, *Pleurotus* spp.

34.1. INTRODUCTION

Mushrooms have achieved significant importance due to their nutritive and medicinal values. Mushrooms, the edible fleshy fungi, represent one of the few valuable sources of nutritive and palatable food. They are rich in proteins (35-48%), vitamins and minerals. They are weight reducers and useful for diabetic patients. Though, these fungi have been naturally occurring and being utilized in India since time immemorial, the commercial cultivation is only of recent times. It was only in 1964 that the first successful crop of mushroom was grown at Solan in Himachal Pradesh. Since then, research on various aspects of mushroom cultivation has been strengthened and today it is one of the important cottage industries, not only in the state of Himachal Pradesh but its commercial cultivation has also been initiated in most of the other states and union territories of India.

34.1.1. Major mushroom producing states in India

34.1.1.1. Bihar: The state produced more than 28,000 tons of mushrooms during 2021-22, which is 10.82 % of the total mushrooms produced in the country.

34.1.1.2. Maharashtra: Maharashtra ranks second after Bihar in mushroom production. The state accounts for 9.89% of the total production throughout the country.

34.1.1.3. Odisha: Odisha is India's 3rd largest producer of mushrooms, with a 9.66 % share in the country. Straw mushroom is the popular mushroom variety in Odisha.

34.1.1.4. Haryana: Sonipat district is popularly known for mushroom cultivation in India. Haryana govt. provides about 40% subsidy to general farmers, while a 90% subsidy to scheduled caste farmers is available.

34.1.1.5. Punjab: The state has around 400 mushroom growers in various parts of Punjab. Moreover, the average annual production of mushrooms is between 45,000 – 48,000 tons per annum.

34.1.1.6. Himachal Pradesh: Solan, Kully, Shimla, and Sirmour are the popular districts of Himachal Pradesh producing mushrooms.

34.1.1.7. Uttar Pradesh: Now the state has over 2000 growers. Milky mushroom cultivation in Uttar Pradesh helps farmers earn profits up to 5 times, compared to 3 times the profit from button mushrooms.

34.1.1.8. West Bengal: More than 300 women from 7 panchayats of West Bengal's Purba Bardhaman district are doing mushroom cultivation. The people in West Bengal focus on the oyster mushroom as it grows well in local climatic conditions from October to February. Moreover, it contains more nutritive value and grows in less space than other varieties.

34.1.1.9. Kerala: Farmers (especially women) cultivate mushrooms across 500 units of 14 districts of Kerala. Furthermore, the program aims to increase earning potential and raise awareness about mushrooms' high nutritional value.

34.2. MUSHROOM NEMATODES

The two commonly grown mushrooms in India include button mushroom, *Agaricus bisporus* and paddy straw mushroom, *Volvariella volvacea*. The first report about nematode damage to mushroom in India was made from district Solan of Himachal Pradesh by Bhardwaj *et al.* (1972) when they reported poor sporophore yield having brownish white, stunted mushrooms from beds infested with *Aphelenchoides* and *Ditylenchus*. Mushroom nematodes comprising mainly *Aphelenchoides agarici, A. composticola, A. myceliophagus, A. neocomposticola, A. sacchari, A. swarupi, Aphelenchus avenae, Ditylenchus myceliophagus* and *Rhabditis* on button mushroom (*Agaricus bisporus*) have now been well recognized as the serious constraint in achieving higher production. *Aphelenchoides composticola* and *Ditylenchus myceliophagus* were most common and were recorded from compost as well as spent compost that is used as casing material; the latter is the main source of contamination in mushroom houses. Population of fungal feeding nematodes, especially of *Aphelenchus avenae* was highest during midcrop season and came down drastically at the end of crop season.

The nematodes associated with mushrooms can be grouped into five categories based on feeding habits. Of these, myceliophagous and saprophagous nematodes are economically important as they are highly pathogenic to common mushroom fungi like *Agaricus* spp. (Sharma and Khanna 1992).

34.2.1. Myceliophagous nematodes

In all, 21 nematode species representing two orders –Aphelenchida and Tylenchida - have been reported to be harmfully associated with mushroom cultivation from various parts of the world. Among these, twenty species belong to four genera (*Aphelenchoides, Aphelenchus, Paraphelenchus* and *Seinura*) of order Aphelenchida and two species belong to two genera (*Ditylenchus myceliophagus* and *Seinura winchesi*) of order Tylenchida. *Aphelenchoides composticola* and *D. myceliophagus* are the major parasitic nematode problems on mushrooms beds in India.

Table 34.1. Pathogenicity of myceliophagous nematodes associated with mushrooms.

Nematode Order	Nematode species	Pathogenicity
Tylenchida	*Ditylenchus myceliophagus**	+++
Aphelenchida	*Aphelencoides agarici**	+++
	A. bicaudatus	+++
	*A. composticola**	+++
	A. dactylocerus	++
	A. limberi	++
	A. neocomposticola	++

	*A. sacchari**	+++
	A. saprophilus	++
	*A. swarupi**	++
	Aphelenchus avenae	++
	Paraphelenchus myceliophthorus	++
	Seinura winchesi	++

* Species reported from India, +++ Highly pathogenic, ++ Pathogenic.

34.2.1.1. *Ditylenchus myceliophagus*: This nematode is widely distributed in almost all mushroom growing belts of the world. In India, it is found in Himachal Pradesh, Jammu and Kashmir and Punjab (Thapa *et al.*, 1981). It is one of the highly destructive species. At an inoculum level of 10-1000 individuals could cause 15-70% mycelial depletion within 60 days (Sharma *et al.*,1985). However, it is less pathogenic than *A. composticola* and *A. sacchari* (Sharma *et al.*, 1981).

34.2.1.2. *Aphelenchoides composticola*: It is one of the most widely distributed species present throughout the mushroom growing countries of the world which is reported from Punjab and Himachal Pradesh, India. This nematode also has a short life cycle of 8 days at 23 °C and multiplies very fast on mushrooms (Cayrol, 1967). An initial inoculum of 10 nematodes/100 g of compost completely depleted the mycelium within 40 days resulting in about 1000 times multiplication rate (Khanna and Sharma, 1988). *A. composticola* was responsible for 35-60% loss in yield of mushroom (Laqman Khan, 2001). The complete destruction of 100 g mycelium was achieved only by 3 nematodes of *D. myceliophagous* within 70 days.

34.2.1.3. *Aphelenchoides agarici*: This species was recorded from mushroom farms in Himachal Pradesh (Seth and Sharma, 1986). It reproduces well on mushroom mycelium which is one of the most destructive nematode species. An initial inoculum level of 10 nematodes/100 g of compost has been shown to cause 100% mycelial depletion within 25 days at 28°C resulting over 800 times multiplication rate under laboratory conditions (Khanna and Sharma, 1988). It completes its life cycle within 8 days at 28 °C and the ratio between female to male was 1.5: 1(Seth, 1984).

34.2.1.4. *Aphelenchoides myceliophagus*: It is recorded from Solan district in Himachal Pradesh. It is also one of the highly destructive nematode species infesting mushrooms showing the pathogenicity status of the level of *A. composticola* (Khanna and Sharma, 1988).

34.2.1.5. *Aphelenchoides neocomposticola*: It was reported from Shimla district of Himachal Pradesh. Khanna and Sharma (1988) found that the nematode caused much severe damage as compared to *A. agarici*, *A. myceliophagus,* and *A. composticola*.

34.2.1.6. *Aphelenchoides sacchari*: It was reported by Sharma *et al.* (1981) and found to be highly pathogenic to white button mushroom, *Agaricus bisporus*. An average of 50-100% yield loss in sporophores in white button mushroom has been reported due to this nematode species (Sharma *et al.*, 1984). It is quite prevalent on mushrooms in India (Thapa *et al.*, 1981). Singh *et al.* (2003) reported that *A. sacchari* was responsible for 40.6 to 100% loss in yield of mushroom.

34.2.1.7. *Aphelenchoides swarupi*: This fungal feeding species was for the first time found in a mushroom farm at Ambala (Haryana), India where though brown wet pinheads were observed initially, no fruiting bodies were produced. An average of about 10,000 individuals/100 g of compost was counted in the beds showing total crop failure.

34.2.1.8. *Aphelenchus avenae*: This nematode is frequently present in low numbers in compost and casing soil samples taken from cropping beds.

34.2.1.9. *Seinura winchesi*: In India, population of *Seinura* sp. were found from the cropping beds of a mushroom farm at Solan, H.P. It reduced sporophore yields even though the initial growth of mycelium was normal.

34.2.2. Saprophagous nematodes

In India and other parts of the world, almost all mushroom farms were severely damaged by the saprophagous nematodes, belonging to the order Rhabditida. *Acrobeloides, Cephalobus, Diplogaster, Caenorhabditis,* and *Panagrolaimus* are the major nematodes occurring normally in mushroom growing areas. The greatest development of *Rhabditis cucumeris* (1500/g) was observed when the nematodes were introduced before spawning; and less (900/g) if the infestation and spawning were simultaneous (Dmowska *et al.*, 1997).

Huge populations of these nematodes have often been recorded from severely damaged mushroom beds on almost all mushroom farms in India and elsewhere.

34.2.2.1. *Rhabditis* spp.: Chandel (1982) demonstrated the pathogenic effects of *Rhabditis* species on A. bisporus mycelium under aseptic condition in India. He reported that the nematode multiplied 500 times on A. bisporus and depleted mycelium by 50% within 30 days.

34.2.2.2. *Caenorhabditis elegans*: Grewal and Richardson (1991) showed that *Caenorhabditis elegans* did not multiply in well prepared pasteurized spawned compost whereas casing material of the cultivated mushroom *A. bisporus* was highly suitable for its reproduction. Mushroom yields were negatively correlated with the initial inoculum level of *C. elegans*. The nematode infection resulted in production of fewer sporophores, absence of usual distinctive flushing patterns, considerable deterioration in mushroom quality and characteristic distortion. Individual sporophores were miss-happen, notched, and had brown- or violet-colored gills.

Grewal and Hand (1992) showed that saprophagous nematodes also cause damage to mushroom indirectly through the bacteria that they supported or associated with viz., *Bacillus cereus, Bacillus* sp., and *Enterobacta amnigenus*. These bacteria caused significant inhibition in mycelial growth of 3 strains of *A. bisporus*: C43, C54 and U3. Similarly, *Pseudomonas aeruginosa, P. fluorescens, Biovar reactans* and *P. meltophilia* resulted in a significant increase in mycelial growth of C54 strain. *B. cereus, E. amnigenus* and *E. cloacae* produced volatile inhibitory substance(s).

34.2.3. Crop losses

Economic losses due to decline in yields are generally very high. Extent of losses mainly depends upon the time of nematode infestation, nematode species involved and population of the nematodes. Yield losses to the tune of 16.3-53.5% were recorded due to *Aphelenchoides composticola*, and *Ditylenchus myceliophagus* inoculated at 100 nematodes/kg compost at the time of spawning and casing in 3 kg plastic bags, respectively. Losses caused by mushroom parasitic nematodes are presented in Table 34.2.

Table 34.2. Yield losses in mushroom due to nematodes

Nematode species	Initial density/ 20 kg compost	Time of inoculation	Yield loss (%)	Reference
Myceliophagous nematodes				
D. myceliophagus	1×10^2	Spawning	89	Khanna (1993)
	5×10^2		90	
	1×10^2	Casing	40	
	5×10^2		50	
A. composticola	5×10^2	Spawning	95	Khanna (1991b)
	5×10^2	Casing	64	
A. sacchari	1×10^1	Spawning	98.2	Sharma *et al.* (1984)
	1×10^2		100.0	
	1×10^3		100.0	
	1×10^4		100.0	
	1×10^1	Casing	90.6	
	1×10^2		92.1	
	1×10^3		87.3	
	1×10^4		100.0	
Saprophagous nematodes				
Rhabditis sp.			76.3	Rao *et al.* (1992a)
C. elegans	4×10^3		27.8	Grewal (1991)
C. elegans + bacterium *Acinebacter calcoaceticus* var. *aintratus*			35.0	

34.2.4. Nature of damage

The mycophagus nematodes thrust their hollow stylet into the hyphal cell. The mycelium bleeds and its sap is sucked upon by to and pro movement of the stylet. After feeding upon one cell the nematode shifts to another cell with the help of the moisture film present in the compost.

The mouth parts of myceliophagous nematodes have the stylet (needle like structure) which is hollow inside and can be moved forward and backward by the contraction and relaxation of the muscles. The nematodes secrete variety of strong enzymes. These enzymes act immediately after ejection and help in penetration of stipe and to convert the cell contents to assimilable forms. The tremendous reproduction potential is in the range of 50-100-fold/week. Rate of multiplication is faster during spawn run period (22-28 ^{0}C) than the cropping period (14-18 ^{0}C). Beyond 30 ^{0}C, these do not reproduce.

34.2.5. Symptoms

Initially whiteness of spawn slowly starts turning brown followed by sparse and patchy appearance of the mycelium. This mycelium is stingy in nature. The compost surface sinks. Alternately, medium, and low quantity of sporophores (fruiting bodies) is obtained in succession of flushes. There are extremely poor yields in the later flushes finally leading to total failure of sporophore production thus reducing the total yields as well as duration of the crop.

Following symptoms of nematode attack appear in infected beds in succession:

» Mycelial growth is sparse, patchy and mycelium turns stingy.
» The compost surface sinks.
» Whiteness of spawn-run slowly changes to brown.
» Mushroom flushes are poor and delayed.
» Alternate high and poor yield in successive flushes.
» Browning of pinheads.
» Decline in yield.
» Complete crop failure.
» White fungal growth over casing in patches.

34.2.6. Biology and life cycle

The life cycle of aphelenchids is very short (8-10 days), while *Ditylenchus* takes about one month.

34.2.7. Spread and survival

The common source of nematode contamination of mushroom beds is through casing soil and compost. However, other common sources responsible for nematode contamination include damp wheat straw, manures, FYM, garden soil, spent compost, platform soil, irrigation water, sciarid flies, and contaminated implements.

When the compost is dried gradually, the nematodes survive as long as two years in a state of anabiosis (a restoring to life from a death like condition), but they die rapidly if the compost is dried quickly.

34.2.8. Management

Prevention of the entry of nematodes into the mushroom beds rather than controlling them afterwards is the key for nematode management in view the nature of the crop, limited availability of safe and potent nematicides, residue problems and other hazards of nematicides. Integrated nematode management using available components is the only option since curative measures are not available.

34.2.8.1. Prophylactic measures

» Purposely built mushroom houses with proper ventilation should be utilized for mushroom cropping using properly pasteurized compost made by short method.
» Throughout the cropping season, strict sanitation and hygiene measures should be followed.
» Avoid the direct contact of compost with the soil by using cemented composting yard disinfected with 2% formalin, 24 hours earlier to compost preparation.
» All the instruments, walls, floors, and galleries should be disinfected with 4% formalin.
» Composting ingredients should always be stored in clean area.
» Cleanliness should be maintained inside and in surroundings of mushroom farms.

- » Casing mixture should be properly pasteurized.
- » Manures used for composting should be thoroughly broken and allowed to decay properly.
- » In long method of composting, covering compost with double PVC sheet for 24 hours after third turning gives effective control of nematodes.
- » Each mushroom unit should be provided with foot dips in the front.
- » Irrigation water should be clean.
- » Proper disinfection of hands and feet is a must for each worker to enter cropping room.
- » All the cropping rooms should be fly proof and only recommended insecticides should be sprayed for the control of flies.
- » In long method, chicken manure may be avoided. Neem cake can be added.

34.2.8.2. Physical methods: Use of heat is the most successful method of nematode control in mushroom cultivation. It is recommended that to make compost nematodes free, air and bed temperature in the pasteurization room must be maintained at 60 ^{0}C for at least 2 hours and cook out of mushroom house at 70 ^{0}C for 5-6 hours or 80 ^{0}C for 30-60 minutes. Complete disinfection of appliances like handling tools and mushroom trays should be carried out by dipping them in formalin or boiling water to kill nematodes. Finally, spent substrate should be disposed-off completely. Destruction of all spent compost is of prime importance as dry infested debris is a potential source of trouble since *D. myceliophagus* can withstand drying for up to 3 years. Lowering down the room temperature to 13 ^{0}C during cropping period retards the pathogenicity of these nematodes.

Seth (1984) tested the efficiency of boiling water against *A. agarici*; complete kill of nematodes was achieved when compost was dipped in boiling water for 15 seconds and above. Steam was most efficient for killing nematodes when used for 4 min or more.

An effective method of casing pasteurization, based on the use of solar heat, was developed wherein the thinly spread casing is exposed to the sun under a polythene cover (150 gauge) for 5-6 hours (Grewal and Grewal, 1988).

Wood and Goodey (1957) found that r-rays in between 48,000 and 96,000 r inactivated *Ditylenchus myceliophagus* in mushroom compost.

34.2.8.3. Cultural methods: Many plants have been found to possess nematicidal properties. Enhancement of mushroom yields by 6.5 to 19% and reduction of the population of *A. composticola* below economic injury levels could be achieved by incorporation of dried leaves of *Azadirachta indica*, *Cannabis sativa*, *Eucalyptus tereticornis* and *Ricinus communis* at 3 kg/100 kg of dry wheat straw, that enhanced the population of thermophilic fungi and mesophilic antibiotic producing fungi (Grewal, 1989). Karanj leaves mixed in compost also reduce the nematode population. Addition of Neem leaf powder @ 2% on w/w basis of compost not only check nematode multiplication in mushroom growing trays but also helps in proper mycelial growth and improved sporophore yields (Khanna and Sharma, 1994). Enhancement of mushroom yields and reduction in nematode reproduction (*A. compostiocola*) was obtained by application of oil cakes like Neem, Karanj (*Pongamia pinnata*), Coconut, Castor and Groundnut @ 5% on w/w basis of compost at spawning (Rao *et al.*, 1992). Karanj cake both at 2 and 4%, coconut, niger and neem cakes each at 4% were effective in reducing *A. composticola* population. Coconut, karanj and neem cakes each at 2 and 4% and sesame cake at 4% were as effective as chemical treatments in reducing *Rhabditis* spp. population (Parvatha Reddy *et al.*, 1999).

The yield of button mushroom was highest when 1% Neem seed kernel extract and Achook at 400 ppm were applied in combination at the time of spawning. The *Aphelenchoides composticola* nematode population was also reduced drastically in the treated boxes compared to inoculated untreated control (Rajesh *et al.*, 2003). Neem cake at 20 g was most efficient in reducing the population of *Aphelenchoides composticola* and Neem leaf at 16 g dose showed manurial effect on *Agaricus bisporus* in terms of yield potential (Gitanjali and Nandal, 2001).

Treatment of compost with Neem seed kernel powder solution at 7.5-liter per100 kg was found effective (Prepare solution by soaking 400 g of Neem seed kernel powder in 10 liters of water overnight, strain through muslin and use).

Neem cake at 5% when mixed in nematode-infested compost at spawning, hindered *A. composticola* multiplication and enhanced the sporophore production significantly (Khanna and Sharma, 1996).

Application of 4% (w/v) Neem seed kernel extract (NSKE) at spawning @ 7.5 L/100 kg compost can be used as prophylactic measure for the management of *Aphelenchoides* spp.

34.2.8.4. Chemical methods: Mushrooms, being indoor crop, provide little scope for pesticidal usage. Short life of the crop and deleterious effect of pesticides on mycelium and residue problems further limits the scope of pesticide application. The nematode population can be managed by using certain safe chemicals especially during long method of composting. Dichlorvos (0.04%) under polythene cover for 3-4 days, Fenamiphos EC at 20 mg/kg incorporated in compost, Thionazin at the rate of 80 ppm have been found to be most effective for control of *A. composticola* and *Rhabditis* sp. without producing any adverse effect on mycelial growth and does not leave any residue in sporophores.

34.2.8.5. Biological methods

i) Antagonistic bacteria: The bacterium, *Bacillus thuringiensis* is found to suppress the population of *Ditylenchus myceliophagus* (Cayrol, 1974).

ii) Antagonistic fungi: Nematode trapping fungus, *Arthrobotrys robusta* commercially available as 'Royal 300' (strain Antipolis) in France has been recommended for the management of *Ditylenchus myceliophagus* on mushroom, *Agaricus bisporus* (Cayrol, 1978). This strain was highly pathogenic to nematodes, a good colonizer of compost and without any antagonism to *A. bisporus.* When added to compost at spawning it traps nematodes by adhesive hyphal networks and has been shown to stimulate mycelial growth and improve cropping. *Arthrobotrys oligospora* and *A. superba* have also been found to reduce nematode populations and increase the yields (Kermarrec, 1973). In India, *Arthrobotrys irregularis* and *Candelatrella musiformis* isolated from spent compost have shown promising results in checking nematode population (Khanna and Sharma, 1990). *A. irregularis* was also found to be highly effective against *Aphelenchoides composticola*. Further studies revealed that *A. irregularis* did not compete with *Agaricus bisporus* mushroom and checked the nematode population significantly and increased the yield.

Simultaneous spawning of compost with *Agaricus bisporus* (0.75%) and *Pleurotus sajor-caju* (0.25%) and inoculation of *A. composticola* (400 larvae/15 kg compost) resulted in remarkable reduction (90%) in nematode population and significant enhancement of yield (Sharma, 1994) (Table 34.3).

Table 34.3. Potentiality of *Pleurotus sajor-caju* for the biocontrol of *Aphelenchoides composticola* in *Agaricus bisporus*

Spawning rate (%)		Nematode inoculation at			% Reduction in nematode population	
A. bisporus	*P. sajor-caju*	Spawning	Casing	Mean	Spawning	Casing
0.50	0.50	202	160	181	94.95	93.27
0.50	0.25	440	300	370	89.00	87.39
0.75	0.25	400	224	312	90.00	90.58
0.50 (Control)	---	4000	2380	3190	---	---
Mean	---	**1260**	**766**	---	---	---

Aphelenchoides composticola and *Ditylenchus myceliophagus* were immobilized after 30-60 min exposure to the culture filtrates of *Pleurotus ostreatus, P. citrinopileatus* and *P. cornucopiae*. After 24-48 hr, the carnivorous *Pleurotus* species grew towards the nematodes and penetrated them through the natural orifices. Subsequently, the body cavity was filled with hyphal contents and later the nematodes were totally digested (Sharma *et al.*, 2002).

Grewal and Sohi (1988) mass produced *Arthrobotrys conoides* on wheat grain and, after incorporating it into compost at spawning, reduced a mixed nematode population of *A. composticola* and *Rhabditis* sp. by 42.5% in 25 days.

iii) Predatory nematodes: Predatory nematodes like *Seinura payeni* have been recovered from mushroom substrates feeding on e.g., *Acrobeloides* and *Bursilla* (Grewal *et al.*, 1991). The widespread distribution of *Seinura* and their feeding on nematodes in mushroom substrates suggest that these predators may also control populations of *Aphelenchoides* parasitic on mushrooms (Grewal *et al.*, 1991).

34.2.8.6. Host resistance: *Agaricus edulis* has been reported to be resistant to *D. myceliophagus*. Thapa *et al.* (1987) reported that *Pleurotus sajor caju* and *Stropharia rugoso annulata* were found resistant to both *A. sacchari* and *D. myceliophagus*. K 30 strain of *A. bitorquis* showed resistance to *A. sacchari* only. Resistance in *Pleurotus sajor caju* against *A. sacchari, A. agarici* and *A. composticola* has been reported by several workers.

34.3. CONCLUSION

The major parasitic nematodes on mushrooms include *Aphelenchoides composticola* and *Ditylenchus myceliophagus* which cause degeneration of the mushroom mycelium in the compost. Nematodes cause heavy losses in mushroom production. Compared to other countries, the productivity of mushrooms is significantly less in India. Oyster mushrooms are relatively resistant to nematodes as compared to the button mushrooms which are generally highly susceptible. Enhancement of yields and effective management of mushroom nematodes should be attempted using plant products (leaves, non-edible oil cakes of neem, pongamia, castor, etc.), prophylactic practices (sterilized production of substrates, containers, implements, gloves, etc.), and maintenance of hygienic production conditions.

Chapter - 35

PROTECTED CULTIVATION OF VEGETABLES

35.1. INTRODUCTION

Due to climate change, population explosion, shrinkage of arable land and reduced water availability, a paradigm shift in global crop production is occurring towards protected agriculture. This input-intensive farming provides year-round production under controlled environment and by occasionally utilizing vertical spaces for soilless cultivation. Unlike open field agriculture, soil-borne plant-parasitic nematodes proliferate heavily inside the protected structure due to continuous monoculture, maintenance of stable microclimate, recycling of nematode-infected growing medium and planting materials by unaware growers. In the current decade, nematodes have emerged as a major yield reducing factor in protected agriculture, especially after phasing out of methyl bromide.

Farmers were compelled to grow some selected vegetable crops under protected cultivation (controlled conditions) to make available horticultural crop produce throughout the year, particularly during off-season; to export high-quality horticultural products; and to get substantial increase in yield (high productivity). Vegetable farming (tomato, capsicum, cucumbers, French bean, crucifers, gherkins, muskmelon, watermelon, *etc.*) inside modern greenhouses is promoted to improve earning potential of farmers especially during harsh winters. The nematode incidence under protected cultivation became severe and led to complete crop losses because of congenial conditions of higher temperature, humidity, and use of high agronomic inputs like fertilizers and plant growth promoters in polyhouses. The proliferation rates of nematodes in polyhouse cultivation reached up to 10 to 30-fold more than in the open field cultivation. Nematode damage has skyrocketed under the protected structures in several parts of the world accounting a heavy toll, and the growers can hardly get rid of this menace.

The high-value crops such as vegetables that are labor-intensive and energy demanding during cooler season are mostly preferred for protected agriculture to optimize the production cost, maximize yield, secure the quality and nutritional value, and simultaneously to minimize abiotic and biotic stresses by exploiting the modern production technologies. The structures that are mainly adopted for covering the crops include:

35.1.1. Low tunnels

Row cover with less than 1 m height, cultural practices performed from outside, mostly made up of conventional polyethylene or ethylene vinyl acetate or polyvinylchloride.

35.1.2. High tunnels

Walk-in tunnels with covers alike of low tunnel, cultural practices can be performed from inside, and moderately tall crops can be grown.

35.1.3. Greenhouses

High and large protection structure made up of glass or plastic, provide improved environmental conditions to the crops against cold, rain, hail, wind *etc.* (Albajes *et a*l., 1999; Maynard and O'Donnell, 2018).

The incident radiation from the sun passes through the transparent or translucent covering/ wall material, falls on the opaque surfaces, i.e., leaves, floor, soil *etc.*, and the heat energy cannot escape *via* convection resulting in inside temperature rise (De Vleeschouwer, 2001). The ubiquitous commonality that occurs inside all these structures is the arbitration of climatic condition making it congenial for nematode development. These covered structures largely maintain monoculture with extended diurnal temperature variation, high humidity, and less ventilation (Dasaeger and Csinos, 2006; Engindeniz and Engindeniz, 2006), aiding in population build-up of humidity loving plant-parasitic nematodes (PPNs).

35.1.4. Advantages

» Higher productivity and higher income
» Quality produce
» Off season or round the year cultivation
» Hardening of tissue culture plants
» Better management of insect pest
» Less use of chemicals
» Efficient use of resources

35.2. REASONS FOR RAPID MULTIPLICATION OF NEMATODES

The survival, reproduction, and population buildup of PPNs are dependent on exposure to certain biotic and abiotic factors, such as continuous prevalence of favorable temperature and moisture, constant availability of suitable hosts with narrow genetic diversity, and minimal exposure to environmental extremities (Greco and Esmenjaud, 2004). The greenhouses and polyhouses mostly have drip/sprinkler irrigation facility, which ensures optimum moisture level in rhizosphere throughout the crop season.

35.2.1. Moisture

Moisture is the foremost important factor for multiplication of nematodes. Continuous moisture availability around root zones through drip irrigation under polyhouses is responsible for fast nematode build-up and movement as compared to open field where flooded and dry conditions prevail. Continuous moisture availability flare-up the nematode population and become more infectious.

35.5.2. Temperature

Temperature affects overall life profile activities of nematode such as hatching, movement, reproduction, development, and survival and the host plant. Optimum temperature range for survival of plant parasitic nematodes is 15-30 °C and become inactive or less active from above and below temperature conditions or may be lethal for nematodes. Under polyhouses, optimum range of temperature for nematode build-up exist. Under optimum moisture and temperature conditions in polyhouses, nematodes can complete several generations in less period of time as compared to open field conditions.

35.2.3. Continuous cultivation of susceptible hosts

Vegetable crops grown under polyhouses are generally susceptible to nematode pest such as tomato, bell pepper, cucumber, etc. due to high economic value monoculture has been adopted by growers. Intensive monoculture of susceptible crops leads to nematode build-up and multiplication rate. All three conditions, susceptible host with favorable microclimate favor fast build-up of nematode population and once it is introduced it is very difficult to get rid from them.

35.3. SOURCES OF NEMATODE INFECTION

The most common and prevalent source of nematode infection under greenhouse and polyhouse conditions arise from the contaminated soil or the soil mixture used as a component of growing medium. Sufficiently fertilized and well-irrigated crops grown under optimum climatic conditions are not an exception to this trend (Albajes *et al.*, 1999). Soil disinfection techniques (e.g., soil solarization, fumigation etc.) are effectively practiced minimizing the initial nematode population level, although these operations create a "biological vacuum" eradicating majority of the beneficial organisms and natural enemies.

Polyhouses were observed to be severely infested with major plant parasitic nematode genera viz., *Meloidogyne, Pratylenchus, Rotylenchulus, Helicotylenchus, Mesocriconema* and *Tylenchorhynchus*. The above-ground symptoms of the most economically important root-knot nematode infection include stunting of plant growth, leaf yellowing and wilting of plants. Root galling is the characteristic under-ground symptom caused by the root-knot nematode which interferes in absorption of water and nutrients from the soil due to disruption of conducting vessels. They feed on the roots devitalizing the plants ultimately affecting both qualitative and quantitative production. Nematode problems on vegetable crops under protected conditions have assumed alarming proportions leading to huge losses (up to 80%) in select crops. The nematode infestations exacerbate severity of fungal and bacterial diseases leading to complete crop failure. In view of damage potential of these tiny hidden enemies of crop plants, an urgent need is felt for developing nematode management strategies.

35.4. NEMATODE MANAGEMENT

Crop rotation, fallowing and summer plowing which are generally employed for nematode management under field conditions are not effective under protected cultivation. Hence, exclusion or avoidance is of prime importance for nematode management under polyhouse. It is very difficult to control nematodes once they are introduced in the polyhouse. Therefore, prevention is better than cure.

Favorable climatic conditions prevail in majority of the protected structures in the regions between 25° and 65° latitude (Berlinger *et al.*, 1999). Solar irradiations, low temperature at higher latitude, maintenance of humidity and temperature and the conditions created favor the nematode incidence. Intense solar irradiation at lower altitude persuades stress in the vegetable crops rendering them susceptible to nematode incidence (Berlinger *et al.*, 1999). Irrespective of the diverse protected structures and materials for buildup, the microclimate inside the protected structures favor the multiplication of plant parasitic nematodes (Heinz *et al.*, 2004). Henceforth, it becomes extremely difficult to manage the nematode pests from single management options. Keeping this in view, integration of all available management techniques/tools is needed for better resolution of the nematode pests. The integrated strategies for control of plant parasitic nematodes can be based upon two basic principles: preventive measures and on-farm techniques. Preventive measures avoid the introduction of pest species in newer areas and on-farm techniques are based upon control measures (cultural, biological, and chemical) adopted by growers, to reduce nematode population below ETL.

The nematodes may be managed by using the following methods:

35.4.1. Preventive measures

Preventive measures are adopted to avoid the introduction of nematode pest in newer areas where nematode problems do not exist before. Some practices have been adopted as preventive measures to control the spread of nematode. Spread of new or emerging species has been checked by regulatory methods to avoid the introduction in newer areas. Soil testing for plant parasitic nematodes is mandatory for all the farmers before erection of polyhouses, green houses and net houses. Entry points for protected conditions should also contain sanitizing stations for hands, shoes, boots, tools, and other equipment. Nematode-free transplanting material is one of the important methods to avoid the nematode infestation under field conditions. Always use nematode-free transplants or plants that build upon soilless substrates from production are increasingly used to exclude soil-borne species of nematodes, but also to promote the plant establishment and crop production. One of the most important methods to prevent spreading of nematodes in nematode- free areas is through growing of nursery crops in soilless media such as organic growing media: peat, coir, bark, sawdust, and compost; and inorganic growing media: rockwool, perlite, pumice, sand, and vermiculite.

35.4.2. Curative measures

Curative measures are used to reduce the nematode population below economic threshold level in nematode infested areas, so growers can get maximum returns.

35.4.2.1. Physical methods: Soil solarization is a most effective method to reduce the nematode population in hot weather areas (temperature around 40-50 °C). In India, conditions are best to adopt this practice to reduce nematode infestation under polyhouses. Transparent polyethylene plastic (25 μm thick LLDP) mulch is used to cover the moist soil for 4–8 weeks in the month of May–June (Katan, 2017a, b; Kumar *et al.*, 2019; Patil *et al.*, 2018b). Green house effects have been created under transparent polyethylene sheet leading to higher temperature lethal to nematodes.

35.4.2.2. Cultural methods

i) Field sanitation: Field sanitation through rapid destruction of infested plant debris and weeds after harvest can minimize the nematode problems from polyhouses.

35.5.4.2. Trap cropping: Growing cowpea (trap crop for root-knot nematode) closely before the main vegetable crop and removing the plants out of polyhouse at 45 days after sowing.

35.5.4.3. Crop rotation and inter cropping: Cultural practices are non-chemical method such as crop rotation with resistant cultivars or non-host crops to reduce pest population. Rotating or inter cropping tomato/ cucumber with non-hosts such as garlic (Allium sativum), marigold (Tagetes sp.), lettuce, radish, cabbage, and cauliflower could reduce root-knot nematode populations in soil. Few options are available in protected conditions to grow non-host crops, so, resistant cultivars are a very good option under protected structures. Raising marigold (antagonistic to nematodes) and incorporating in situ after plucking the flowers.

ii) Organic amendments: Enormous organic amendments are used for suppression of plant parasitic nematodes in protected cultivation. Suppression efficacy of organic amendments depends on the active ingredient and their concentration. Non-edible oil cakes of neem (Azadirachta indica), castor (Ricinus communis), karanj (Pongamia glabra), mahua (Madhuca latifolia) etc. are used for management of root-knot nematode in protected cultivation (Patil et al., 2020). Other organic substances like FYM, vermicompost, slurry, green manure etc. are also effective for suppression of PPNs.

35.4.2.3. Chemical methods: Application of Carbofuran 3G at 1 kg a.i./ha at planting and repeated after 45 days.

35.4.2.4. Biological methods: Higher efficiency, targeted results, environmentally sound and local acceptability among the growers gain much popularity of the bio-agents in recent era. Egg parasitic fungus- *Paecilomyces lilacinus, Pochonia clamydosporia,* antagonistic fungus- *Trichoderma viride, T. harzianum,* VAM fungus- *Glomus* spp., bacterial parasite- *Pasteuria penetrans* and PGPR bacteria- *Pseudomonas fluorescence* is used as potential bio-agents against plant parasitic nematodes (Patil *et al.*, 2017b, c; Patil *et al.*, 2018b; Patil *et al.*, 2021). Treatment of seed with biopesticide - *Pseudomonas fluorescens* at 20 g per kg seeds was found effective. Bio-agents enriched organic amendments are very effective strategy to control nematodes in protected cultivation.

35.4.2.5. Host resistance: Resistant cultivars are one of the convenient options against plant parasitic nematodes. Grafting of commercially desired susceptible cultivars on resistant rootstock is a trending method among vegetable crops under protected conditions (Rivard *et al.*, 2010). Resistant rootstock of brinjal wild relatives, *Solanum toxicarium*, *Solanum sisymbriifolium* and *S. torvum* have been grafted by commercial tomatoes, noticeable reduction in galling was observed (Black *et al.*, 2003). Various grafted rootstock of melon and capsicum were produced that confirmed extraordinary results in minimize root galling in the greenhouses (Kokalis-Burelle and Rosskopf, 2011).

Genetic engineering approaches has made natural resistance with synthetic resistance may be the auspicious tools for management of nematode in tomato production (Gheysen *et al.*, 1996; Jung *et al.*, 1998; Opperman *et al.*, 1998). RNA Interference (RNAi) has emerging tool to downregulate gene activity and recognized efficient tactic against root-knot nematode (Elling, 2013). RNAi first performed for *Caenorhabditis elegans,* and it was used for gene silencing by overwhelming their expression in a plant parasitic nematode (Ali *et al.*, 2017). Nematode feeding site formation gene has been silenced by using dsRNA or siRNA that elicit a systemic RNAi response (Lilley *et al.*, 2017). Root-knot nematode produces effector proteins determined by parasitism genes, and these effectors epitomize the molecular interface between the nematode and host (Elling, 2013). Effectors secreted in nematode esophageal glands play perilous role in parasitism (Lilley *et al.*, 2012).

The advancement of the Clustered Regularly Interspaced Short Palindromic Repeats (CRISPR) technology has become a commanding alternative to gene silencing (Ali *et al.*, 2019). Foreign DNA sequences has incorporated host loci to produce short crRNAs (CRISPR RNAs) that direct sequence-specific cleavage of homologous target dsDNA by Cas endonucleases (Jinek *et al.*, 2012). Recently, documentation of pathogen and host novel genes responsible for infection help in developing the CRISPR technique for improving the resistance to Meloidogyne spp. under protected systems.

35.4.2.6. Integrated methods

i) Bioagents and botanicals: The seedlings can be raised in the coco-peat or any other substrates such as FYM, Vermicompost, and Neem/Pongamia cakes enriched with biopesticides (2 kg each of *Purpureocillium lilacinum*, *Trichoderma harzianum* and *P. fluorescens* per 2000 kg of FYM/500 kg of Vermicompost/200 kg of Neem/Pongamia cakes).

The nursery beds can be treated with Neem or Pongamia cake at 50 g, *T. harzianum* at 25 g or *Pochonia chlamydosporia* + *P. lilacinum* at 25 g/ m^2 at an interval of 2 months.

Soil application of 2 tons of farmyard manure enriched with bio-pesticide-*T. harzianum* per acre before planting, along with 100-200 kg of Neem or Pongamia cake.

ii) Botanicals and chemicals: If the beds are highly infested with the nematodes, the population can be brought down by applying Neem/Pongamia cake at 500 g + Carbofuran 3G at 50 g/ m^2.

35.5. TOMATO, *Solanum lycopersicum*

Productivity of tomato in greenhouse is increased manifold in comparison to open field. Some of the other advantages of growing tomato in polyhouse include high economic returns, availability of fresh vegetables in winter, and improved taste, flavor and shelf life. Unlike field-grown tomatoes which require harvesting and marketing over a short period of time (3-4 weeks), greenhouse tomatoes are harvested and marketed over a 30-week period.

35.5.1. Root-knot nematodes, *Meloidogyne incognita, M. javanica*

35.5.1.1. Crop losses: *M. incognita* was responsible for 30.57 to 46.92% loss in fruit yield of tomato, while *M. javanica* caused 77.5% loss in yield.

35.5.1.2. Symptoms: Root-knot nematodes cause root galls on the feeder roots and sometimes affect the entire root system showing heavy galling. This affects the uptake of nutrients and water, and the plants show wilting during warmer part of the day and results in stunting of plant growth, leaf chlorosis, and reduced yields. In tomato, dynamics of root-knot nematode showed enhanced population build-up from 1 to 30 juveniles per g soil within a period of 6–12 months, which is comparatively higher in contrast to the open cultivation.

35.5.1.3. Management

i) Physical methods: The effectiveness of solarization, a soil disinfection technique that uses passive solar heating, to control the incidence of root-knot nematode under greenhouse condition was studied.

Solarisation of moist soil with polythene mulching for four weeks during peak summer in closed polyhouses increase the soil temperature. Maximum soil temperatures achieved under solarization were 51°C in the alluvial soil and 52.4 °C in the heavy clay soil type at 20 cm. depth. In the non-solarized unheated greenhouse, soil temperature achieved was 39 °C.

The effect of soil solarization treatments on the population densities of the root-knot nematode *M. incognita* was very clear. The larval density per 100 ml of soil was markedly reduced compared with the untreated control. The soil solarization method had a significant influence to reduce the severity of root damage. Root galling index was dramatically reduced in both solarized greenhouses compared with untreated control where the highest root galling index was noticed.

The growth of tomato plants, as indicated by visual assessment and plant height measurements was significantly improved by the soil solarization. Tomato fruit yields in solarized greenhouses were four-fold higher than those obtained in untreated control. Such significant increases were consistent with effective control of nematode provided by solarization.

Solarization is also known to improve plant growth and yield through the release of nutrient induced by high soil temperatures. This technique at the same time showed good results to reduce the infection of a severe tomato disease like corky root disease (*Pyrenochaeta lycopersici*).

ii) Cultural methods: Crop rotation with marigold, wherever possible was found effective. The tomato-cabbage-cowpea and tomato-cucumber-mustard rotations were found to have maximum suppressive effect on *M. incognita* (Sharma *et al.*, 2015).

Soil application of Neem cake at 320 g/m^2 proved to be promising for root-knot nematode management with benefit: cost ratio of 5.65 in tomato (Anon, 2015; Chandel *et al.*, 2014).

iii) Chemical methods: Application of Carbofuran 3G at 1 kg a.i. /ha at transplanting proved beneficial.

Addition of Dazomet at 1 kg a.i. /ha to the nursery beds gave effective control of root-knot nematodes.

Soil application of Metham sodium (40 ml/m^2) and Dazomet (40 g/m^2) proved promising in polyhouse tomato. Benefit: cost ratio in Metham sodium and Dazomet was found to be 5.69 and 4.38, respectively in tomato (Anon, 2015; Chandel *et al.*, 2014).

Metham Sodium at 40 ml/m^2 reduced nematode reproduction viz. no. of galls/plant, no. of egg masses/plant, no. of eggs and larvae/egg mass and nematode population/200 ml soil and enhanced plant growth characters. Highest yield was also recorded with Metham sodium (10.32 kg/plant) over control (5.25 kg/plant) (Giriraj *et al.*, 2018).

iv) Biological methods: Seed treatment with biopesticide, *Pseudomonas fluorescens* at 10g/kg seed was found effective.

Nursery bed treatment with *T. harzianum* at 50 g/ m^2 proved beneficial.

v) Integrated methods: Addition of enriched FYM/ Compost/ Vermicompost enriched with *Trichoderma harzianum/T. viride* + *Purpureocillium lilacinum* + *Pseudomonas fluorescens* each at 2 kg/ ton of organic amendments at 5 t/ha gave effective management of root-knot nematodes.

After 15 to 30 days of planting, addition of enriched Neem cake (with *T. harzianum/T. viride* + *P. lilacinum* + *P. fluorescens* each at 2 kg/ton) at 50 g/m^2 gave effective management of nematodes under polyhouse conditions. Nematode management in polyhouse beds can be achieved by application of Neem cake suspension enriched with bioagents (20 kg Neem cake + *T. harzianum/T. viride* + *P. lilacinum* + *P. fluorescens* each at 2 kg) through drip irrigation.

Application of *P. lilacinum* (2 x 10^6 cfu/g) as seed treatment @ 10 g/kg seed followed by its soil application @ 50 g/m^2 was found effective for the management of root-knot nematode, *M. incognita.*

35.6. BELL PEPPER, *Capsicum annuum*

Greenhouses can be a means to economically maintain a warm environment during cool season, to protect bell pepper plants from rain, wind, and high solar radiation, and to retain pollinators and beneficial insects while excluding unwanted insect pests. In greenhouses, capsicum fruits are harvested with full maturation color and high-quality fruit yields are greater and usually produced at a time of the year when production in the field is not possible and market prices for capsicum are highest.

35.6.1. Root-knot nematode, *Meloidogyne incognita*

35.6.1.1. Symptoms: Root-knot nematodes cause root galls on the feeder roots which sometimes affect the entire root system showing heavy galling. Root galls caused by root-knot nematodes on sweet pepper are frequently small. This affects the uptake of nutrition and water, and the plants also show wilting during warmer part of the day resulting in stunting of plant growth, leaf chlorosis, and reduced yields. In capsicum, a pathogenic bacterium, *Ralstonia solanacearum* gets entry into the roots infested by root-knot nematode and together they produce wilting disease complex that reduces yield by 60–70%.

35.6.1.2. Management

i) Physical methods: Solarization using plastic sheets for covering soil during peak summer in greenhouse reduces nematodes, bacteria, fungi, and weed population.

ii) Cultural methods: Several cultural methods like selection of nematode-free nursery sites; destruction of infested roots after crop harvest; either alone or in combination proved to be reasonably effective, cheap and prevent nematode reproduction on bell pepper.

iii) Chemical methods: Bare-root dip treatment of capsicum seedlings in 0.1% Carbosulfan/ Monocrotophos for 6 hr. eliminates root-knot infection.

iv) Biological methods

a) Antagonistic bacteria: Zuckermann *et al.* (1993) reported that significant enhancement of yield and reduction in root galling due to *M. incognita* was obtained by soil incorporation of *Bacillus thuringiensis* strain to bell pepper.

b) Antagonistic fungi: Application of 'Royal 350' (*Arthrobotrys irregularis* cultured on oat seed medium) at 140 g/m^2 a month before transplantation of capsicum resulted in good protection against root-knot nematodes. Among the egg parasites, efficacy of *Purpureocillium lilacinum* (commercially formulated as 'Biocon' in Philippines) has been found to be comparatively higher in suppressing the population of *Meloidogyne* spp. on capsicum.

Treatment of the greenhouse beds with the formulation of *Pochonia chlamydosporia* at 50g/m^2 was significantly effective in reducing the galling index, number of nematodes in roots and soil, increasing the per cent parasitization of eggs by bioagent and enhancing yields. Effective management of root-knot nematode can also be achieved by treatment of nursery with *P. chlamydosporia* and treatment of seed with *P. fluorescens* (Naik, 2004).

v) Host resistance: Capsicum cvs. Mississipi-68, Santanka, Anaheim Chile and Italian Pickling are reported to be resistant to root-knot nematodes. Black Indica, Naharia and Pant C1 were reported to be resistant to *M. incognita*, while All Big as moderately resistant. California Wonder and Naharia were resistant to *M. javanica*, while Early California Wonder as moderately resistant. Naharia was reported to be resistant to *M. arenaria.*

The southern root-knot nematode (*M. incognita*) resistant 'Carolina Wonder' and 'Charleston Belle' are recommended for use by greenhouse growers.

vi) Integrated methods: Integration of *P. fluorescens* and *P. chlamydosporia* in the greenhouse bed has proved significantly effective in reducing the root-galling index (*M. incognita*), number of nematodes in the roots and soil and increasing the yield of the capsicum crop (Naik, 2004).

Integration of soil incorporation of Neem based *P. fluorescens* and *T. harzianum* followed by seed treatment with *P. fluorescens* + soil incorporation of *T. harzianum* increased the seedling stand in bell pepper (Naik, 2004).

35.7. CUCUMBER, *Cucumis sativus*

Cucumber and gherkins are being extensively grown in greenhouses in many countries. The Continental or Burpless cucumber is the main type of cucumber and has long fruit, whereas the Lebanese (mini) cucumber has shorter fruit (parthenocarpic). Compared with field cucumbers, these two types are seedless and do not require pollination to produce fruits. Compared with field crops, protected cultivated cucumbers have a finer texture, greener in color, smoother, flesh that has fewer total acids (0.11 per cent versus 0.17 per cent) and more sugars (3 per cent versus 2.8 per cent). The skin is more tender, not peeled and are easier to digest. The crops produced in greenhouses fetch higher income due to good quality produce and higher productivity per unit area as also due to being available in off season. The cropping period gets extended, so regular supply could be maintained for a long time.

35.7.1. Root-knot nematode, *Meloidogyne incognita*

35.7.1.1. Symptoms: Root-knot nematodes *(Meloidogyne* spp., especially *M. incognita)* cause stunting of plant growth, leaf yellowing, wilting of plants, and severe galls on the root system. The disease is more common and serious on sandy soils than on clay soils; crop injury is more serious during dry seasons than seasons with adequate rainfall. Root-knot is one of the most common and serious diseases of cucumbers and all growers should practice control measures.

35.7.1.2. Management

i) Physical methods: Soil solarization in the month of May-July by light irrigation and covering of soil with 25 micron transparent polythene sheet for 30 days was found effective for the management of root-knot nematodes (*Meloidogyne* spp.).

ii) Cultural methods: Rising of nurseries in plug trays proved beneficial. Addition of Neem cake in (40 g per pit) about 15 days before the transplantation of the crop is beneficial.

iii) Chemical methods: If nematode infestation is high, mixing soil with Carbofuran @ 1 g per plant at the time of transplanting is beneficial.

Sterilization of severely infested polyhouses with Dazomet (at 40 g/m^2) or Metham sodium (at 40 ml/m^2) was found effective forthe management of root-knot nematodes.

iv) Biological methods: *Purpureocillium lilacinum* was found most effective in increasing plant growth parameters like plant height, fresh and dry shoot weight, and root growth. The reason for increased plant growth and other parameters observed could be attributed to the release of growth promoting substances by bio-agents (Baker *et al.*, 1986) or by production of toxic metabolites which inhibited nematodes and other deleterious microorganisms.

P. lilacinum caused maximum reduction in soil nematode population, root galling and egg mass production (due to parasitic activity of *P. lilacinum* on eggs and other nematode stages) and was superior over *P. fluorescens, P. chlamydosporia*, and marigold intercropping.

v) Host resistance: Cucumber cv. Vertina was moderately resistant to *M. incognita* (11-30 galls per root systems) at harvest.

vi) Integrated methods: Soil fumigation with Metham sodium at 30 ml/m^2 with polythene mulch with 15 days waiting period prior to sowing/planting + Neem cake 200 g/m^2 enriched with *P. lilacinum* @ 50 g/m^2 (2x10^6 cfu/g) mixed 15 days prior to sowing/planting was effective.

Soil application of Neem cake @ 200 g/m^2 three weeks prior to sowing + soil application of *P. lilacinum* @ 50 g/m^2 at sowing reduced soil nematode population and increased the yield of salad cucumber under polyhouse conditions.

Soil solarization and application of partially decomposed farmyard manure enriched with *P. lilacinum* reduced crop loss in salad cucumber under polyhouse condition. In the next crop the population was reduced to 180 nematodes per 200 cc soil. The longevity of the crop also extended.

35.8. LETTUCE, *Lactuca sativa*

Greenhouse-grown lettuce cvs. 'Kwiek', 'Emerald', and 'Kloek' are grown because of their good quality, availability in off-season (November to May), and high productivity of 313 t/ha/year.

35.8.1. Root-knot nematodes, *Meloidogyne* spp.

35.8.1.1. Symptoms: Root-knot nematodes (*Meloidogyne incognita, M. hapla, M. arenaria, M. javanica*) cause root galls on the feeder roots which sometimes affect the entire root system showing heavy galling. This affects the uptake of nutrition and water, and the plants also show wilting during warmer part of the day resulting in stunting of plant growth, leaf chlorosis, and reduced yields.

35.8.1.2. Management: The root-knot nematodes infecting lettuce can be managed by the use of soil solarisation, nematode-free planting stock, pre-planting soil treatment with Oxamyl and Dazomet, and soil incorporation of oil cakes enriched with biological control agents.

35.9. FUTURE THRUSTS

The future lines of research work recommended include the following practices:

» At the time of new polyhouse establishment, ensuring nematode-free soil should be strictly adhered by testing soil for the presence nematodes or otherwise in ICAR Institutes/State Agricultural Universities.

» Region-wise dedicated referral labs should be established at State Agricultural Universities/ ICAR Research Institutes for analysing samples for the presence of nematodes.

» Sites for erecting new polyhouses should be properly treated with recommended chemicals and followed by supplementation with biological control agents enriched farmyard manure/ oil cakes/ vermicompost.

» There is an urgent need to grant provisional registration for new chemicals having proven efficacy against root-knot and other nematodes.

» In case of heavy population of plant parasitic nematodes species, soil fumigation with permissible chemicals should be undertaken.

» Evaluation of rootstocks of crop genotypes against target nematode species (even up to races) should be carried out and resistant rootstocks should be employed for effective nematode management.

» Efforts on breeding vegetable varieties resistant to nematodes and suitable for polyhouse conditions should be intensified.

» Cluster approach may be followed to tackle the emergent nematode and other pest problems under polyhouse conditions.

35.10. CONCLUSION

The major plant parasitic nematodes associated with vegetable crops under polyhouse cultivations include species of *Meloidogyne*, *Rotylenchulus*, *Pratylenchus*, and *Helicotylenchus*. Nematodes assume new proposition in protected cultivation of vegetable crops due to moderate climate and intensive cultivation. As the accrued losses due to nematode pests are tangible, proper attention is a must for their management. Various chemicals for the management of nematodes have been used under protected cultivation but fumigants like dazomet and metham sodium used as pre-plant under tarp, have filled the gap along with other methods of management. There is further needed to develop viable options for nematode management, including use of biological control agents and antagonistic crops like crotalaria, sun hemp, and marigold (as green manure crops). There is a need to assess the benefits of integrating practices such as cover crops, resistant cultivars, and environmentally compatible chemicals for effective management of root-knot and other plant-parasitic nematodes in vegetable crop production systems. Since the choice of crops and nematode incidence in protected agriculture differ according to locations, futuristic management studies should run in a holistic approach considering the interactions between nematode communities, host plants and biotic/abiotic environmental factors.

Chapter - 36

FUTURE CHALLENGES AND OPPORTUNITIES

36.1. INTRODUCTION

The world needs to increase food production by at least 2% every year across all crops to ensure an adequate food supply for the world's growing population which is estimated to reach 9.5 billion by 2050. With the biotic and abiotic factors impacting food production now and, in the future, all forms of crop production improvement will be relevant for food security for the generations to come. Those working in nematology have an important role to play in securing an adequate food supply for the world.

Plant-parasitic nematodes cause colossal problems to major crops throughout the world, including vegetables (Shapiro-Ilan *et al.*, 2015). Further, these nematodes signify serious risk to the world economy and are responsible for excessive losses in crop production systems worldwide (Abd-Elgawad and Askary, 2015). Several vegetable crops, including bitter gourd, cabbage, carrot, cucumber, okra, chilli, lettuce, melons, mustard, pumpkin, sponge gourd, squash, eggplant, potato, and tomato are affected by different species of plant-parasitic nematodes. Many nematodes have been identified in vegetable crops; however, only a few of them are economically important. They include root-knot nematodes (*Meloidogyne* spp.), cyst nematodes (*Heterodera* and *Globodera* spp.), reniform nematodes (*Rotylenchulus reniformis*), root-lesion nematodes (*Pratylenchus* spp.), stunt nematodes, *Tylenchorhynchus* spp., and stem and bulb nematode (*Ditylenchus dipsaci*). Our inability to adequately control many nematodes makes them a significant part of the 'big giveaway'. It is to some extent incomprehensible that phytonematology, which evolved as a science to reduce nematode losses to food crops through research and management in the late 1800s, has still not reached this goal for many crop–nematode interactions after 20 decades of intense research.

Plant parasitic nematodes are major constraints and part of the solution to sustainable food production. Of course, major strides have been made in reducing crop losses caused by several nematodes, but there is still much to do to shift from what could be called a direct control mode to a true INM mode in many crops. INM has contributed extensively to world food security by ensuring higher yields on a broad array of crops. Nematology has made major advancements in improving food production and will remain important as the above constraints affect crop health and risks to the farmers we serve.

36.2. INTEGRATED NEMATODE MANAGEMENT (INM)

INM was defined by Bird (1981) as the integration of research, development, technology transfer and implementation needed to integrate two or more control procedures to manage one or more nematode species. INM is often expanded to include specific crops, control methods, production practices and one or more plant parasitic nematodes either alone or present concomitantly within a cropping system. Nematode management, however, is not static but changes as external drivers and constraints impact crop growth, as nematode problems evolve, and as grower needs shift in importance with the development of new technologies (Sikora and Roberts, 2018).

Any grower that has dealt with a serious nematode problem can attest to the difficulty of managing them, and while chemical control was the preferred option in the past, the future will require a far more integrated approach. The consequences of non-action regarding improving INM will be severe, including further proliferation of nematode problems and deterioration of soil health.

While it was clear that there is no one-size-fits-all approach to INM, employing a diversity of management practices and focusing on key concepts such as targeted rotations, intercropping, advanced genetics for resistance breeding, remote sensing to monitor nematode distribution and densities, precision agriculture to target control treatments and molecular tools to measure soil suppressiveness, were repeatedly mentioned by most authors.

36.3. PILLERS OF INM

The pillars of INM and the wide array of tools they encompass are outlined in Table 36.1 highlighting future anticipated developments in nematode management. The five major pillars of INM used here are: crop rotation, cultivar choice and establishment, soil management, targeted control, and monitoring and evaluation. Each pillar is constructed of numerous control tools that prevent, reduce populations, or enhance plant tolerance to nematode damage as well as provide tactical tools to make INM more effective (Desaeger *et al.* 2022).

Table 36.1. Integrated nematode management (INM) in the decades to come (Tabular form).

INM				
Monitoring & evaluation	**Targetted control**	**Crop rotation**	**Cultivar choice & establishment**	**Soil management**
Quarantine explained[1]	Novel sanitation techniques[1]	Seed certification expansion [1]	Seed certification expansion [1]	Soil-free production [1]
Robotics for targetted treatment[2]	Novel chemistry[2]	New resistant cover, trap, rotation crops [2]	New sources of resistance [2]	Soil antagonistic potential [2]
Integrated field management[3]	Biological soil infestation[2]	Mixed and relay cropping[3]	Grafting on resistant rootstock[2]	Volunteer weed control[2]
Nematode soil analysis molecular/ deep learning[4]	Improved biocontrol strains[2]	Tolerant crops[3]	Seed and seedling treatments[3]	Integrated crop management[3]
Satellite imaging for monitoring & mapping[4]	Seed treatments[3]	DSS nematode schemes[4]	High yielding tolerant cultivars[3]	Microbe manipulation engineering[3]

Genomics for holistic soil evaluation[4]	Biochemicals/ stimulants[3]		DSS cultivar schemes[4]	Field maps from geo platforms[4]
	Variable rate application [4]			Population dynamics modelling[4]

[1]Prevention [2]Reducing populations [3]Reducing impact of nematodes [4]Supportive tactics

The building blocks of INM are discussed below with their advantages and limitations (Sikora *et al.* 2022).

36.3.1. Prevention

The most effective nematode control tool is prevention of introduction and spread of plant parasitic nematodes. Certified seed and planting material is a highly effective tool used to prevent invasive nematode species from expanding their range. Quarantine policies are aimed at keeping new species out of an area for as long as possible. A better approach to ensure maximum delay of nematode dispersal would be a system whereby growers detect infestations early through intensive soil sampling so that they can take immediate action. This approach is not part of most governmental plant protection organizations' policies.

36.3.2. Crop rotation

Policy makers often live under the assumption that nematode problems are preventable by expanding the length of rotations with non-host crops. This is true only for nematode species with narrow host ranges such as cyst nematode species. For polyphagous species such as those of the genera *Meloidogyne* and *Pratylenchus*, it is not the frequency of cultivation of a crop that is decisive but the makeup of the cropping sequence. Farmers also tend to overlook the fact that some green manure crops are good hosts. This fact needs to be taken into account when constructing rotation schemes.

36.3.3. Cultivar selection

Crop cultivars with resistant or tolerant traits are available and offering a good option to build an effective INM system. However, most breeders search for resistance in crops in a rotation with known nematode problems. They usually overlook potential resistance in other crops in the rotation sequence planted prior to the main susceptible crop. The use of resistant/ tolerant cultivars should be coupled with resistance management to prevent the development of virulent races or pathotypes. Sometimes damage can be prevented by choosing a sowing and harvesting date to escape early root infection of a nematode species or to avoid development of late season generations, through early host removal.

36.3.4. Soil management

Optimal growing conditions help crops to endure nematode damage. Therefore, proper soil management is an important pillar of INM. Nutrient and water management, proper tillage and organic matter content are all basic conditions for a healthy crop, a condition that is not standard around the world. A strong plant is more resilient to pest and disease impacts.

36.3.5. Targeted control

In the past, synthetic nematicides were the starting point of targeted control in INM. This view is no longer tenable when we aim to build a sustainable agricultural system. Nematicides are still an important measure within INM, but they are no longer the starting point but rather the safety net in difficult situations. Nematicides should no longer be used preventatively, but only used in situations where nematode populations have exceeded damage threshold levels as determined through sampling or crop damage observations. Targeted variable rate application within the field where necessary is increasingly becoming the favoured approach. In addition to chemical and biological agents, there are also physical techniques such as solarization, inundation and anaerobic soil disinfestation that can be utilized.

36.3.6. Monitoring and evaluation

The monitoring and evaluation pillars are at the same time the first and final step in an INM system. Knowing what species are present at what infestation levels make it possible to begin to develop an INM plan and to check whether the results of INM are satisfactory or need restructuring. Historical information of a field is a good starting point. Once a nematode species has been introduced into a field, it will typically stay there below or above detection levels.

36.3.7. Sampling

Evaluation of the soil and observation on the health of root systems are other important sources of information. During harvest of root crops like potato, carrots and sugar beet, a great deal of information can be gathered by looking for symptoms on the harvested product as well as on the root system. During the growing season, plant canopy development can tell the farmer that there is a problem developing under the soil surface. Examination of the root system of randomly selected plants for nematode damage is therefore a valuable tool in INM. Canopy surveillance by satellite, drones or other techniques for remote sensing is becoming increasingly important in farming operations.

36.3.8. Decision support systems (DSS)

There are very few DSS on the market or in the public domain that can be used by farmer extension agents for keeping track of the large amount of data accumulated for proper selection of the tools needed at any specific time. Useable DSS programs and trained extension agents for complex cropping programs will be needed.

36.3.9. Communication

Integrated nematode management and knowledge transfer go together. INM is knowledge intensive so effective knowledge exchange is crucial. It is important to share 'success stories' within the farm community. Communication should be interactive and traditional methods such as on-farm demonstrations and field training should be supplemented with websites, blogs, and webinars. Data visualization tools and easy to use DSS such as smartphone apps help to transfer validated information and discover knowledge gaps or misunderstandings. The use of geographic information systems to visualize the available information on a farm map improves the farmer's understanding of the situation. Economic analysis of different measures in different cropping systems will help to judge if a given measure fits in the economic goals of the farm. Communicating with farmers without a realistic outlook on cost/ benefits is a waste of time. By presenting the right data and combining it with relevant knowledge, a farmer can make well-founded decisions.

36.4. MODIFICATIONS AND EXPANSIONS TO INM FOR FUTURE

The INM building blocks that make up the pillars will need to be modified and expanded upon to be able to deal with future changes impacting soil and root health. The future impact of climate change was considered a major factor in INM. The innovative technologies and some of the solutions that will influence INM and fill the knowledge gap in the future are presented in Table 36.2 (Desaeger *et al.* 2022).

Table 36.2. Current and anticipated progress in technologies that will have major impacts on integrated nematode management and soil and plant health.

Technologies	2000–2020	2020–2050
Sanitation	Hot water treatment of plant material. Certified seed and planting material but limited equipment cleaning.	New sanitation techniques (e.g., controlled atmosphere temperature treatment, steam, accelerated electrons). Seed certification more widespread Quarantine expanding.
Rotation and cover crops	Standard but often narrow, decreasing in importance as INM tool. Often limited information on specific nematode host status.	Clever rotations, intercropping, mixed and relay cropping increasingly designed for nematode management. More nematode-targeted crop and cover crop cultivars available.
Biological soil disinfestation	Anaerobic soil disinfestation, solarization, biofumigation not widely used and results often inconsistent.	Integrated programs including a biological oil component more common. More effective agents and methods (e.g., microbial volatiles).
Soil free production	Substrate production for vegetables (mostly EU and Canada). Ultraviolet radiation treatment of irrigation water.	Expanded to include more crops and regions. Ozone generators for decontamination of irrigation water in closed systems
Cultivar selection	Resistant cultivars and rootstocks used where available, relatively few crops with nematode resistance. Tolerant cultivars often preferred over resistant.	New sources of resistance, increasing use of resistant and tolerant cultivars and rootstocks. Expanding the success and acceptance of modern genetic editing (e.g., CRISPR)
Seed-based control	Combined fungicide/nematicide seed coating for row crops introduced.	Increasing availability of chemical and biological seed treatment options combined with nematode resistance/tolerance.
Chemical control	Old chemistry is still widely used but being replaced. First new reduced risk nematicides becoming available.	Older nematicides and fumigants removed from market. New nematicides becoming available (including basipetal treatments) and replacing old nematicides, especially high-value crops
Biological control	Several products with limited or scattered levels of success, and relatively few field data. Effective seed-based treatments are still lacking. Potential use of biochemical treatments still untested.	More competitive agents and strains, including plant- and microbial-derived products. Seed and seedling-based treatments are becoming standard for many crops. Increased use of biologicals for targeted post-plant treatment and partnered with chemicals and bio-chemicals.

Soil antagonistic potential	Presence acknowledged but impact unsatisfactory and not used on a practical scale. Organic agriculture is still leading promoter. Soil microbiome research expanding.	Soil microbiome advances lead to targeted methods to stimulate soil antagonistic potential. Metagenomics and holistic approaches available for monitoring. Holobiont concept used to help monitor nematode populations and improve soil antagonistic potential.
Soil sampling	Field scouting, manual soil, and root sampling.	Remote sensing, drone driven mapping, use of robotics to aid sampling. Analysis of samples using molecular and machine-learning tools, including holistic soil analysis approaches.
Precision agriculture	Technology developed; field application still limited.	Site-specific treatment using sensors and robotics adopted in field and horticultural crops.
Modelling population dynamics	Lack of data on population dynamics limits use. Initiation of remote sensing for detection of nematode distribution tested.	Mathematical modelling is available to calculate population development over rotations and Pi estimation. Remote sensing coupled with practical models used in precision INM. Satellite imaging allows detection of nematode hotspots and nematode densities.
Other supportive tactics	Limited number of decision support tools/systems (DST/ DSS) with low level of implementation.	DST with IT and artificial intelligence expanding especially in horticultural crops. High-resolution data platforms providing the farmer/adviser with detailed information and INM recommendations, facilitating integrated crop management.

The table includes the suggestions and projections made by many of the authors when anticipating how INM will need to change in the coming decades. We need to adjust our approaches to offset the drivers of change: human population growth, food insecurity, soil degradation, climate change and volatility, and the anticipated increase in nematode impact on soil and crop health.

36.5. NEW BUILDING BLOCKS OF INM

The new building blocks of INM as summarized in Table 1 that could improve the future of nematode management are as follows (Desaeger *et al.* 2022):

36.5.1. Chemical control in future needs a new look

Nematicides can be credited for having put the science of nematology firmly on the map. The chemical control of nematodes started changing in the 1960s and 1970s, when some of the less desirable side effects of pesticides and nematicides started to become public. In the past decades, regulatory pressure on broad-spectrum fumigant, carbamate and organophosphate nematicides has led to their removal from the market, and as a result renewed pressure on crop protection companies to develop new and safer nematicides. New and more selective nematicides are entering the market and are expected to be more compatible with INM programs and will replace older nematicides. However, for many resource-poor farmers, the new generation of nematicides may not be an option due to the high cost. The same goes for large acreage field crops, unless nematicides can be applied as seed treatments and be cost effective. The main initial use of these new nematicidal compounds will be in high-value crops like fruits, vegetables, and cash crops.

36.5.2. Biological control receiving prominence

While biological products may show good effects when tested under highly controlled conditions, they often fail to show high levels of control and consistent results under field conditions. Furthermore, biopesticides require suitable formulations and application approaches, which are often difficult to develop or are cost prohibitive. The use of biological nematicides, however, has been growing and will continue to do so, but their success will depend on the efficacy and cost of the products. Growers are always looking for 'quick fix' or 'risk-limiting' solutions to prevent potential crop losses when faced with an urgent or unexpected nematode problem. Efficacy for growers is measured in terms of yield, not in terms of reduced nematode populations. Organic growers, having more limited options, may be willing to pay more, but conventional growers that use biological control agents are much less likely to do so. Public concern about chemical pesticides and governmental regulatory controls will further benefit the growth of biological nematicides, especially when more effective and affordable chemical options become unavailable.

There has been little research in understanding the long-term impacts of nematode biocontrol organisms. Little is known of the side effects of fermentation products and shifts in microbial population structure due to botanicals. In addition, not much is known about the active ingredient(s) themselves. Increased scrutiny could slow down the growth of biological nematicides, but as they are in high demand, the expectation is that many more products will become available in the future.

36.5.3. Plant holobiont concept opens a new perspective

Finally, the holobiont concept opens an exciting and daunting new perspective on the future use of biological agents and INM in general. In brief, a holobiont is defined as the host and all its microbial symbionts, including transient and stable members (Margulis and Fester, 1991; Bordenstein and Thies, 2015). Therefore, growth and development of living organisms, be it the nematode or its plant host, is influenced by the native microbiota that are naturally associated with them. These associated microbiotas may or may not provide protection against introduced or native biological antagonists. The holobiont concept is just developing for PPNs and their hosts. Understanding the interactions between microorganisms from the plant host and microorganisms associated with PPN is a monumental task, but one that will greatly aid INM in the future (Sikora *et al.*, 2008).

36.5.3.1. Plant innate immunity: Plants rely on innate immunity which exist in two forms (Jones and Dangl, 2006):

- Pathogen - associated molecular pattern (PAMP)-triggered immunity (PTI), formerly called horizontal resistance.
- Effector - triggered immunity (ETI), also called R-gene-based immunity and previously known as vertical resistance.

Plant pathogens can use effectors, which are tools or weapons to manipulate their hosts by suppressing PTI mechanisms (Kamoun, 2006). Reciprocally, plants have developed the perception of danger signals through the recognition of effectors. ETI corresponds to a direct or indirect interaction between a pathogen effector and the product of a specific plant resistance gene, which leads to a strong response (i.e., hypersensitive response) consisting of cell apoptosis and local necrosis. In this case, nuclear binding sites (NBSs) interact with a variety of LRR domains to initiate cell death (Boller and Felix, 2009). It must be underlined that the knowledge corpus of plant innate immunity is mostly based on reductionist analyses focusing on one host and one microbe. Photoprotection is also a consequence of biotic interactions where the plant microbiota is suggested to play an important role.

It has been suggested that, upon attack by a pathogen, plants are able to recruit protective microorganisms from their ectosphere reservoir (Berendsen *et al.*, 2012), procuring a clear modulation of host plant immunity by beneficial microbes (Zamioudis and Pieterse, 2012), which induces systemic resistance primed for accelerated activation (Berendsen *et al.*, 2012). To date, the recruitment mechanisms of microorganisms from the ectosphere remain poorly understood, while understanding of the mechanisms by which the plant microbiota avoid the plant's defenses is growing.

36.5.3.2. Microbiota and plant defense 'manipulation': Hormonal modulation of plant immunity (Pieterse *et al.*, 2012) is especially effective through salicylic acid (SA) and jasmonic acid (JA) acting as backbone cellular signal molecules interplaying with complex networks of regulation (Pieterse *et al.*, 2012). Like some successful plant pathogens, certain root endophytes have been shown to be able to 'hijack' or manipulate plant hormone signaling in different ways, by:

- » Synthesizing auxins and auxin analogs along with gibberellins (GAs) (Sirrenberg *et al.*, 2007; Ben Lugtenberg and Kamilova, 2009), which probably attenuate SA signaling (Pieterse *et al.*, 2012);
- » Using effectors which modify the hormonal signaling pathways, as is the case for mycorrhizal fungi (Kloppholz *et al.*, 2011; Plett *et al.*, 2011);
- » Transient accumulation of JA at an early stage of mycorrhiza formation and root-nodule formation, supposedly to 'bypass' the SA-triggered response (Gutjahr and Paszkowski, 2009).

36.5.4. Use of seed treatment will increase

Due to the increased desire to protect seeds from infection, seed treatments using chemical and biological agents is one of the fastest growing market segments. Most of the major seed companies now offer seed not just coated with fungicides and insecticides, but also with a nematicide. Farmers prefer the convenience of seed treatments and for that reason their use will continue to increase. In particular, the availability of biological nematicide and microbial seed treatments is expected to grow significantly in the next decades.

Seed treatments offer many advantages to the farmer by saving time and reducing the overall amount of product applied per field. However, seed treatment efficacy is generally lower as compared to soil applications. Seeds have and will continue to become more and more valuable, equipped internally with proprietary traits and externally with protectants and stimulants. Furthermore, seed treatments will only reduce early root infection processes and increase yield, they will not lead to a reduction in overall population densities.

36.5.5. Resistance breeding needs more emphasis

Breeding for nematode resistance is probably the most desirable and urgent future need. The problem is that except for potato and beet cyst nematodes, and in limited cases root-knot on tomato and some green manure crops, nematodes are rarely perceived as the number one priority for resistance breeding. Resistance breeding is a lengthy and expensive process, and requires a significant and long-term investment. The availability of large and diverse germplasm collections is also important, especially traditional locally adapted cultivars. There is some mild optimism that interest in nematode resistance is increasing among seed companies, but for nematode resistance to become more widely available, a much more concerted effort from public and private entities will be needed. For large acreage crops, susceptible rotational crops should be targeted for resistance breeding as well, as this would help to reduce the overall nematode pressure on the main cash crop in the rotation.

In some instances, tolerance could be an attractive strategy as well. The detection of genes for tolerance to beet cyst nematode led to the development of tolerant, tolerant-resistant and resistant-intolerant beet cultivars. Tolerant cultivars are less costly to the grower than resistant cultivars and yield higher.

Genome-editing technologies, especially CRISPR/Cas9, provide new and faster avenues for resistance breeding, but societal acceptance for these technologies will remain a major hurdle in much of the world. The emergence of genome-editing tools combined with our progress in understanding plant–nematode interactions could provide many new opportunities for nematode management using biotechnology. RNA interference and other genetic approaches could bring broad spectrum and long-lasting new solutions but predicting the time scales for development of these approaches is about as uncertain as the weather.

Resistance is certainly not a single shot solution, as nematodes will find a way to break the resistance. This is especially true when nematode management relies solely on the use of resistant cultivars without resistance management, again stressing the need for effective resistance management tactics as part of an INM approach.

36.5.6. Antagonistic potential of suppressive soil

Stimulation of the soil antagonistic potential to improve suppressiveness is a future resource for nematode management. There is a need for a holistic approach in which the agroecological system is considered rather than looking at one pest or pathogen. Significant advances are being made in our knowledge of how complex microbial communities influence root health and plant growth. The important functions the microbial communities provide include outcompeting and antagonizing nematodes and other diseases and in stimulating root defences and plant growth. A better understanding of how these communities interact with one another, with crops, and with the surrounding soil environment will provide predictive power to our understanding of nematode suppressiveness and how it will respond to changes in climate, cropping system and soil management.

The key feature of the antagonistic potential in agricultural soils is the reduction of root damage from soil-borne diseases and nematodes. We need to be able to use INM management tools and modern technologies to monitor and then manipulate this natural biocontrol phenomenon in favor of stable root health (Sikora, 1992). One piece of the puzzle may be the emerging understanding of the role of soil and plant microbiomes and of organisms as holobionts. Such research has just started for PPN, but one day it may be possible to predict which functions are mediated by a certain microbiome and which functions must be provided externally by (for example) organic compounds, plant extracts or microorganisms. It is well known that low levels of organic matter reduce natural nematode population regulation. Especially in naturally poor soils, insufficient return of organic matter to agricultural land has led to widespread and severe soil degradation across the world.

Microbiome-based solutions for crop protection are now actively being pursued by many companies. Some of these novel methods to help manipulate whole soil communities include whole soil microbiome plus microfauna inoculation, microbiome engineering and *ex situ* cultivation of soil microbiomes (Ke *et al.*, 2021). As most soil processes are regulated not only by soil microbial communities, but also by the whole soil food web. The survival and successful establishment of newly introduced bacteria or fungi to stimulate nematode suppressiveness will depend in large part on the regulation of their abundance by the entire soil community and soil food web after introduction.

36.5.7. Adaption to climate change

In many areas of the world soil biodiversity will be strongly affected by changes in soil temperature and soil moisture. Increases in temperature will cause a significant loss in organic matter which will have negative effects on microbial communities and the antagonistic buffering capacity of ecosystem services. Rain-fed crops in climate change hotspots will become more stressed and require new management approaches, which offset heat and drought stress as well as nematode root infection that synergistically increases the impact on plant growth of these two abiotic constraints. Root health management combined with development of drought resistant crops will be even more critical in the future.

The high-tech Netherlands-style greenhouse farming (protected cultivation) has made this country the second largest vegetable exporter in the world. There can be little doubt that this type of vegetable farming will expand in many regions. We may someday see skyscrapers of vertical vegetable farms where farming will be completely independent of the outside climate. Nematode management in such systems where artificial growing media are used, will be mostly a sanitation issue. However, in many places this technology will not be economically feasible, and more traditional soil-based farming will continue to prevail.

36.5.8. Developing solutions on a regional basis

The face of agriculture is changing rapidly all over the world with corporate farming and high-tech greenhouses transforming the way food is being produced, and how nematodes are managed. At the same time, while agriculture is changing rapidly in some regions, this is certainly not the case everywhere. Access to new technology, seed and credit is not available to many smallholder farmers who represent 98% of the farmers in the world. These farmers probably suffer more loss from nematodes than anyone else, and it should be clear that future technological advances in nematode management are of no value to them, at least in the short to mid-term. This emphasizes the importance of developing solutions on a regional basis and having nematologists working closely with farmers across the world. So, the answer to the question 'what should we expect in the future?' will very much depend on where this future happens to be. There is a dire need for nematology centers to link more closely with nematologists in food-insecure countries around the world to improve their expertise and access to modern technology that made and are making major strides in improving nematology around the world (Cortada *et al.*, 2019) [e.g. International Meloidogyne Project (IMP) at North Carolina State University USA, The Nematology Initiative for Eastern and Southern Africa (NIESA) at Rothamsted Experimental Station UK, and Nematology Education in Sub-Saharan Africa (NEMEDUSSA) at Ghent University, Belgium].

36.5.9. Loss of nematology positions

The past decades have experienced a consistent loss of applied nematology positions at universities and plant protection agencies and the scope of research has been narrowed by focusing on advanced technologies, e.g., remote sensing, genetics, and molecular aspects. Applied nematology is a time-consuming discipline with no quick solutions. This is especially true for INM which is a systems approach and highly data driven. Therefore, for INM to become successful, more applied nematology positions are needed, and a re-alignment of priorities in nematology science is needed, especially regarding the training of students in INM.

Furthermore, nematologists will have to become more effective and focused at exchanging ideas and findings. One way to do this is through global collaborative projects that are specifically targeted at nematode management, as was done in the IMP mentioned above and now within NEMEDUSSA. Most collaborative projects tend to have a regional focus, combining different competencies that may or may not include a nematologist. This means that nematologists can easily become isolated and deprived of new insights. Our knowledge and understanding of nematodes and their environment has grown tremendously in recent years, but at the same time there seems to be a massive disconnect between fundamental and applied research. A new global nematology projects could help bridge this gap.

Funding is the ultimate bottleneck, and it is sad but true that researchers have to spend an exorbitant amount of time writing and managing grant proposals. Nematodes are rarely on top of funding agencies priority lists, which is why it is imperative that nematologists continue to raise nematode awareness among the public. Such activities are the domain of the extension nematologist, which is unfortunately becoming an increasingly rare breed. While social media can help to popularize nematodes and increase awareness, it is no substitute for the 'boots on the ground' approach of traditional extension nematology.

36.5.10. Whether farmers are adopting INM technologies recommended?

The recommended nematode management practices should be practical and economically attractive to the farmer. With nematodes being only one of the many constraints that farmers face, INM recommendations will have a much greater chance of being adopted if they can be easily integrated with other management practices. This is the reason why soil fumigants became popular as they provided weed, disease, and nematode control all at once. Similarly, the use of cover crops has been a traditional practice for farmers as it has many benefits in terms of soil fertility and quality, in addition to the potential benefits in terms of nematode management. We should keep in mind that ultimately INM is not a goal in itself, but just one of the many components of a broader soil, crop and farm management plan.

If INM programs are to be fully accepted by farmers, it will be necessary for nematologists to work closely with agronomists when developing INM programs, so they become a component of larger integrated crop management programs. Undoubtedly, more and more INM extension will move to online sites, and we need to ensure that such information is easily accessible and understandable. Decision support tools for INM therefore need to be developed on a regional basis and with input from growers, only then will they be acceptable and implemented by farmers and lead to more effective and sustainable nematode management.

36.6. CONCLUSION

INM is not a set of fixed practices, but rather a philosophy, and will have many different shapes and forms depending on the cropping system and region (Sikora and Roberts, 2018). Only when INM is fully integrated in crop and farm management plans will it be acceptable to farmers and able to improve food safety, the livelihood and well-being of farmers, and strengthen the perception of farmers as good environmental stewards.

Nematode management has acquired a new urgency in recent years, not in the least because applied nematology has been vastly neglected and underfunded for decades. This has created a huge gap with few nematology graduates trained in applied nematology and able to connect with industry and farmers. For INM to become a reality, applied nematology needs to be at the forefront of the science of nematology again, and funded accordingly. Only then will the advances in molecular methods, artificial intelligence, and precision agriculture lead to tangible, sustainable solutions for the world's food production (Desaeger *et al.* 2022).

References

Abawi, G.S. and Barker, K.R. 1984. Effects of cultivar, soil temperature, and population levels of *Meloidogyne incognita* on root necrosis and *Fusarium* wilt of tomatoes. *Phytopathology* 74: 433–438.

Abawi, G.S., Widmer, T.L., Ludwig, J.W. and Mitkowski, N.A. 1999. Biology and management of *Meloidogyne hapla* on carrots, onions and lettuce in New York. *Journal of Nematology* 31(4): 521-585.

Abd El-Aal, E.M., Shahen, M., Samy Sayed, S. *et al.* 2021. *In vivo* and *in vitro* management of *Meloidogyne incognita* (Tylenchida: Heteroderidae) using rhizosphere bacteria, *Pseudomonas* spp. and *Serratia* spp. compared with oxamyl. *Saudi J Biol Sci.* 28(9): 4876–4883.

Abd-Elgawad, M.M.M. 2020. Optimizing biological control agents for controlling nematodes of tomato in Egypt. *Egypt J Biol Pest Cont* 30, 58. https://doi.org/10.1186/s41938-020 -00252-x.

Abd-Elgawad M.M.M. and Askary T.H. 2015. Impact of phytonematodes on agriculture economy. In: T.H. Askary and P.R.P. Martinelli (eds.) *Biocontrol Agents of Phytonematodes*, pp. 3–49. CABI, Wallingford.

Abuzar, S. 2012. Response of field pea (*Pisum sativum*) selections/ cultivars to root-knot nematode (*Meloidogyne incognita*) – wilt fungus (*Fusarium oxysporum*) disease complex under greenhouse conditions. *Int. J. Appl. Res. Studies* 1(2): 1-9.

Abuzar, S. and Haseeb, A. 2006. Efficacy of carbofuran, bavistin, neem, *Trichoderma harzianum* and *Aspergillus niger* against *Meloidogyne incognita* and *Fusarium oxysporum* f. sp. *vasinfectum* disease complex on okra. *Indian J. Nematol.* 36: 299-301.

Afriyani, Maulidia, V., Alfizar. and Sriwati, R. 2019. Endophytic bacteria (genus: *Pseudomonas* spp.) isolated from Aceh bamboo root as biological agent againts nematode *Meloidogyne* spp. *Conf. Ser.: Earth Environ. Sci.* 425 012074. doi:10.1088/1755-1315/ 425/1/012074.

Agbaglo, S.Y., Nyaku, S.T., Vigbedor, H.D. and Cornelius, E.W. 2020. Pathogenicity of *Meloidogyne incognita* and *Fusarium oxysporum* f. sp. *vasinfectum* on growth and yield of two okra varieties cultivated in Ghana. *International Journal of Agronomy* Article ID 8824165, https://doi.org/10.1155/2020/8824165.

Ahmed, F.A and Choudhury, B.N. 2004a. Management of *Meloidogyne incognita* in French bean through crop rotation. *Ann. Plant Prot. Sci.* 12: 118-120.

Ahmed, F.A. and Choudhury, B.N. 2004b. Management of root-knot nematode, *Meloidogyne incognita* using organic amendments in French bean. *Indian J. Nematol.* 24: 137-139.

Ahmed, F.A and Choudhury, B.N. 2004c. Effect of rabbing on *Meloidogyne incognita* in French bean. *Ann. Plant Prot. Sci.* 12: 156-158.

Ahuja, S. 1983a. Studies on the variability in infection of root-knot nematode *Meloidogyne incognita* on potato. *Third All India Nematol. Symp.*, Y.S. Parmar Univ. Hort. Forest., Solan, p. 4.

Ahuja, S. 1983b. Comparative efficacy of some nematicides against the root-knot nematode, *Meloidogyne incognita* in tomato, okra, brinjal and cauliflower. *Third Nematological Symposium,* Y.S. Parmar Univ. Hort. Forest., Solan, P. 43.

Ajirloo, A.R., Shaaban, M. and Motlagh, Z.R. 2015. Effect of K nano-fertilizer and N bio-fertilizer on yield and yield components of tomato (*Lycopersicon esculentum* L.). *Int. J. Adv. Biol. Biom. Res.* 3(1): 138-143.

Alam, M.M., Ali, Q.G., Masood, A. and Khan, A.M. 1976. Studies on the chemical changes by the infection of the stunt nematode (*Tylenchorhynchus brassicae*) in cabbage and cauliflower roots. *Indian J. Expt. Biol.* 14: 517-518.

Alam, M.M., Khan, A.M. and Saxena, S.K. 1973. Efficacy of 'Vydate' oxamyl for the control of root-knot nematode, *Meloidogyne incognita* on eggplant and okra. *Indian Journal of Nematology* 3: 148-152.

Alam, M.M., Khan, A.M. and Saxena, S.K. 1974. Reaction of some cultivated varieties of eggplant, pepper and okra to the root-knot nematode, *Meloidogyne incognita. Indian Journal of Nematology* 4: 64-68.

Alam, M.M., Saxena, S.K. and Khan, A.M. 1973. Control of root-knot nematode, *Meloidogyne incognita* (Kofoid & White, 1919) Chitwood, 1949 on tomato and eggplant with V. C. -13 and Basamid liquid as bare-root dip. *Indian Journal of Nematology* 3: 154-156.

Alam, M.M., Saxena, S.K. and Khan, A.M. 1977. Influence of interculture of marigold and margosa with some vegetable crops on plant growth and nematode population. *Acta Botanica Indica* 5: 33-39.

Alam, M.M., Siddiqi, Z.A., Khan, A.M. and Saxena, S.K. 1980. Effect of different cropping sequences on the population of plant parasitic nematodes. *Indian Journal of Nematology* 10: 35-39.

Albajes, R., Gullino, M.L., Lenteren, J.C. and Elad, Y. (eds.). 1999. *Integrated Pest and Disease Management in Greenhouse Crops.* Kluwer Academic Publishers. The Netherlands.

Al-Hazmi, A.S. 1985. Interaction of *Meloidogyne incognita* and *Macrophomina phaseolina* in a root-rot disease complex of French bean. *J. Phytopathol.* https://doi.org/10.1111/j. 1439-0434.1985. tb04831.x.

Ali, M.A., Azeem, F., Abbas, A. *et al.* 2017. Transgenic strategies for enhancement of nematode resistance in plants. *Frontiers in Plant Science* 8: 750.

Ali, M.A., Shahzadi, M., Zahoor, A. *et al.* 2019. Resistance to cereal cyst nematodes in wheat and barley: an emphasis on classical and modern approaches. *International Journal of Molecular Sciences* 20: 432.

Ali, S.S. 1995. *Nematode Problems in Chickpea.* Indian Institute of Pulses Research, Kanpur, India.

Ali, S.S. and Sharma, S.B. 2003. Nematode survey of chickpea production areas in Rajasthan, India. *Nematol. Mediterr.* 31: 147–149.

Althaf, M., Widayati, W. and Purnawati, A. 2022. Endophytic bacteria from wet soil of South Kalimantan as biological control agent for root nematodes (NPA) in celery (*Apium graveolens*). *Seminar Nasional Magsiter Agroteknologi 2022.* NST Proceedings, pp. 19-23. doi: 10.11594/ nstp.2023.3203.

Amalendu Chatterjee, 1995. Management of Phytonematodes by bioagents. *Indian Sci. Congress*, Calcutta.

Amalraj, S.F.A. and Rao, G.R. 1992. Potato cyst nematode pathogens in the southern hills of India. *Indian J. Nematol.* 22: 82-85.

Anita, B. 2012. Crucifer vegetable leaf wastes as biofumigants for the management of root knot nematode (*Meloidogyne hapla* Chitwood) in celery (*Apium graveolens* L.). *J. Biopest.* 5 (Suppl.): 111-114.

Anita, B. and Selvaraj, N. 2011. Biology, yield loss and integrated management of root-knot nematode, *Meloidogyne hapla* infecting carrot in Nilgiris. *Indian J. Nematol.* 41(2): 144-149.

Anon, 1985. *Kufri Swarna, A High Yielding Late Blight and Cyst Nematode Resistant Potato Cultivar suitable for Cultivation in Tamil Nadu Hills.* No. 21(E), Central Potato Research Institute, Shimla, 4 pp.

Anon. 1993a. *QRT Report of AICRP on Nematodes (1989-93)*, Dept. of Nematol., Haryana Agri. Univ., Hisar, India, 149 pp.

Anon. 1993b. *Biennial Report (1991-93) of AICRP on Plant Parasitic Nematodes with Integrated Approach for Their Control*, Dept. of Nematology, Haryana Agri. Univ., Hisar, 80 pp.

Anon. 1994. *Biennial Report for AICRP on Nematode Pests of Crops and their Management.* Dept. of Entomol., Y.S. Parmar Univ. of Hort. & Forestry, Solan, Himachal Pradesh, India.

Anon. 2011. *Quinquennial Report.* AICRP on Nematode pests of crops & their Control, Palampur centre, Himachal Pradesh, 98 pp.

Anon. 2012a. *Annual Report 2011-12.* Indian Inst. of Hort. Res., Bangalore, pp. 23-24.

Anon. 2012b. *Biennial Report (2010-2011).* AICRP on Nematode Pests of Crops & their Control, Jorhat Centre, AAU, Jorhat.

Anon. 2015. *Biennial Report*, AICRP on Nematode Pests of Crops & their Control, Palampur centre, Himachal Pradesh, 47 pp.

Anon. 2017. *Biennial Report (2015-2016)*, AICRP on Nematode Pests of Crops & their Control, Assam Agri. Univ., Jorhat.

Anon. 2019. Ministry of agriculture and Farmers welfare of Agriculture and Co-operation. Directorate of Economics and Statistics (http://dacnet.nic.in Retrieved on 2019).

Ansari, S., Ahmad, G., Elhakem, A. *et al.* 2021. Effects of organic and inorganic fertilization with bio-inoculants on the sustainable management of plant-parasitic nematodes infesting okra (*Abelmoschus esculentus*). *Notulae Botanicae Horti Agrobotanici Cluj-Napoca* 49(4): 12544. https://doi.org/10.15835/nbha 49412544.

Anver, S. 2020. Biomanagement of root-knot and reniform nematodes on okra. *Int. Res. J. Pharm.* 11(12): 53-55.

Anwar, A. and Verma, A.C. 1994. How to control root-knot nematode in parwal. *Farm Digest* (NDUAT Supplement): 25.

Anwar, A., Mughal, N.M., Shahnaz, E. *et al.* 2021. Management of root-knot nematode, *Meloidogyne incognita* dreaded invading in pointed gourd (*Trichosanthes dioica* Roxb.) crop prone to eastern U.P. of India. In C. Bello and T.E. Kaspary (eds.) *Nematodes - Recent Advances, Management and New Perspectives.* http://dx.doi.org/10.5772/intechopen. 94719.

Atwal, A.S. and Mangar, A. 1969. Repellent action of root exudates of *Sesamum orientale* against the root-knot nematode, *Meloidogyne incognita* (Heteroderidae: Nematoda). *Indian Journal of Entomology* 31: 286.

Ayyar, P.N.K. 1926. A preliminary note on the root nematode, *Heterodera radicicola* Mitter and its economic importance in South India. *Sci. Cult.* 22: 391-393.

Ayyar, R.K. 1933. Some experiments on the control of root-gall nematode (*Heterodera radicicola* Greeff) in South India. *Madras Agricultural Journal* 21: 97-107.

Baghel, K.S. and Singh, R. 2013. Alarming population of *Pratylenchus* spp. in chickpea growing areas in Rewa and its vicinity and its effect on plant growth and nodulation. *Environ. Ecol.* 31: 328–333.

Baheti B.L., Yadav P.K., Nama C.P. and Khandelwal S.K. 2018. Environment friendly management of root-knot nematode, *Meloidogyne incognita* infecting cluster bean (*Cyamopsis tetragonoloba* L.). *Indian J. Nematol.* 48: 56-61

Baheti, B.L., Yadav, P.K., Samdani, K.L. and Nama, C.P. 2017. Efficacy of pesticides as seed treatment against root-knot nematode, *Meloidogyne incognita* on cluster bean (*Cyamopsis tetragonoloba* L.). *Curr. Nematol.* 28(1): 89–94.

Bajaj, K.L. and Mahajan, R. 1977. Phenolic constituents of tomato resistant and susceptible to *Meloidogyne incognita* (Kofoid & White) Chitwood. *Nematologia Mediterranea* 5: 329-333.

Baker, R., Paulitz, T., Windham, M.H. and Elad, Y. 1986. Enhancement of growth of ornamentals by a biological control agent. *Colorado Greenhouse Grow. Assoc. Res. Bull.* 431:1.

Ban, S.G., Žanić, K., Dumičić, G. *et al.* 2014. Growth and yield of grafted cucumbers in soil infested with root-knot nematodes. *Chilean J. Agri. Res.* 74(1): 29-34.

Barua, L. and Bora, B.C. 2009. Compatibility of *Trichoderma harzianum* and *Pseudomonas fluorescens* against *Meloidogyne incognita* and *Ralstonia solanacearum* complex on brinjal. *Indian J. Nematol.* 39: 29-34.

Bawa, J.A., Sani, Z.R., Mohammed, I. *et al.* 2016. Incidence and population of plant parasitic nematodes in green amaranth (*Amaranthus hybridus* L.) (Caryophyllales: Amaranthaceae) from three selected areas in Dutsin-Ma Town, Katsina State, Nigeria. *Brazilian Journal of Biological Sciences* 3(6): 319-330.

Becker, W.F., 1999. The effect of abamectin on garlic infected by *Ditylenchus dipsaci. Nematologia Brasileira* 23: 1-8.

Bello A. 1998. Biofumigation and integrated pest management. In A. Bello, J.A. González, M. Arias and R. Rodriguez-Kábana (eds.) *Alternatives to Methyl Bromide for The Southern European Countries*, pp. 99-126. DG XI EU, CSIC, Valencia.

Berendsen, R.L., Pieterse, C.M.J. and Bakker, P.A.H.M. 2012. The rhizosphere microbiome and plant health. *Trends in Plant Science* 17: 478– 486.

Berlinger, M.J., Jarvis, W.R., Jewett, T.J. and Lebiush-Mordechi, S. 1999. Managing the greenhouse crop and crop environment. In: R. Albajes, M.L. Gullino, J.C. van Lenteren and Y. Elad (eds.) *Integrated Pest and Disease Management in Greenhouse Crops*, pp. 97-123. Kluwer Academic Publishers, Netherlands.

Bhagawati, B., Choudhury, B.N and Sinha, A.K. 2009. Management of *Meloidogyne incognita-Rhizoctonia solani* complex on okra through bioagents. *Indian J. Nematol.* 39: 156-161.

Bhagawati, B., Das, B.C. and Sinha, A.K. 2007. Interaction of *Meloidogyne incognita* and *Rhizoctonia solani* on okra. *Ann. Plant Prot. Sci.* 15: 533-535.

Bhagawati, B and Goswami, B.K. 2000. Interaction of *M. incognita* and *Fusarium oxysporum* f. sp. *lycopersici* on tomato. *Indian J. Nematol.* 30: 93-94.

Bhagawati, B., Goswami, B.K. and Singh, C.S. 2000. Management of disease complex of tomato caused by *M. incognita* and *Fusarium oxysporum* f. sp. *lycopersici* through bioagents. *Indian J. Nematol.* 30: 16-22.

Bhagawati, B. and Phukan, P.N. 1990. Chemical control of *Meloidogyne incognita* on pea. *Indian J. Nematol.* 20: 73-83.

Bhardwaj, A.K., Sharma, P.L. and Thakur, J.R. 1972. Mushroom nematodes. *Indian Bot. Reptr.* 42: 257–258.

Bhati, S.S., Baheti, B.L. and Kumhar, R.N. 2019. Bio-agents: An effective method for suppression of root-knot nematode, *Meloidogyne incognita* infecting bitter gourd (*Momordica charantia* L.) as seed treatment. *Indian J. Ent. Zool. Studies* 7(5):1197-1201.

Bhattacharya. and Swarup, G, 1989. *Pasteuria pe*netrans a pathogen of the genus *Heterodera*, its effect on nematode biology and control. *Indian Journal of Nematology* 18: 61-70.

Bhatti, D.S. and Jain, R.K. 1977. Estimation of loss in okra, tomato and brinjal yield due to *Meloidogyne incognita. Indian J. Nematol.* 7: 37-41.

Bhosle, B.B., Sehgal, M., Puri, S.N. *et al.* 2006. Efficacy of organic amendment in management of root-knot nematode, *Meloidogyne incognita* on okra. *Indian Journal of Nematology* 36: 37-40.

Bird, G.W. 1981. Integrated nematode management for plant protection. In B.M. Zuckerman *et al.* (eds.) *Plant Parasitic Nematodes* Vol. III, pp. 355–375. Academic Press, New York.

Bishnoi, S.P. and Yadav, B.S. 1989. Effect of some host plants on biology of *Rotylenchulus reniformis* Linford and Oliveira, 1940. *International Nematology Network Newsletter* 6: 10-11.

Biswas, B.C. 2010. Fertigation in hi-tech agriculture a success story of a lady farmer. *Fertil. Mark. News.* 41(10): 4- 8.

Black, L.L., Wu, D.L., Wans, J.F. *et al.* 2003. *Grafting Tomato for Production of in Hot–Wet Season.* AVRDC Publication, Taiwan, pp. 1-6: 03-551.

Boag, B. 1979. Nematodes associated with carrots in Scotland. *Annals of Applied Biology* 93: 199-204.

Boller, T. and Felix, G. 2009. A renaissance of elicitors: perception of microbe-associated molecular patterns and danger signals by pattern-recognition receptors. *Annual Review of Plant Biology* 60: 379– 406.

Bordenstein, S.R. and Theis, K.R. 2015. Host biology in light of the microbiome: Ten principles of holobionts and hologenomes. *PLoS Biology* 13: e1002226.

Bozkurt, S., Uygur, V., Agca, N. and Yalcin, M. 2011. Yield responses of cauliflower (*Brassica oleracea* var. *botrytis* L.) to different water and nitrogen levels in a Mediterranean coastal area. *Acta Agriculturae Scandinavia Section B-Soil and Plant Science* 61: 183-194.

Calvache, M. and Reichardt, K. 1999. Effects of water stress imposed at different plant growth stages of common bean (*Pahaseolus vulgaris*) on yield and N_2 fixation. In C. Kirda, P. Moutonnet, C. Hera and D.R. Nielsen (eds.) *Crop Yield Response to Deficit Irrigation*. Kluwer Academic Publishers, Dordrecht, The Netherlands.

Castagnone-Sereno, P. and Kermarrec, A. 1991. Invasion of tomato roots and reproduction of

Meloidogyne incognita as affected by raw sewage sludge. *Supplement to Journal of Nematology* 23: 724-728.

Castillo, P., Mora-Rodríguez, M.P., Navas-Cortés, J.A. and Jiménez-Díaz, R.M. 1998. Interactions between *Pratylenchus thornei* and *Fusarium oxysporum* f. sp. *ciceris* on chickpea. *Phytopathology* 88: 836–844.

Castillo, P. and Vovlas, N. 2007. Pratylenchus *(Nematoda: Pratylenchidae): Diagnosis, Biology, Pathogenicity and Management*. Brill, Leiden, Netherlands.

Castle, L.A., Wu, G. and McElroy, D. 2006. Agricultural input traits: Past, present and future. *Curr. Opin. Biotechnol.* 17: 105-112.

Caswell, E.P., deFrank, J., Apt, W.J. and Tang, C.S. 1991. Influence of nonhost plants on population decline of *Rotylenchulus reniformis*. *J. Nematol.* 23(1): 91-98.

Caubel, G., Curvale, J.P., Hemery, F. and Le Bohec, J. 1981. Observations écologiques et experimentations des méthodes de lutte contre *Pratylenchus penetrans*, nématode nuisible à l'"artichaut en Bretagne. In: *3 Congresso Internazionale di Studi sul Carciofo*, 1979, Bari, Italy, pp. 985-1002.

Caubel, G., Hemery, J. and Le Bohec, F. 1978. Observations sur l'"effet de *Tagetes patula* L. et *T. erecta* L. sur *Pratylenchus penetrans* Cobb. *Extrait de la Revue La Defense des Vegetaux* 194: 1-10.

Cayrol, J.C. 1967. E'tude de cycle e'volutif d' *Aphelenchoides composticola*. *Nematologica* 13: 23-32.

Cayrol, J.C. 1974. *Proc. XII Intnl. Symp. De Nematologia*, Society of European Nematologists, Grenada, Spain, pp. 21-22.

Celetti, M. 2009. *Bulb and Stem Nematodes in Garlic*. Ontario ministry of Agriculture Food and Rural Affairs (OMAFRA). Ontario, Canada.

Celetti, M. 2012. *Fusarium Basal Plate Rot of Onion and Garlic*. Accessed online March 2017. https://onvegetables.com/2012/07/24/fusarium-basal-plate-rot-of-onion-and garlic/.

Chahal, P.P.K. and Chahal, V.P.S. 1988. Biological control of root-knot nematode with *Azotobacter chroococcum* on brinjal. In M.A. Maqbool, A. Morgan Golden, A. Gaffar and L.R. Krusberg (eds.) *Advances in Plant Nematology*, pp. 257-263. Natl. Nematol. Centre, Univ. of Karachi, Karachi, Pakistan.

Chahal, P.P.K. and Chhabra, H.K. 1984. Interaction of *Meloidogyne incognita* with *Rhizoctonia solani* on tomato. *Indian Journal of Nematology* 14: 56-57.

Chahal, V.P.S. and Chahal, P.P.K. 1991. Control of *Meloidogyne incognita* with *Bacillus thuringiensis*. In R.J. Wright, V.C. Baligar and R.P. Newman (eds.) *Plant and Soil Interaction*, pp. 677-680. Klower Academic Publishers.

Chaitali, L., Singh, S. and Goswami, B.K. 2003. Effect of cakes with *Trichoderma viride* for the management of disease complex caused by *Rhizoctonia bataticola* and *Meloidogyne incognita* on okra. *Ann. Plant Prot. Sci.* 11: 178-180.

Chakraborti, S. 2000. Management approaches for root-knot nematode in pointed gourd. *Indian J. Nematol.* 30: 136–140.

Chandel, Y.S., Ravichandra, N.G., Mhase, N.L. *et al.* 2014. Integrated management of root-knot nematodes (*Meloidogyne spp.*) under protected cultivation. *Indian J. Nematol.* 44(1): 92-96.

Chandel, Y.S. and Sharma, N.K. 1991. Relative susceptibility of some commonly grown mushroom species to *Aphelenchoides sacchari*. *Indian Phytopath.* 44: 143-144.

Chauhan, J.S., Rajendra Prasad, S. and Pal, S. 2013. Quality seed and productivity enhancement in major crops in India. In: A.L. Singh *et al.* (eds.) *Current Trends in Plant Biology Research.* National Conference of Plant Physiology. Directorate of Ground nut Research, Junagadh, India, pp. 160-168.

Chaurasiya, P.C. and Sahu, G.D. 2016. Effect of different type of mulch, fertigation and drip irrigation in chilli on open field condition. *International Journal of Advanced Multidisciplinary Research* 1(1): 1-6.

Chaya, M.K., Rao, M.S. and Ramachandran, N. 2010. Biomanagement of disease complex in okra (*Abelmoschus esculentus* L. Moench) using microbial antagonists. *National Conf. on Innovations in Nematol. Res. for Agri. Sustainability – Challenges and A Roadmap Ahead*, Tamil Nadu Agri. Univ., Coimbatore, pp. 96-97.

Chen, J., Abawi, G.S. and Zuckerman, B.M. 2000. Efficacy of *Bacillus thuringiensis, Paecilomyces marguandii* and *Streptomyces costaricanus* with and without organic amendments against *Meloidogyne hapla* infecting lettuce. *J. Nematol.* 32: 70-77.

Chhabra, H.K. and Sharma, J.K. 1981. Combined effect of *Meloidogyne incognita* and *Rhizoctonia bataticola* on pre-emergence damping-off of okra and brinjal. *Science and Culture* 47: 256-257.

Chhabra, H.K., Sidhu, A.S. and Singh, I. 1977. *Meloidogyne incognita* and *Rhizoctonia solani* interaction on okra. *Indian Journal of Nematology* 7: 54-57.

Choudhury, B.N., Bhagawati, B. and Bora, B.C. 2011. *Three decades of Nematology in Assam (1977-2011).* AICRP on Plant Parasitic Nematodes with Integrated Approach for their Control, Dept. of Nematol., Assam Agri. Univ., Jorhat, 49 pp.

Cortada, L., Inge Dehennin I., Bert, W. and Coyne, D. 2019. Integration of nematology as a training and research discipline in sub-Saharan Africa: progress and prospects. *Nematology* 2: 1–21.

Curto, G., Lazzeri, L., Dallavalle, E. and Santi, R. 2008. Management of *Meloidogyne incognita* (Kofoid *et* White) Chitw. in organic horticulture. *Third Int. Biofumigation Symp.*, Canberra, Australia.

Curto, G., Lazzeri, L., Santi, R. and Dallavalle, E. 2006. Evaluation of alternative strategies for the control of southern root-knot nematode *Meloidogyne incognita* (Kofoid *et* White) Chitw., on tomato crop, in plastic greenhouse. *Proceedings of the Second International Biofumigation Symposium*, Moscow, Idaho. Springer.

Dahiya, R.S., Bansal, R.K. and Paruthi, I.J. 1998. Efficacy of mustard cake and poultry manure for the management of *Meloidogyne javanica* in bottle gourd and okra. *Proc. of National Symp. on Rational Approaches in Nematode Management for Sustainable Agri.*, Nematol. Soc. of India, New Delhi, pp. 100-102.

Dalal, M.R. and Vats, R. 1998. Estimation of loss in pea (*Pisum sativum* L.) due to *Rotylenchulus reniformis.* In Usha K. Mehta (ed.) *Nematology – Challenges and Opportunities in 21st Century,* pp. 8-9. Sugarcane Breeding Institute, Coimbatore.

Dalla Costa, L. and Giovanardi, R. 2000. Nitrogen use efficiency in tomato and potato as affected by water regime and N fertilization. *In* G. Parente and J. Frame *(eds.) Proc. COST 814 Final Conf., pp.* 443–447. Pordenone, Italy.

Darekar, K.S. and Mahse, N.L. 1988. Assessment of yield losses due to root-knot nematode, *Meloidogyne incognita* Race 3 in tomato, brinjal and bitter gourd. *Int. Nematol. Network Newsl.* 5(4): 7-9.

Darekar, K.S. and Patil, B.D. 1981. Reaction of some cowpea varieties to root-knot nematodes. *Indian Journal of Nematology* 11: 83-84.

Das, N. and Sinha, A.K. 2005. Integrated management of root-knot nematode (*Meloidogyne incognita*) on okra (*Abelmoschus esculentus* (L.) Moench). *Indian Journal of Nematology* **35**: 175-182.

Das, A.K. 1994. *Assessment of Yield Loss due to* Meloidogyne incognita *on French Bean and its Management.* M. Sc. (Agri.) thesis, Assam Agri. Univ., Jorhat.

Dasgupta, D.R. and Seshadri, A.R. 1971. Races of the reniform nematode, *Rotylenchulus reniformis* Linford and Oliveira, 1940. *Indian J. Nematol.* 1(1): 21-24.

Davide, R.G. and Zorilla, R.A. 1986. Evaluation of a fungus, *Paecilomyces lilacinus* for the biological control of root-knot nematodes *Meloidogyne incognita* on okra. *International Nematology Network Newsletter* 3(3): 32-33.

Dawar, S., Tariq, M. and Zaki, M.J. 2008. Application of *Bacillus* species in control of *Meloidogyne javanica* (Treub) Chitwood on cowpea and mash bean. *Pak. J. Bot.* 40(1): 439-444.

Dayal, S. and Sharma, M.K. 2007. Management of reniform nematode, *Rotylenchulus reniformis* on mung bean through seed dressing. *Indian J. Nematol.* 37: 89–91.

De Mastro, G., D'Addabbo, T., Lazzeri, L., Verdini, L. and Grassano, N. 2008a. Effectiveness of biofumigating *Brassica* treatments for the control of root-knot nematode *Meloidogyne incognita* on melon and watermelon. *Third Int. Biofumigation Symp.*, Canberra, Australia.

De Mastro, G., D'Addabbo, T., Verdini, L. and Radicci, V. 2008b. Biofumigation in greenhouse for the control of root-knot nematodes. *Third Int. Biofumigation Symp.*, Canberra, Australia.

De Mastro G., D'Addabbo T., Verdini L. and Radicci V. 2008c. Control of carrot root-knot and cyst nematodes by biofumigating treatments. *Third Int. Biofumigation Symp.*, Canberra, Australia.

De Vleeschouwer, O. 2001. Greenhouses and Conservatories. Flammarion.

Decraemer, D. 1995. *The Family Trichodoridae: Stubby-root and Virus Vector Nematodes.* Kluwer Academic Publishers.

Deepthi, N.V. 1993. *Control of Collar Rot and Root Knot of Cowpea with Vesicular Mycorrhizal Fungi.* M. Sc. (Ag.) thesis, Kerala Agri. Univ. Thrissur, 114 pp.

Desaeger, J. and Csinos, A. 2006. Root-knot nematode management in double crop plasticulture vegetables. *Journal of Nematology* 38: 59-67.

Desaeger, J., Sikora, R.A. and Molendijk, L.P.G. 2022. Outlook: A vision of the future of integrated nematode management. In R.A. Sikora, J. Desaeger and L.P.G. Molendijk (eds.) *Integrated Nematode Management: State-of-The-Art and Visions for The Future*, pp. 475-484. CAB International. Wallingford, UK.

Devi, T.P. and Goswami, B.K. 1992. Effect of VA mycorrhiza on the disease incidence due to *Macrophomina phaseolina* and *Meloidogyne incognita* on cowpea. *Ann. Agric. Res.* 13: 253-265.

Devi, Geetanjali. 1993. *Pathogenicity, Crop Loss Assessment and Management of* Meloidogyne incognita *(Kofoid and White, 1919) Chitwood, 1949 on Carrot (*Dacus carota *L.).* M. Sc. (Agri.) thesis, Assam Agri. Univ., Jorhat.

Dhawan, S.C. and Sethi, C.L. 1977. Inter-relationship between root-knot nematode, *Meloidogyne incognita* and little leaf of brinjal. *Indian Phytopathology* 30: 55-63.

Dhawan, S.C. and Singh, S. 2009. Compatibility of *Pochonia chlamydosporia* with nematicide and neem cake against root-knot nematode, *Meloidogyne incognita* infesting okra. *Indian J. Nematol.* 39(1): 85-89.

Dhawan, S.C., Singh, S. and Kamra, A. 2008. Biomanagement of root-knot nematode, *Meloidogyne incognita* by *Pochonia chlamydosporia. Indian J. Nematol.* 38: 119-121.

Di Vito, M., Greco, N. and Saxena, M.C. 1992. Pathogenicity of *Pratylenchus thornei* on chickpea in Syria. *Nematol. Mediterr.* 20: 71-73.

Diamond, J., Kimpinski, J. and Gallant, C.E. 1991. Root-lesion and root-knot nematodes associated with crops grown in rotation with carrots on Prince Edward Island. *Canadian Plant Disease Survey* 71: 13-15.

Dmowska, E., Dmoch, J. and Ilieva, K. 1997. Early interaction of the bacterivorous nematode *Rhabditis cucumeris* and the edible fungus *Agaricus bisporus* in relation to time and quantity of nematode inoculation. *Int. J. Mush. Sci.* 2: 15-23.

Dingre, S.K., Pawar, D.D., Kale, K.D. and Kadam, M.M. 2016. Onion seed productivity, nutrient use and quality response to drip NPK fertigation in Semi-arid India. *Journal of Plant Nutrition* 39: 1391-1403.

Echdvez-Badel, R. 1989. Performance of cucumber varieties in soil infested with root-knot and reniform nematodes. *J. Agric. Univ. P.R.* 73(4): 321-325.

Egunjobi, O.A. 1985. Effect of cocoa pod husks soil amendment on cowpea infestation by *Meloidogyne* spp. *Pakistan Journal of Nematology* 3: 99-103.

Elling, A.A. 2013. Major emerging problems with minor *Meloidogyne* species. *Phytopathology* 103: 1092-1102.

El-Nagdi1, W.M.A., Haggag, K.H.E., Abd-El-Fattah, A.I. and Abd-El-Khair, H. 2011. Biological control of *Meloidogyne incognita* and *Fusarium solani* in sugar beet. *Nematol. Medit.* 39: 59-71.

El-Nagdi, W.M.A., Youssef, M.M.A., El-Khair, H.A. and Abd-Elgawad, M.M.M. 2019. Effect of certain organic amendments and *Trichoderma* species on the root-knot nematode, *Meloidogyne incognita*, infecting pea (*Pisum sativum* L.) plants. *Egyptian Journal of Biological Pest Control* 29:75.

El-Remaly, E., Osman, A.A., El-Gawad, H.G.A. *et al.* 2022. Bio-management of root-knot nematodes on cucumber using biocidal effects of some brassicaceae crops. *Horticulturae* 8: 699. https://doi.org/10.3390/ horticulturae8080699.

Engindeniz, S. and Engindeniz, D.Y. 2006. Economic analysis of pesticide use on greenhouse cucumber growing: a case study for Turkey. *Journal of Plant Disease and Protection* 113: 193-198.

Escobar-Ávila, I.M. 2021. Population dynamics of the carrot cyst nematode *Heterodera carotae* under greenhouse condition. *Rev. Mex. Fitopatol.* 39(1): https://doi.org/10. 18781/r. mex.fit.2007-1.

Farina, A. and Ghorbani, A. 2014. Effect of K Nano-fertilizer and N-bio fertilizer on yield and yield components of red bean (*Phaseolus vulgaris* L.). *Intl. J. of Biosci.* 5(12): 296-303.

Faruk, M.I. 2021. Development of bio-rational based management packages against root-knot (*Meloidogyne incognita*) of bottle gourd. *Bangladesh J. Agril. Res.* 46(4): 373-383.

Faske, T.R. and Hurd, K. 2015. Sensitivity of *Meloidogyne incognita* and *Rotylenchulus reniformis* to fluopyram. *Journal of Nematology* 47: 316-321.

Fassuliotis, G. 1970. Resistance of *Cucumis* spp. to the root-knot nematode, *Meloidogyne incognita*. *J. Nematol* 2: 174-178.

Fatma, G.R. and Singh, S.N. 2020. Survey of root knot disease of sponge gourd (*Luffa cylindrica*) at Muzaffarpur in city areas. *Indian J. Sci. Res.* 10(2): 27-32.

Francl, L.J. and Wheeler T.A. 1993. Interaction of plant parasitic nematodes with wilt inducing fungi. In: M.W. Khan (ed.) *Nematode Interactions*, pp. 79-103. Chapman and Hall, London, UK.

Fulton, J.P. 1962. Transmission of tobacco ringspot virus by *Xiphinema americanum*. *Phytopathology* 52: 375.

Gandariasbeitia, M., López-Pérez, J.A., Juaristi, B. *et al.* 2021. Biodisinfestation with agricultural by-products developed long-term suppressive soils against *Meloidogyne incognita* in lettuce crop. *Front. Sustain. Food Syst.* 5: 663248. doi: 10.3389/fsufs.2021.663248.

Ganguly, S. and Dasgupta, D.R. 1981. Protein pattern in resistant and susceptible tomato varieties inoculated with the root-knot nematode, *Meloidogyne incognita*. *Indian Journal of Nematology* 14: 189-196.

Ganguly, S., Singh, M. and Ganguly, A.K. 2010. Record of potato cyst nematode, *Globodera rostochiensis* and *G. pallida* in Shimla, Himachal Pradesh, India. *Indian J Nematol.* 40: 96-102.

Gaur, H.S., Beane, J, and Perry, R.N. 1996. Effect of storage *in vitro* and in soil on the hatch from cysts of the pigeon pea cyst nematode, *Heterodera cajani*. *Fundamental and Applied Nematology* 19(6): 537-544.

Gaur, H.S. and Singh, I. 1977. Pigeon-pea cyst nematode, *Heterodera cajani*, associated with the moong crop in the Punjab State. *Journal of Research, Punjab Agricultural University* 14: 509.

Gheysen, G., van der Eycken, W., Barthels, N. *et al.* 1996. The exploitation of nematode-responsive plant genes in novel nematode control methods. *Pesticide Science* 47: 95-101.

Giriraj, Sharma, M.K. and Nama, C.P. 2018. Efficacy of fumigants for the management of root-knot nematode, *Meloidogyne incognita* on tomato (*Solanum esculentum* Mill.). *Int. J. Curr. Microbiol. Appl. Sci.* (Special Issue-7): 3198-3201.

Gitanjali, D. and Nandal, S.N. 2001. Nematodes associated with white button mushroom (*Agaricus bisporus*) in Sonipat district of Haryana. *Indian J. Nematol.* 31: 95-96.

Golden, J.K. and Van Gundy, S.D. 1975. A disease complex of okra and tomato involving the nematode, *Meloidogyne incognita* and the soil inhabiting fungus, *Rhizoctonia solani*. *Phytopathology* 65: 265-273.

Goswami, B.K., Pandey, R.K., Goswami, J. and Tiwari, D.D. 2007. Management of disease complex caused by root knot nematode and root wilt fungus on pigeon pea through soil organically enriched with Vesicular Arbuscular Mycorrhiza, karanj (*Pongamia pinnata*) oilseed cake and farmyard manure. *J. Environ. Sci. Health* 42: 899-904.

Goswami, B.K., Uma Rao and Singh, S. 1998. Potentiality of some fungal bioagents against root-knot nematode, *Meloidogyne incognita* infecting tomato. *Proc. First National Symp. Pest Mangmt. Hort. Crops*, Bangalore, pp. 304-307.

Goswami, S.K. and Chenulu, V.V. 1974. Interaction of root-knot nematode, *Meloidogyne incognita* and tobacco mosaic virus in tomato. *Indian Journal of Nematology* 4: 69-80.

Gowda, M.T., Rai, A.B. and Singh, B. 2017. *Root-Knot Nematode: A Threat to Vegetable Production and its Management*. Tech. Bull. No. 76, Indian Inst. of Veg. Res., Varanasi, 32 pp.

Gratwick, M. (ed). 1992. Docking disorder of sugar beet. In: *Crop Pests in the UK*. Collected edition of MAFF leaflets. Chapman & Hall.

Greco, N. 1987. Heterodera carotae*: Destructive Nematode of Carrot*. Nematology Circular No. 140, Florida Department of Agriculture and Consumer Services, Division of Plant Industry, Gainesville, FL, USA.

Greco, N., Brandonisio, A. and Elia, F. 1985. Control of *Ditylenchus dipsaci, Heterodera carotae* and *Meloidogyne javanica* by solarization. *Nematol. Medit*. 13:191-197.

Greco, N. and Esmenjaud, D. 2004. Management strategies for nematode control in Europe. In: R. Cook and D.J. Hunt (eds.) *Nematology Monographs and Prospective*, *Proceedings of the Fourth International Congress of Nematology*, Spain, pp. 33-43.

Greco, N., Brandonisio, A. and Elia, F. 1985. Control of *Ditylenchus dipsaci, Heterodera carotae* and *Meloidogyne javanica* by solarization. *Nematol. Medit*. 13: 191-197.

Greco, N., Di Vito, M., Saxena, M.C. and Reddy, M.V. 1988a. Effect of *Heterodera ciceri* on yield of chickpea and lentil and development of this nematode on chickpea in Syria. *Nematologica* 34: 98-114.

Greco, N., Di Vito, M., Saxena, M.C. and Reddy, M.V. 1988b. Investigation on the root lesion nematode *Pratylenchus thornei* in Syria. *Nematol. Mediterr*. 16: 101-105.

Greco, N. and Grandoision, A. 1990. Effect of soil solarization and SIP 5561 on *Heterodera carotae* and *Ditylenchus dipsaci* and on yield of carrot and onion. *Nematol. Medit*. 18: 189-193.

Greco, N., Vovlas, N. and Inserra, R.N. 1992. *The Chickpea Cyst Nematode*, Heterodera ciceri. Nematology Circular 198, Florida Department Agriculture.

Grewal, P.S. 1989. Effect of leaf-matter incorporation on *Aphelenchoides composticola* (Nematode), mycofloral composition, mushroom compost quality and yield of *Agaricus bisporus*. *Ann. Appl. Biol.* 115: 299-312.

Grewal, P.S. 1991. Relative contribution of nematodes (*Caenorhabditis elegans*) and bacteria towards the disruption of flushing patterns and losses in yield and quality of mushrooms (*Agaricus bisporus*). *Ann. Appl. Biol.* 119: 483–499.

Grewal, P.S. and Sohi, H.S. 1988. A new and cheaper technique for the rapid multiplication of *Arthrobotrys conoides* and its potential as a bio-nematicide in mushroom culture. *Curr. Sci.* 57: 44-46.

GRIN. 2008. *Germplasm Resources Information Network*. Retrieved 27/10, 2008, from http://www.ars-grin.gov/.

Guan, W., Zhao, X. and Dickson, D.W. 2014. Root-knot nematode resistance, yield, and fruit quality of specialty melons grafted onto *Cucumis metulifer*. *HortScience* 49(8): 1046-1051.

Gupta, D.C. and Mukhopadhyaya, M.C. 1971. Effect of N, P and K on the root-knot nematode, *Meloidogyne javanica* (Treub) Chitwood. *Science and Culture* 37: 246-247.

Gupta, D.C. and Verma, K.K. 1990. Studies on avoidable losses in mung bean (*Vigna radiata*) due to root-knot nematode, *Meloidogyne javanica* and its control under field conditions. *Indian J. Nematol.* 20: 148-151.

Gutjahr, C. and Paszkowski, U. 2009. Weights in the balance: jasmonic acid and salicylic acid signaling in root–biotroph interactions. *Molecular Plant–Microbe Interactions* 22: 763– 772.

Haegeman, A., Mantelin, S., Jones, J.T. and Gheysen, G. 2012. Functional roles of effectors of plant-parasitic nematodes. *Gene* 492: 19-31.

Haidar, M.G., Askary, T.H. and Nath, R.P. 2001. Nematode population as influenced by paddy based cropping sequences. *Indian J. Nematol.* 31: 68-71.

Haider, M.G., Nath, R.P., Thakur, S.C. and Ojha, K.L. 1987. Interaction of *Meloidogyne incognita* and *Pseudomonas solanacearum* on tomato plants. *Indian J. Nematol.* 17: 174-176.

Hajihassani, A., Tyson, C. and Shirley, A.C. 2020. *Biology and Management of Stubby-Root Nematodes on Onion*. University of Georgia Extension Circular 1211, Georgia, USA, 3 pp.

Hajra, N., Shahina, F., Firoza1, K. and Maria, R. 2015. Damage induced by root-knot nematodes and its alleviation by vesicular arbuscular mycorrhizal fungi in roots of *Luffa cylindrica*. *Pakistan Journal of Nematology* 33(1): 71-78.

Handa, D.K. and Mathur, B.N. 1981. Control of root-knot of chillies with nematicides. *Second Nematology Symposium*, Tamil Nadu Agricultural University, Coimbatore, p. 48.

Hanna, H.Y. 2004. *Improving Cucumber Yields Following Nematode-Resistant Tomatoes*. LSU Ag Center and LSU College of Agriculture.

Haque, M.M. and Gaur, H.S. 1985. Effect of multiple cropping sequences on the dynamics of nematode population and crop performance. *Indian J. Nematol.* 15: 262-263.

Haris, M., Ahmad, G., Shakeel, A. and Khan, A.A. 2019. Utilization of fly ash to improve the growth and the management of root-knot nematode on carrot. *Haya Saudi J. Life Sci.* 4(7): 221-226.

Haroutunian, G. 2013. *The Use of Biofumigation Crops as an Alternative to Methyl Bromide for The Management of The Root-Knot Nematode in Greenhouse Cucumber Production*. Doctoral Thesis, L'Institut des Sciences et Industries du Vivant et de l'Environnement (AgroParisTech), 16 rue Claude Bernard, 75231. Paris Cedex 05.

Hasan, A. 1988. Interaction between *Pratylenchus coffeae* and *Pythium aphanidermatum* and/ or *Rhizoctonia solani* on chrysanthemum. *Phytopathol. Zeist* 123: 227-232.

Hasan, A. 1989. Efficacy of certain non-fumigant nematicides on the control of pigeon pea wilt involving *Heterodera cajani* and *Fusarium udum*. *Phytopathol. Zeist* 126: 335-342.

Hasan, M., Singh, B., Singh, M.C. *et al.* (eds.). 2010. *Fertigation Scheduling for Horticultural Crops*. Technical Bulletin, IARI, New Delhi, 44 pp.

Hasan, N. and Jain, R.K. 1992. Field application of *Paecilomyces lilacinus* in combination with certain organic matters for controlling *Meloidogyne incognita* infecting cowpea followed by lucerne. *First Afro-Asian Nematol. Symp.*, Aligarh Muslim Univ., Aligarh, pp. 10-11.

Hasan, N. and Jain, R.K. 1998. Nematological research in Uttar Pradesh: An overview. In P.C. Trivedi (ed.) *Phytonematology in India*, pp. 150-169. C.B.S. Publishers and Distributors, New Delhi.

Hasaneen, M.N.A.G., Abdel-aziz, H.M.M. and Omr, A.M. 2016. Effect of foliar application of engineered nanomaterials: carbon nanotubes NPK and chitosan nanoparticles NPK fertilizer on the growth of French bean plant. *Bioch. Biot. Res.* 4(4): 68-76.

Haseeb, A. and Shukla, P.K. 2005. Wilt disease complex of pigeon pea and its management. In S. Nehra (ed.) *Plant Diseases: Biocontrol Management*, pp. 84-96. Avishkar Publishers, Distributors, Jaipur.

Haseeb, A., Kumar, V., Shukla, P.K. and Ahmad, A. 2006. Effect of different bioinoculants, organic amendments and pesticides on the management of *Meloidogyne incognita – Fusarium solani* disease complex on tomato cv. K-25. *Indian J. Nematol.* 36: 65-69.

Haseeb, A., Sharma, A. and Shukla P.K. 2008. Studies on the management of root-knot nematode, *Meloidogyne incognita*-wilt fungus, *Fusarium oxysporum* disease complex of green gram, *Vigna radiata* cv. ML-1108. *Journal of Zhejiang University Science B* 6: 736-742.

Hassan, M.G. and Devi, S. 2004. Evaluation of mung bean, urd bean, pea, lentil and chickpea germplasm for resistance to root-knot nematode (*Meloidogyne incognita*). *Indian J. Nematol.* 34: 228–229.

Hazarika, B.P. and Roy, A.K. 1974. Effect of *Rhizoctonia solani* on the reproduction of *Meloidogyne incognita* on eggplant. *Indian Journal of Nematology* 4: 246-248.

Hazarika, K. 1990. *Pathogenicity and Management of* Meloidogyne incognita *on Brinjal* (Solanum melongena). M. Sc. (Ag) thesis, Assam Agri. Univ., Jorhat.

Heald, C.M. and Thomas, C.E. 1983. Nematode control by soil solarization. *J. Nematol.* 13: 114-115.

Heffes, T.P., Coates-Beckford, P.L. and Robotham, H. *et al.* 1991. Effects of *Meloidogyne incognita* on growth and nutrient content of *Amaranthus viridis* and two cultivars *of Hibiscus sabdariffa*. *Nematropica* 2: 7-18.

Heinz, K.M., van Driesche, R.G. and Parella, M.P. 2004. Biocontrol in protected culture. *Ball Hort Technology* 5: 6-23.

Hema. 2014. *Eco-friendly Management of Nematodes in Cucumber (*Cucumis sativus *L.) under Protected Conditions*. M. Sc. Thesis. Dr. Y.S. Parmar Univ. of Hort. and Forestry, Solan, Himachal Pradesh, 57 pp.

Hickman, G.W. 2011. *A Review of Current Data on International Production of Vegetables in Greenhouses*. www.cuestaroble.com., 73 pp.

Hnoosh, H.J.H. and Aljuaifari, W.A.R. 2020. Examination of biological seed treatments to management of root-knot nematode (*Meliodogyne* sp.) on okra (*Abelmoschus esculentus*). *Plant Archives* 20 (Supplement 1): 1039-1045.

Hockland, S. 2000. Control of nematodes in root vegetable crops. In: *Report on The 25th International Symposium on Nematology*, Herzliya, Israel, 2-7 April 2000. CSL Internal Report for MAFF (Plant Health Division).

Huang, S.W., Wang, L., Liu, L.M. *et al.* 2014. Nonchemical pest control in China rice: A review. *Agron. Sustain. Dev.* 34: 275–291.

Hussain, Z. and Bora, B.C. 2008. Integrated management of *Meloidogyne incognita* and *Ralstonia solanacearum* complex in brinjal. *Indian J. Nematol.* 38: 159-164.

Ibrahim, I.K.A. and Rezk, M.A. 1988. The root-knot nematodes – a major problem in crop production in Egypt. In M.A. Maqbool, A.M. Golden, A. Gaffar and L.R. Krusberg, (eds.) *Advances in Plant Nematology*, pp. 81-98. Natl. Nematol. Res. Centre, Univ. of Karachi, Karachi, Pakistan.

Ibrahim, A., Shahid, A.A., Shafiq, M. and Haider, M.S. 2012. Management of root-knot nematodes on the turnip plant (*Brassica rapa*) by using fungus (*Trichoderma harzianum*) and neem (*Azadirachta indica*) and effect on the growth rate. *Pak. J. Phytopathol.* 24(2): 101-105.

INCID. 1994. *Drip Irrigation in India.* Indian National Committee on Irrigation and Drainage, New Delhi.

Ingham, R.E., Hamm, P.B., Williams, R.E. and Swanson, W.H. 2000. Control of *Paratrichodorus allius* and corky ringspot disease of potato in the Columbia Basin of Oregon. *Journal of Nematology (Supplement)* 32(4S): 566-575.

Inglis, D.A. 1998. *Pea Cyst Nematode: Biology and Prevention.* WSU Cooperative Extension Bulletin 1872.

Inglis, D.A. 2001. Root-lesion nematode. In: J.M. Kraft and F. L. Pfleger (eds.) *Compedium of Pea Diseases and Pests*, 2nd edition, pp. 21-22. APS Press, St. Paul, MN.

Iqbal, Z., Nasir, H., Hiradate, S. and Fujii, Y. 2006. Plant growth inhibitory activity of *Lycoris radiata* Herb. and the possible involvement of lycorine as an allelochemical. *Weed Biol. Manag.* 6: 221–227.

Ismael, J.H.S. and Mahamood, A.A. 2020. Integrated management of root-knot nematode (*Meloidogyne* spp.) In cucumber (*Cucumis sativus* L.) and its effect on nematode population density, plant growth and yield in Sulaimani Governorate, Kurdistan, Iraq. *Applied Ecology and Environmental Research* 18(3): 4709-4741.

Ives, L. 2019. *Epidemiology and Management of Stem and Bulb Nematode.* M. Sc. thesis, The University of Guelph, Guelph, Ontario, Canada.

Jadhav D.K., Fugro P.A., Kadam J.J. 2010. Management of anthracnose and root-knot nematode complex and evolution of resistance source in bitter gourd. *J. Plant Dis. Sci.* 5(2): 238-241.

Jagadeeswara, R. and Singh, R.V. 2011. Management of reniform nematode, *Rotylenchulus reniformis* on okra hybrid by organic amendments. *Indian Journal of Nematology* 41(1): 9-13.

Jain, R.K. and Bhatti, D.S. 1978. Bare root treatment with systemics for controlling root-knot nematodes in tomato transplants. *Indian Journal of Nematology* 8: 19-24.

Jain, R.K. and Bhatti, D.S. 1981. Systemics as seed treatment for the control of *Meloidogyne javanica* in okra. *Second Nematology Symposium,* Tamil Nadu Agricultural University, Coimbatore, p. 44.

Jain, R.K. and Bhatti, D.S. 1983. Chemical control of *Meloidogyne javanica* in tomato through nursery bed treatment. *Third Nematology Symposium*, Dr. Y.S. Paramar Univ. Hort. and Forestry, Solan, pp. 37-38.

Jain, R.K. and Bhatti, D.S. 1985. Effect of summer ploughing alone and in combination with some other effective practices on the incidence of root-knot nematode (*Meloidogyne javanica*) on tomato (Cultivar HS 101). *Indian Journal of Nematology* 15: 262.

Jain, R.K., Bhatti, D.S., Bhutani, R.D. and Kalloo, G.1983. Screening of germplasm of some vegetable crops for resistance to root-knot nematode *Meloidogyne incognita. Indian Journal of Nematology* 13: 212-215.

Jain, R.K. and Hasan, N. 1984. Increased seed production of cowpea with low dose of phenamiphos. *Pesticides* 18: 49.

Jain, R.K. and Hasan, N. 1986. Efficacy of "Neem" cake on fodder production, photosynthetic pigments and nematodes associated with oats and its residual effect on cowpea. *Indian Journal of Nematology* 16: 98-100.

Jain, R.K. and Hasan, N. 1988. Role of vesicular arbuscular mycorrhizal (VAM) fungi and nematode activities in forage production. *Acta Botanica Indica* 16: 84-88.

Jain, R.K. and Sethi, C.L. 1987. Pathogenicity of *Heterodera cajani* on cowpea as influenced by the presence of VAM fungi, *Glomus fasciculatum* and *G. epigaeus*. *Indian J. Nematol.* 17: 89-93.

Jain, R.K. and Sethi, C.L. 1988a. Influence of endomycorrhizal fungi *Glomus fasciculatum* and *G. epigaeus* on penetration and development of *Heterodera cajani* on cowpea. *Indian J. Nematol.* 18: 89-93.

Jain, R.K. and Hasan, N. 1995. Nematode problems in forage crops. In G. Swarup, D.R. Dasgupta and J.S. Gill, (eds.) *Nematode Pest Management- Appraisal of Eco-Friendly Approaches*. Nematol. Soci. of India, Indian Agri. Res. Inst., New Delhi, India, pp. 217-227.

Jain, R.K., Paruthi, I.J. and Gupta, D.C. 1990. Influence of intercropping on the incidence of root-knot nematode (*Meloidogyne javanica*) in brinjal. *Indian J. Nematol.* 20: 236-238.

Jain, R.K., Paruthi, I.J., Gupta, D.C. and Dhankar, B.S. 1986. Appraisal of losses due to root-knot nematode, *Meloidogyne javanica* on okra under field conditions. *Trop. Pest Mangmt.* 32: 341-342.

Jakobsen, I. 1999. Effects of *Pseudomonas fluorescens* DF 57 on growth and P uptake of two arbuscular mycorrhizal fungi in symbiosis with cucumber. *Mycorrhiza* 8: 319-334.

Jatala, P. 1985. Biological control of nematodes. *Ann. Rev. Phytopath.* 24: 453-489.

Jayakumar J., Rajendran G. and Ramakrishna, S. 2005. Management of reniform nematode, *Rotylenchulus reniformis* on okra through *Streptomyces avermitilis*. *Indian Journal of Nematology* 35(1): 59-62.

Jinek, M., Chylinski, K., Fonfara, I. *et al.* 2012. A programmable dual-RNA-guided DNA endonuclease in adaptive bacterial immunity. *Science* 337: 816-821.

Johnson, A.W., Rohde, W.A., Dowler, C.C. *et al.* 1981. Influence of water and soil temperature on the concentration and efficacy of phenamiphos for control of root-knot nematodes. *J. Nematol.* 13(2): 148-153.

Jones, F.C.E. 1961. The potato root eelworm *Heterodera rostochiensis* Woll. in India. *Current Science* 30: 187.

Jones, J.D.G, and Dangl, J.L. 2006. The plant immune system. *Nature* 444: 323– 329.

Jones, J.G., Kleczewski, N.M., Desaeger, J. and Meyer, S.L.F. 2017. Evaluation of nematicides for southern root-knot nematode management in lima bean. *Crop Protection* 96: 151-157.

Jumder V.M, Pankaj, S., Chawla G. and Singh J.2004. Effect of urea coated with neem formulations on root-knot and reniform nematodes in Okra. *Indian Journal of Nematology* 34(1): 60-63.

Jung, C., Cai, D. and Kleine, M. 1998. Engineering nematode resistance in crop species. *Trends Plant Science* 3: 266-271.

Kadim, A.A.Y.A.M. 2018. Impact of nano chelated iron, GA_3 and organic fertilizer (acadian) in moringa leaves content of α-tocopherol and phytosterols. *Res. J. Pharmacy and Technol.* 11(5): 1840-1846.

Kafkafi, U. and S. Kant. 2004. Fertigation. In: D. Hillel, C. Rosenzweiz, D. Powlson, K. Scow, M. Singer and D. Sparks (eds.) *Encyclopaedia of Soils in The Environment*. Elsevier Ltd., Oxford, UK, Vol. 2: 1-9.

Kalanjiyam, S. and Manickam, K. 2015. Fertigation studies in tomato. *Journal of Plant and Agriculture Research* 1(1):1-5.

Kamoun, S. 2006. A catalogue of the effector secretome of plant pathogenic oomycetes. *Annual Reviews of Phytopathology* 44: 41– 60.

Kandasamy, K. and Sivakumar, C.V. 1981. Effect of *Meloidogyne arenaria* on the germination of chilli (*Capsicum annuum* Linn.) seeds. *Second Nematology Symposium*, Tamil Nadu Agricultural University, Coimbatore, p.16.

Kannan, C. and Lingaraju, S. 2002. Interaction of *Cladosporium herbarum* (Pers.) Link ex. Gray, *Heterodera cajani* Koshy and few organic amendments on pigeon pea (L.) Mill. *Indian J. Nematol.* 32(2): 193-196.

Kanwar, R.S. 1990. *Studies on Cropping Systems for Management of Root-knot Nematode,* Meloidogyne javanica *(Treub, 1985) Chitwood, 1949*. Ph. D. thesis, Haryana Agri. Univ., Hisar, 118 pp.

Kanwar, R.S. and Bhatti, D.S. 1990. Yield performance of resistant and susceptible tomato cultivars in monoculture and related reproduction of *Meloidogyne javanica. International Nematology Network Newsletter* 7(3): 16-18.

Kapkavalci, G., Tuzel, Y., Dura, O. and Oztekyn, G.B. 2009. Effects of alternative control methods against *Meloidogyne incognita* in organic tomato production. *Ekoloji* 18(72): 23-31.

Kaspary, T.E., García, M.A. and Bellé, C. 2022. Agricultural weeds maintain populations of *Meloidogyne* spp. root-knot nematodes. *Rhizosphere* 23. 100576.

Katan, J. 2017. Diseases caused by soil-borne pathogens: biology, management and challenges. *J. Plant Pathol.* 99(2): 305-315.

Ke, J., Wang, B. and Yoshikuni, Y. 2021. Microbiome engineering: synthetic biology of plant-associated microbiomes in sustainable agriculture. *Trends in Biotechnology* 39: 244–261.

Khan, A., Khan, F., Shariq, M. *et al.* 2022. Screening for resistance and susceptibility in some beet root cultivars to root-knot nematode, *Meloidogyne javanica*. *Journal of Environmental Biology* 43(4): 571-577.

Khan, A.H., Mahmood, I. and Saxena, S.K. 1986. Interaction of certain crucifers to *Tylenchorhynchus brassicae*. *Int. Nematol. Network Newsl.* 3: 12.

Khan, A.M., Saxena, S.K. and Khan, M.W. 1971. Interaction of *Rhizoctonia solani* Kuhn and *Tylenchorhynchus brassicae* Siddiqi, 1961 in pre-emergence damping-off of cauliflower seedlings. *Indian J. Nematol.* 1: 85-86.

Khan, A.M., Saxena, S.K. and Siddiqi, Z.A. 1971a. Efficacy of *Tagetes erecta* in reducing root infesting nematodes of tomato and okra. *Indian Phytopathology* 24: 166-169.

Khan, A.M., Saxena, S.K. and Siddiqi, Z.A. 1971b. Reaction of some cultivated and wild species of *Lycopersicon* and cucurbits to *Meloidogyne incognita* (Kofoid and White) Chitwood. *Indian Phytopathology* 24: 197-199.

Khan, A.M., Saxena, S.K., Siddiqi, Z.A. and Upadhyay, R.S. 1975. Control of nematodes by crop rotation. *Indian J. Nematol.* 5: 214-221.

Khan, F.A. and Khan, A.M. 1973. Studies on the reniform nematode, *Rotylenchulus reniformis* I. Host range and population changes. *Indian Journal of Nematology* 3: 24-30.

Khan, M.M., Khan, M.R. and Reshu. 2010. Evaluation of effectiveness of seed treatment with pesticides and biocontrol agents against the disease complex of pigeon pea in pots. *National Conf. on Innovations in Nematological Research for Agricultural Sustainability – Challenges and A Roadmap Ahead*, Tamil Nadu Agri. Univ., Coimbatore, p. 101.

Khan, M.R. and Ghadipur, M.H. 2004. Management of root-knot disease of some solanaceous vegetables by soil application of fly ash. *National Symp. on Paradigms in Nematol. Res. for Biodynamic Farming*, Univ. of Agri. Sci., Bangalore, p. 92.

Khan, M.R., Bhattacharya, I., Chattopadhyay, S.B. and Ghosh, S. 2009. Integrated management of root-knot (*Meloidogyne incognita*) and other nematodes in pointed gourd (*Trichosanthes dioica*). *Indian Journal of Nematology* 39(1): 25-28.

Khan, M.R., Kounsar, K. and Hamid, A. 2002. Effect of certain rhizobacteria and antagonistic fungi on root nodulation and root-knot nematode disease of green gram. *Nematol. Mediterr.* 30: 85–89.

Khan, M.W., Khan, A.M., Haider, S.R. and Ashraf, S.S. 1988. Identity of races of *Meloidogyne incognita* in Western region of Uttar Pradesh, India. *Nematologica* 34: 115-116.

Khan, T.A. and Husain, S.I. 1988. Studies on the efficacy of *Paecilomyces lilacinus* as biocontrol agent against disease complex caused by the interaction of *Rotylenchulus reniformis, Meloidogyne incognita* and *Rhizoctonia solani* on cowpea. *Nematol. Medit.* 16: 229-231.

Khan, T.A. and Hussain, S.I. 1989. Relative resistance of six cowpea cultivars as affected by concomitance of two nematodes and a fungus. *Nematol. Medit.* 17: 39-41.

Khan, T.A. and Husain, S.I. 1990. Biological control of root-knot and reniform nematodes and root rot fungus on cowpea. *Bioved.* 1: 19-24.

Khan, T.A., Khan, S.T. and Saxena S.K. 1992. Effect of interaction of *Meloidogyne incognita* and *Fusarium solani* on the relative response of papaya cvs. *First Afro-Asian Nematology Symposium*, Aligarh Muslim Univ., Aligarh, p. 14.

Khanna, A.S. 1991. Yield losses in white button mushrooms due to nematode infestation. *Nematol. Medit.* 19(1): 35–36.

Khanna, A.S. 1993. Estimation of yield losses in white button mushrooms due to nematode infestation. *Curr. Nematol.* 4(2): 173–176.

Khanna, A.S. and Sharma, N.K. 1988. Pathogenicity of various *Aphelenchoides* species on *Agaricus bisporus* (Lange) Singer. *Indian Phytopath.* 41(3): 472–473.

Khanna, A.S. and Sharma, N.K. 1989. Resistance of *Pleurotus sajor caju* against myceliophagous nematode *Aphelenchoides agarici. Nematol. Medit.* 17: 61.

Khanna, A.S. and Sharma, N.K. 1990. Efficacy of nematophagous fungi against myceliophagous nematodes *in vitro*. In *Proc. of National Seminar on Bioagents in Nematode Management*, pp. 83–84. Indian Agri. Res. Inst., New Delhi.

Khanna, A.S. and Sharma, N.K. 1994. Management of *Aphelenchoides composticola* infesting white button mushroom by neem cake. *Indian J. Hill Fmg.* 7: 114-115.

Khanna, A.S. and Sharma, V.P. 1996. Management of *Aphelenchoides composticola* infesting white button mushrooms by neem cake. *Indian J. Hill Fmg.* 1(2): 41–44.

Kimenju, J.W., Karanja, N.K. and Macharia, I. 1999. Plant parasitic nematodes associated with common bean in Kenya and the effect of *Meloidogyne* infection on bean nodulation. *African Crop Sci. J.* 7: 503-510.

Kloppholz, S., Kuhn, H. and Requena, N. 2011. A secreted fungal effector of *Glomus intraradices* promotes symbiotic biotrophy. *Current Biology* 21: 1204– 1209.

Koenning, S. R., Overstreet, C., Noling, J. W *et al.* 1999. Survey of crop losses in response to phytoparasitic nematodes in the United States for 1994. *Journal of Nematology* 31: 587-618.

Kokalis-Burelle, N and Rosskopf, E.N. 2011. Microplot evaluation of rootstocks for control of *Meloidogyne incognita* on grafted tomato, muskmelon, and watermelon. *Journal of Nematology* 43: 166-171.

Kovacs, T., Kovacs, G. and Szito, J. 1999. Crop yield response to deficit irrigation imposed at different plant growth stages. In: C. Kirda, P. Moutonnet, C. Hera and D.R. Nielsen (eds.) *Crop Yield Response to Deficit Irrigation*. Kluwer Academic Publishers, Dordrecht, The Netherlands.

Krishna Prasad, K.S. 1986 Potato nematodes. In G. Swarup and D.R. Dasgupta (eds.) *Plant Parasitic Nematodes in India*. Indian Agri. Res. Inst., New Delhi.

Krishna Rao, A.B.V., Padhi, N.N. and Acharya, A. 1987. Effect of different nematicides, oilcakes and urea in the control of *Rotylenchulus* in okra. *Indian Journal of Nematology* 17: 171-173.

Kuiper. 1977. Introductie en vestiging van planteparasitaire aaltjes in nieuwe polders, in het bijzonder van *Trichodorus teres*. *Verslagen en Mededelingenvan de Plantenziektenkunndige Dienst te Wageningen* 136: 193-200.

Kumar, S. and Prabhu, S. 2008. Biological control of *Heterodera cajani* in pigeon pea by *Trichoderma harzianum* and *Pochonia chlamydosporia*. *Indian J. Nematol.* 38: 65-67.

Kumar, S. and Sivakumar, C.V. 1981. Disease complex involving *Rotylenchulus reniformis* and *Rhizoctonia solani* in okra. *Second Nematology Symposium*, Tamil Nadu Agricultural University, Coimbatore, p.12.

Kumar, A., Patil, J.A, Verma, K.K. 2019. Management of root-knot nematode, *Meloidogyne* spp. in vegetable crops grown under protected cultivation through fumigants. *Indian Journal of Nematology* 49(2): 125-130.

Kumar, A., Patil, J.A., Yadav, S. and Verma, K.K. 2020. Response of organic amendment and bio-agents against root-knot nematode, *Meloidogyne incognita* infesting cluster bean. *Plant Pathology Journal* 19(4): 221-225.

Kumar, S. and Sivakumar, C.V. 1981. Disease complex involving *Rotylenchulus reniformis* and *Rhizoctonia solani* in okra. *Second Nematology Symposium*, Tamil Nadu Agricultural University, Coimbatore, p.12.

Labuy'ere, R.E., Den Ouden, H. and Seinhorst, J.W. 1959. Experiments on the interaction of *Hoplolaimus uniformis* and *Fusarium oxysporum* f. *pisi* race 3 and its importance in" early yellowing" of peas. *Nematologica* 4: 336-343.

Ladan Moghadam, A., Vattani, H., Baghaei, N. and Keshavarz, N. 2012. Effect of different levels of fertilizer nano iron chelates on growth and yield characteristics of two varieties of spinach (*Spinacia oleracea* L.): Varamin 88 and viroflay. *Res. J. Appl. Sci. Eng. Technol.* 4(12): 4813-4818.

Lakshmanan, P.L. and Sivakumar, C.V. 1981. Control of *Rotylenchulus reniformis* in bhendi by integrated method. *Second Nematology Symposium*, Tamil Nadu Agricultural University, Coimbatore, p. 54.

Lal, B.S. and Ansari, M.N.A. 1960. Field studies on the root-knot nematodes (*Meloidogyne* spp.) (Nematoda: Heteroderidae). *Sci. Cult.* 26: 279-281.

Lalhruaitluanga, Pc. and Hazarika, K. 2020. Bio-management of *Meloidogyne incognita* in pumpkin. *Int. J. Curr. Microbiol. App. Sci.* 9(11): 762-770.

Lamers, J.G. and Roosjen, J. 1993. *De waardplantgeschiktheid van groenbemestingsgewassen voor het Noordelijk wortl-knobbelaaltje*. Verlag-Proefstation voor de Akkerbouw en de Groenteteelt in de Vollegrond No. 163, 40 pp.

Laqman Khan, M. 2001. Dr. Y.S. Paramar University of Horticulture and Forestry, Solan, Himachal Pradesh. In S.C. Dhawan *et al.* (eds.) *Indian Nematology-Progress and Prospectives*. Divn. of Nematol., Indian Agri. Res. Inst., New Delhi, pp. 153-158.

Lee, W.Y. 1992. Use of *Sicyos angulatus* cv. Andong as rootstock for cucurbits (in Korean). *Rural Development Admin., Suwon, Korea Res & Extension* 33(4): 69-73.

Lee, W.Y., Hwang, G.S., Shin, C.G. *et al.* 1992. Yield performance test of watermelon (*Citrullus vulgaris* Schrad) grafted onto the rootstock, *Sicyos angulatus* L in the farm field (Abstr.) (in Korean with tables in English). *Kor Soc Hort Sci* 10(2): 38-39.

Lilley, C.J., Davies, L.J. and Urwin, P.E. 2012. RNA interference in plant parasitic nematodes: a summary of the current status. *Parasitology* 139: 630– 640.

Liu, B., Ren, J., Zhang, Y. *et al.* 2014. A new grafted rootstock against root-knot nematode for cucumber, melon, and watermelon. *Agr. Sustain. Dev.* 35: 251 259.

Loganathan, M., Swarnakumari, N., Sivakumar, M. *et al.* 2001. Biological suppression of fungal nematode complex diseases of major cruciferous vegetables. *South Indian Horticulture* 49: Special issue.

Lone, J.A., Parveen, B.G. and Khan T.A. 2014. Comparison of concomitant and sequential inoculation of *Steinernema* sp. in the management of reniform (*Rotylenchulus reniformis*) nematode infecting eggplant. *e-J Sci & Tech* 2(9): 97–111.

Machal, M., Borah, A. and Mahanta, B. 2022. Screening of bitter gourd germplasms against root-knot nematode, *Meloiodogyne incognita*. *The Pharma Innovation Journal* 11(5): 1617-1620.\

Mahalik, J.K. and Sahu, A. 2018. Ecofriendly management of root-knot nematode (*Meloidogyne incognita*) infecting okra in Odisha, India. *Int. J. Curr. Microbiol. App. Sci.* (Special Issue) 6: 1999-2004.

Mahalik, j., Sahu, A. and Behera, S. 2020. Ecofriendly management of root knot nematode, *Meloidogyne incognita* in okra (*Abelomoschus esculentus* (L.) Moench) through integrated application of oil cakes and bioagents. *Journal of Entomology and Zoology Studies* 8: 1781-1784.

Mahapatra, B.C. and Pahdi, N.N. 1986. Pathogenicity and control of *Rotylenchulus reniformis* on *Cicer arietinum*. *Nematol. Mediterr.* 14: 287-290.

Mahapatra, S.N., Swain, P.K., 2001. Interaction between *Meloidogyne incognita* and *Fusarium oxysporum* on black gram. *Annals of Plant Protection Sciences* 9: 92–94.

Mahmood, K., Qumar, S., Naqvi, A. and Alam, M.M. 1974. The influence of bottle gourd mosaic virus and brinjal mosaic virus on the development of root-knot caused by *Meloidogyne incognita* (Kofoid and White, 1919) Chitwood, 1949. *Indian Phytopathology* 27: 610-611.

Maluf, L.E.J., Okada, A.T., Gomes, L.A.A. *et al.* 2003. Reação de cultivares de alface à infecção por *Meloidogyne incognita. Horticultura Brasileira, Recife, PE* 21(2): Suplemento 2. 1CD-ROOM. Trabalho apresentado no 43° Congresso Brasileiro de Olericultura, 2003.

Mani, M.P., Rajeswari, S. and Sivakumar, C.V. 1998. Management of the potato cyst nematodes, *Globodera* spp. through plant rhizosphere bacterium *Pseudomonas fluorescens* Migula. *J. Biol. Control* 12: 131-134.

Mankau, R. 1980. Biological control of nematode pests by natural enemies. *Ann. Rev. Phytopathol.* 18: 415-440.

Margulis, L. and Fester, R. 1991. *Symbiosis as a Source of Evolutionary Innovation.* MIT Press, University of Massachusetts, Amherst, Massachusetts, 454 pp.

Masood, A. and Husain, S.I. 1976. Phenolic and ortho-dihydroxy phenolic changes and their role in the resistance and susceptibility of three tomato varieties to *Meloidogyne incognita. Indian Journal of Nematology* 6: 86-93.

Mastievich, N.V., Shelpetene-Yu, A. and Kucheryavaya, O.A. 1991. Nematophagous predatory fungi against gall nematode. *Zashchita-Rastenii-Moskava* 7: 16-17.

Maynard, E. and O'Donnell, M. 2018. Managing the environment in high tunnels for cool season vegetable production. https://mdc.itap.purdue.edu/item.asp?Item_Number= HO-297- W.

McGuire, J.M. 1964. Efficiency of *Xiphinema americanum* as vector of tobacco ringspot virus. *Phytopathology* 54: 799-801.

McGuire, J.M. and Good, M.J. 1970. The effect of benomyl on *Xiphinema americanum* and tobacco ring spot virus infection. *Phytopathology* 60: 1150-1151.

McSorley, R. and Parrado, J.L. 1986. Application of soil solarization to Rockdale soils in a subtropical environment. *Nematropica* 16: 125-140.

Melakeberhan, H., Xu L., Kravchenko, A. *et al.* 2006. Potential use of arugula (*Eruca sativa* L.) as a trap crop for *Meloidogyne hapla*. Nematology 8(5): 793-799.

Melek, E., Dursun, A., Yildirim, E. and Parlakova, F. 2014. Effects of nanotechnology liquid fertilizers on the plant growth and yield of cucumber (*Cucumis sativus* L.). *Acta Sci. Pol. Hortorum Cultus.* 13(3): 135-141.

Mhase, N.L. 2012. *Studies on the Root-Knot Nematode,* Meloidogyne incognita *(Kofoid & White, 1919) Chitwood, 1949 Infesting French Bean,* Phaseolus vulgaris *L.* M. Sc. Thesis, Mahatma Phule Agri. Univ, Rahuri, Maharashtra.

Midha, S.K. 1985. Efficacy of *Paecilomyces lilacinus* in controlling root-knot infestation on cowpea and mung bean. *Indian Journal of Nematology* 15: 267.

Mishra, S.D. and Chakrabarti, U. 2001. Distribution and intensity of nematode problems in oilseeds and pulses. *National Congress on Centenary of Nematology in India: Appraisal and Future Plans.* 5-7 December, pp. 25-26.

Mohamed, S.A., El-Ghamry, A.M., Selim, E.M.M. *et al.* 2014. Influence of composting of rice straw with effective microorganisms and humic acid on quality and quantity of potato plants (*Solanum tuberosum* L.) through fertigation system. *Middle East Journal of Applied Sciences* 4(3): 484-493.

Mohammed, R.K.A. and Khan, M.R. 2021. Management of root-knot nematode in cucumber through seed treatment with multifarious beneficial microbes in polyhouse under protected cultivation. *Indian Phytopathology* 74: 1035–1043.

Mohapatra, S. 2015. *Pathogenicity and Avoidable Yield Loss due to* Meloidogyne incognita *in French Bean*. M. Sc. (Nematol.) thesis, Orissa Univ. of Agri. and Technol., Bhubaneswar, Orissa.

Mohler, C.M. and Johnson, S.E. 2009. *Crop Rotation on Organic Farms: A Planning Manual.* Sustainable Agriculture Research and Education (SARE) Plant and Life Sciences Publishing (PALS) Cooperative Extension NRAES 177, Cornell University, Ithaca, New York, 154 pp.

Montasser, S.A., Mahmoud, N.A., El-Mesalamy, A.F. and Abdel-Mageed, M.A. 2016. Influence of *Meloidogyne javanica* and *Rotylenchulus reniformis* alone on nodulation of some leguminous crops. *Egypt. J. Agronematol.* 15(1): 1-11.

Morra, L. and Billoto, M. 2006. Evaluation of new rootstocks for resistance to soil borne pathogens and productive behavior of pepper (*Capsicum annuum* L.). *J. Hort. Sci. Biotechnol.* 81: 518-524.

Morison, J.I., Baker, N.R., Mullineaux, P.M. and Davies, W.J. 2008. Improving water use in crop production. *Philo. Trans. R. Soc. London B. Biol. Sci.* 12: 639-658.

Mukhopadhyay, A.K., Roy, K. and Panda, T.R. 2006. Evaluation of nematicides and neem cake for managing root-knot nematodes in pointed gourd, *Trichosanthes dioica*. *Nematol. Medit.* 34: 27-31.

Mukhtar, T., Hussain, M.A. and Kayani, M.Z. 2017. Yield responses of 12 okra cultivars to southern root-knot nematode (*Meloidogyne incognita*). Bragantia 76(1) https://doi.org/ 10.1590/1678-4499.005.

Mweke, A.N., Kimenju, J.W., Seif, A.A. *et al.* 2008. Potential of sequential cropping in the management of root-knot nematodes in okra. *Asian Journal of Plant Sciences* 7(4): 399-403.

Nageswari, S.; Mishra, S.D. 2005. Integrated nematode management schedule incorporating neem products, VAM and soil solarization against *Heterodera caja* infecting pigeon pea. *Indian J Nematol* 35: 68-71.

Naik, D. 2004. *Biotechnological Approaches for The Management of Wilt Disease Complex in Capsicum (*Capsicum annum *L.) and Eggplant (*Solanum melongena*) with Special Emphasis on Biological Control.* Ph. D. thesis, Kuvempu Univ., Shimoga, Karnataka, India.

Naik, D., Rao, M.S., Shylaja, M. *et al.* 2003. Management of wilt disease complex in capsicum using *Bacillus pumilus* and *Paecilomyces lilacinus. Proc. of Int. Plant Growth Promoting Rhizobacteria Workshop*, Calicut, India, pp. 276-281.

Nama, C.P. and Baheti, B.L. 2021. Assessment of losses due to root-knot nematode, *Meloidogyne incognita* on cluster bean (*Cyamopsis tetragonoloba* L.). *Journal of Entomology and Zoology Studies* 9(1): 855-858.

Nanjappa, C.K., Panda, M. and Seshadri, A.R. 1978. Effects of *Meloidogyne incognita* and *Rotylenchulus reniformis* on the emergence and survival of seedlings. *Third Int. Cong. on Plant Pathol.*, Germany.

Narayanamoorthy, A. 2004. Drip irrigation in India: Can it solve water scarcity? *Water Policy* 6(2): 117-130.

Nath, R., Khan, M.N., Wanshi, R.S.K. and Dwivedi, R.P. 1984. Influence of root-knot nematode, *Meloidogyne javanica* on pre- and post-emergence damping-off of tomato. *Indian Journal of Nematology* 14: 135-140.

Nath, R.P., Sinha, B.K. and Haider, M.G. 1976. Studies on the nematodes of vegetables in Bihar-II. Combined effect of *Meloidogyne incognita* and *Ozonium texanum* var. *parasiticum* on germination of eggplant. *Indian Journal of Nematology* 6: 177-179.

Nchore, S.B. 2012. *Incidence, Prevalence and Management of Root-Knot Nematodes (Meloidogyne spp.) on Selected Indigenous Leafy Vegetables in Kisii and Trans-Mara Counties, Kenya*. M. Sc. thesis (Plant Pathology), Kenyatta University, Kenya.

Netscher, C. 1983. Control of *Meloidogyne incognita* in vegetable production by crop rotation in Ivory Coast. *Acta Hort.* 152: 219-225.

NCPA. 1990. *Status, Potential and Approach for Adoption of Drip and Sprinkler Irrigation Systems.* National Committee on the Use of Plastics in Agriculture, Pune, India.

Ngundo, B.W. 1977. Screening of bean cultivars for resistance to *Meloidogyne* spp. in Kenia. *Plant Disease Reporter* 61: 991-993.

Nibin, P.M., Ushakumari, K. and Ishrath, P.K. 2019. Organic nano NPK formulations on soil microbial and enzymatic activities on post-harvest soil of bhindi. *Int. J. Curr. Microbiol. App. Sci.* 8(4): 1814-1819.

Nicewarner-Pena, S.R., Freeman, R.G., Reiss, B.D. *et al.* 2001. Sub-micrometer metallic barcodes. *Sci.* 294: 137-141.

Nirula, K.K. and Bassi, K.K. 1965. Thermotherapy for root-knot nematode, *Meloidogyne incognita* in potato tubers. *Indian Potato J.* 7: 9-11.

Ntidi, K.N., Daneel, M.S. and Fourie, H. 2019. The host status of *Amaranthus* spp. genotypes to *Meloidogyne* spp. and verification of resistance in genotype Thohoyandou. *Crop Protection* 115: 64-74.

Ntidi, K.N., Daneel, M.S. and Fourie, H. 2019. The host status of *Amaranthus* spp. genotypes to *Meloidogyne* spp. and verification of resistance in genotype Thohoyandou. *Crop Protection* 115: 64-74.

Ogundele, R.A., Oyedele, R.A. and Adekunle, O.K. 2016. Management of *Meloidogyne incognita* and other phytonematodes infecting *Amaranthus cruentus* and *Telfairia occidentalis* with African marigold (*Tagetes erecta*) and Siam weed (*Chromolaena odorata*). *Australasian Plant Pathol.* DOI 10.1007/s13313-016-0438-z.

Oostenbrink, M. 1972. Evaluation and integration of nematode control methods. In J.M. Webster (ed.) *Economic Nematology*. Academic Press, New York, pp. 497-514.

Oostenbrink, M., Kuiper, K. and S'Jacob, J.J. 1957. *Tagetes* als feindpflanzen von *Pratylenchus arten. Nematologica Suppl.* 2: 424-433.

Opperman, C.H., Acedo, G.N., Saravitz, D.M. et al. 1998. Bioengineering resistance to sedentary endoparasitic nematodes. In F. Lamberti, C. de Giorgi, and D.M. Bird (eds.) *Advances in Molecular Plant Nematology*, pp. 221-232. Plenum Press, New York.

Palanisamy, S. and Sivakumar, C.V. 1981. Assessment of avoidable yield loss in cowpea, black gram, maize and finger millet. *Second Nematology Symposium,* Tamil Nadu Agricultural University, Coimbatore, p. 58.

Pandey, R.K. and Nayak, D.K. 2018a. Assessment of avoidable yield loss due to root-knot nematode, *Meloidogyne incognita* infesting *Luffa acutangul*a (L.) Roxb. *Journal of Entomology and Zoology Studies* 6(5): 911-914.

Pandey, R.K. and Nayak, D.K. 2018b. Screening and evaluation of ridge gourd varieties /cultivars against induced population of root-knot nematode, *Meloidogyne incognita*. *Journal of Entomology and Zoology Studies* 6(5): 1954-1959.

Pani, A.K. and Das, S.N. 1972. Studies on etiological complexes in plant disease. I. Association of root-knot nematodes in bacterial wilt of tomato. *Journal of Research Orissa University of Agriculture & Technology* 2: 54-59.

Park, Y.L., Krel, R.K. and Carroll, M. 2007. Theory, technology, and practice of site-specific insect pest management. *Journal of Asia-Pacific Entomology* 10(2): 89–101.

Paruthi, L.J., Jain, R.K. and Gupta, D.C. 1987. Effect of different periods of degradation of subabool leaves alone and in combination with nematicides in okra. *Indian Journal of Nematology* **17**: 30-32.

Paruthi, I.J., Nandal, S.N. and Verma, K.K. 2010. Field evaluation of mustard and neem cake against rot-knot nematode (*Meloidogyne* spp) in bottle gourd. *National Conf. on Innovations in Nematological Research for Agricultural Sustainability – Challenges and A Roadmap Ahead.* Tamil Nadu Agri. Univ., Coimbatore, p. 84.

Parvatha Reddy, P. 1984. Efficacy of seed treatment with three nematicides for the control of *Meloidogyne incognita* infecting cowpea, French bean and peas. *Indian Journal of Nematology* 14: 39-40.

Parvatha Reddy, P. 1985a. Chemical control of root-knot nematodes infecting peas. *Indian Journal of Nematology* 15: 226.

Parvatha Reddy, P. 1985b. Estimation of crop loss in peas due to *Meloidogyne incognita. Indian Journal of Nematology 15*: 226.

Parvatha Reddy, P. 2008. *Disease of Horticultures Crops, Nematode Problems and their Management.* Scientific Publishers, Jodhpur, p. 131.

Parvatha Reddy, P. and Khan, R.M. 1988. Evaluation of *Paecilomyces lilacinus* for the biological control of *Rotylenchulus reniformis* infecting tomato as compared with carbofuran. *Nematologia Mediterranea* 16: 113-116.

Parvatha Reddy, P. and Khan, R.M. 1991. Integrated management of root-knot nematodes infecting okra. *Current Nematology* 2: 115-116.

Parvatha Reddy, P., Nagesh, M. and Devappa, V. 1997. Effect of integration of *Pasteuria penetrans, Paecilomyces lilacinus* and neem cake for the management of root-knot nematodes infecting tomato. *Pest Mangmt. Hort. Ecosystems* 3: 100-104.

Parvatha Reddy, P., Rao, M.S., Tewari, R.P. and Nagesh, M. 1999. Evaluation of oil cakes for their efficacy against mushroom nematodes in chicken manure. *Pest Mangmt. Hort. Ecosystems* 5: 50-53.

Parvatha Reddy, P. and Singh, D.B. 1979. Control of root-knot nematode, *Meloidogyne incognita* on tomato by chemical bare-root dips. *Indian Journal of Nematology* 9: 40-43.

Parvatha Reddy, P. and Singh, D.B. 1981. Assessment of avoidable yield loss in okra, brinjal, French bean and cowpea due to root-knot nematodes. *Third International Symposium on Plant Pathology*, New Delhi, pp. 93-94.

Parvatha Reddy, P., Singh, D.B. and Ram Kishun. 1979. Effect of root-knot nematodes on the susceptibility of Pusa Purple cluster brinjal to bacterial wilt. *Current Science* 48: 915-916.

Parvatha Reddy, P., Singh, D.B. and Sharma, S.R. 1979. Interaction of *Meloidogyne incognita* and

Rhizoctonia solani in a root rot disease complex of French beans. *Indian Phytopathology* 32: 651-652.

Passioura, J.B. 2002. Environmental biology and crop improvement. *Funct. Plant Biol.* 29: 537–546.

Patel, G.J., Shah, H.M. and Patel, D.J. 1979. Reaction of some tomato cultivars to root-knot nematodes. *Indian Journal of Nematology* 9: 60-61.

Patel, M. and Patel, A.D. 2017. Management of *Meloidogyne incognita* in mung bean by seed soaking in different chemicals under pot condition. *Int. J. Curr. Microbiol. App. Sci.* 6(10): 1080-1084.

Pathak, K.N. and Keshari, N. 2004. Interaction of *Meloidogyne incognita* with *Fusarium oxysporum* f. sp. *conglutinans* on cauliflower. *Indian J. Nematol.* 34: 85-87.

Patil, J., Goel, S.R. and Yadav, S. 2017b. Bio-management of cucumber wilt complex caused by root-knot nematode, *Meloidogyne incognita* and *Fusarium oxysporum* f. sp. *cucumerinum* in polyhouse under protected cultivation. *Journal of Pure and Applied Microbiology* 11 (4): 1909-1917.

Patil, J., Kumar, A., Yadav, S. and Goel S.R. 2018b. Nematicidal effect of fumigants on the *Meloidogyne incognita* and *Fusarium oxysporum* f. sp. *cucumerinum* on cucumber in polyhouse. *Plant Pathol. Journal* 17(1): 25-32.

Patil, J., Kumar, A., Yadav, S. *et al.* 2018c. Bio-efficacy of phytotherapeutic substances against *Meloidogyne incognita* and *Fusarium oxysporum* f. sp. *cucumerinum* affecting cucumber in polyhouse under protected cultivation. *Indian Journal Nematogy* 48(2): 190-197.

Patil, J.A., Kumar, A., Yadav, S. and Verma, K.K. 2020. Nematicidal effect of cruciferous bio-fumigants against the root-knot nematode, *Meloidogyne incognita* infesting okra. *J. Nematol.* 52. DOI: 10.21307/jofnem-2020-080.

Patil, J.A., Yadav, S. and Kumar, A. 2020. Evaluation of organic oils for the management of root-knot nematode, *Meloidogyne incognita* and fungus infesting cucumber under polyhouse conditions. *Indian Journal of Nematology* 50(2): 79-86.

Patil, J.A., Yadav, S. and Kumar, A. 2021. Management of root-knot nematode, *Meloidogyne incognita* and soil borne fungus, *Fusarium oxysporum* in cucumber using three bioagents under polyhouse conditions. *Saudi Journal of Biological Sciences* 28(12): 7006-7011.

Pedroche, N.B., Villanueva, L.M. and De Waele, D. 2009. Management of root-knot nematode, *Meloidogyne incognita* in carrot. *Commun. Agric. Appl. Biol. Sci.* 74(2): 605-615.

Pedroche, N.B., Villanueva, L.M. and De Dirk, W. 2007. Response of five lettuce cultivars to root-knot nematode, *Meloidogyne incognita*. *Commun Agric Appl Biol Sci.* 72(3): 659-666.

Perry, R.N. and Moens, M. 2006. *Plant Nematology*. CABI, Wallingford, England.

Perveen, S., Haque, E. and Gaffar, A. 1998. Efficacy of *Pseudomonas aeruginosa* and *Paecilomyces lilacinus* in the control of root rot-root knot disease complex in some vegetable crops. *Nematol. Medit.* 26: 209-212.

Perwez, M.S., Rehman, M.F. and Haider, S.R. 1988. Effect of *Tagetes erecta* on *Meloidogyne javanica* infecting lettuce. *Int. Nematol. Network Newsl.* 5(3): 18-19.

Pieterse, C.M.J., Van der Does, D., Zamioudis, C. *et al.* 2012. Hormonal modulation of plant immunity. *Annual Review of Cell and Developmental Biology* 28: 489– 521

Plett, J.M., Kemppainen, M., Kale, S.D. *et al.* 2011. A secreted effector protein of *Laccaria bicolor* is required for symbiosis development. *Current Biology* 21: 1197– 1203.

Ploeg, A. 2012. *Integrated Management of Root-knot Nematodes in Carrot*. Univ. of Calif., Riverside, USA.

Pofu, K.M., Mashela, P.W. and Oelofse, D. 2016. Nematode resistance in bitter gourd to *Meloidogyne incognita*. *Acta Agriculturae Scandinavica, Section B — Soil & Plant Science* 65(1): 1–5.

Potter, J.W. and Olthof, T.H.A. 1993. Nematode pests of vegetable crops. In: K. Evans, D.L. Trudgill and J.M. Webster (eds.) *Plant Parasitic Nematodes in Temperate Agriculture*. CAB International.

Prasad, K.S.K. 1989. Nematological problems and progress of research on potato in India. *Fourth Group Meeting on Nematological Problems of Plantation Crops*, University of Agricultural Sciences, Bengaluru.

Prasad, K.S.K. and Krishnappa, K. 1981. Effect of bare-root dip treatment with pesticides on *Rotylenchulus reniformis* affecting brinjal. *Indian Journal of Nematology* 11: 154-158.

Prasad, K.S.K., Raj, D. and Sharma, R.K. 1983. Effect of planting time on *Meloidogyne incognita* infesting potato at Shimla Hills. *Third Nematology Symposium*, Dr. Y.S. Paramar Univ. of Hort. and Forestry, Solan, pp. 5-6.

Qu L, Dai K, Wang J, *et al.* 2023. Microbial landscapes of the rhizosphere soils and roots of *Luffa cylindrica* plant associated with *Meloidogyne incognita*. *Front. Microbiol.* 14 1168179. doi: 10.3389/fmicb.2023.1168179.

Raj, D. and Gill, J.S. 1983. Reaction of some potato varieties to root-knot nematode *Meloidogyne incognita*. *Third Nematology Symposium,* Dr. Y.S. Paramar Univ. of Hort. and Forestry, Solan, pp. 27-28.

Rajagopalan, P. and Naganathan, T.B. 1977. Studies on nematode parasites of banana. *Tamil Nadu Agricultural University Annual Report* 6: 131.

Rajaraman, G. and Paramaguru, P. 2011. Influence of fertilizer levels on yield and economics of leafy types of coriander (*Coriandrum sativum* L.). *Crop Research* 42: 210-214.

Rajendran, G., Vadivelu, S. and Muthukrishnan, T.S. 1976. Pathogenicity of *Meloidogyne incognita* on crossandra. *Indian Journal of Nematology* 6: 115-116.

Rajesh, V., Bajaj, H.K. and Mann, S.S. 2003. Integrated management of *Aphelenchoides composticola* in button mushroom with neem seed kernel water extract and Achook. In R.V. Singh, S. Pankaj, S.C. Dhawan and H.S. Gaur (eds.) *Proc. National Symp. Biodiversity and Mangmt. of Nematodes in Cropping Systems for Sust. Agri.*, Jaipur, India, pp. 228-231.

Rajvaniya, D.K., Sharma, H.K., Sharma M.K. and Jangir, B. 2021. Eco-friendly management of reniform nematode (*Rotylenchulus reniformis*) using botanicals on cluster bean (*Cyamopsis tetragonoloba* L.). *Journal of Entomology and Zoology Studies* 9(5): 459-460.

Ram, B. and Baheti, B.L. 2004. Management of reniform nematode, *Rotylenchulus reniformis* on cowpea through seed and soil treatment with plant products. *Indian J. Nematol.* 34: 193-195.

Ramakrishnan, C. and Balasubramanian, P. 1981. Control of *Meloidogyne incognita* in chilli *(Capsicum annum* L.) nursery with nematicides. *Second Nematology Symposium*, Tamil Nadu Agricultural University, Coimbatore, p. 54.

Ramana, K.V. and Mohandas, C. 1988. Occurrence of potato cyst nematode, *Globodera pallida* (Stone, 1973) in Kerala. *Indian Journal of Nematology* 18: 141.

Ramappa, H.K. 1988. *Studies on Nematode Parasites of French Bean and their Control.* M. Sc. (Agri.) thesis, Univ. of Agri. Sci., Bangalore.

Rani, P., Singh, M., Prashad, H. and Sharma, M. 2022. Evaluation of bacterial formulations as potential biocontrol agents against the southern root-knot nematode, *Meloidogyne incognita*. *Egypt J Biol Pest Control* 32, 29. https://doi.org/10.1186/s41938-022- 00529-3.

Rao, M.S., Gowen, S.R., Pembroke, B. and Parvatha Reddy, P. 1997f. Relationship of *Pasteuria penetrans* spore encumbrance on juveniles of *Meloidogyne incognita* and their infection in adults. *Nematol. Medit.* 25: 129-131.

Rao, M.S., Kamalnath, M., Umamaheswari, R. *et al.* 2017. *Bacillus subtilis* IIHR BS-2 enriched vermicompost controls root-knot nematode and soft rot disease complex in carrot. *Scientia Horticulturae* 218: 56-62.

Rao, M.S., Kerry, B.R., Gowen, S.R. *et al.* 1997e. Management of *Meloidogyne incognita* in tomato nurseries by integration of *Glomus deserticola* with *Verticillium chlamydosporium*. *J. Pl. Dis. Prot.* 104: 419-422.

Rao, M.S., Mohandas, S. and Parvatha Reddy, P. 1993c. Management of reniform nematode on eggplant by integrating botanical (castor cake), mycorrhiza (*Glomus fasciculatum*) and bioagent (*Paecilomyces lilacinus*). *Golden Jubilee Symposium on Horticultural Research - Changing Scenario,* Bengaluru, pp. 259-260.

Rao, M.S., Naik, D. and Shylaja, M. 2004. Bio-intensive management of root-knot nematodes on bell pepper using *Pochonia chlamydosporia* and *Pseudomonas fluorescens*. *Nematol. Medit.* 32: 159-163.

Rao, M.S., Naik, D., Shylaja, M. and Parvatha Reddy, P. 2002. Prospects for the management of the root-knot nematode and bacterial wilt disease complex in capsicum by integration of fungal and bacterial biocontrol agents. *Proc. of Int. Conf. on Vegetables,* Bangalore, pp. 347-351.

Rao, M.S., Pankaj, Sadashiva, A.T. and Parvatha Reddy, P. 1998e. Effects of manifestation of *Mi* gene in a tomato hybrid and its parental line on the penetration and development of *Meloidogyne incognita*. *Nematol. Medit.* 26: 213-216.

Rao, M.S. and Parvatha Reddy, P. 1992. Prospects of management of root-knot nematode on tomato through the integration of biocontrol agents and botanicals. *Seminar on Current Trends in Diagnosis and Management of Plant Diseases,* Bangalore, p.10.

Rao, M.S. and Parvatha Reddy, P. 1993a. Effective use of 'Welgro' and 'RD-9 Repelin' for the management of root-knot nematodes on tomato. In R.P. Singh, M.S. Chari, A.K. Raheja and W. Kraus (eds.) *Neem and Environment* 2: 651-656. Oxford & IBH Publishing Co. Pvt. Ltd., New Delhi & Calcutta.

Rao, M.S. and Parvatha Reddy, P. 1993b. Interactive effect of *Verticillium chlamydosporium* and castor cake on the control of root-knot nematode on eggplant. In D. Rajagopal, R.D. Kale and K. Bano (eds.) *Soil Organisms and Sustainability*, pp. 114-118. Indian Society of Soil Biology and Ecology, Univ. of Agri. Sci., Bengaluru.

Rao, M.S. and Parvatha Reddy, P. 1994. A method for conveying *Paecilomyces lilacinus* to soil for the management of root-knot nematodes on eggplant. *Nematol. Medit.* 22: 265-267.

Rao, M.S. and Parvatha Reddy, P. 2001. Control of *Meloidogyne incognita* on eggplant using *Glomus mosseae* integrated with *Paecilomyces lilacinus* and neem cake. *Nematol. Medit.* 29: 153-157.

Rao, M.S., Parvatha Reddy, P., Archana Mittal, Chandravadana, M.V. and Nagesh, M. 1996a. Effect of some secondary plant metabolites as seed treatment agents against *Meloidogyne incognita* on tomato. *Nematol. Medit.* 24: 49-51.

Rao, M.S., Parvatha Reddy, P. and Chawla, G. 1996b. Potential use of botanicals for the development of formulation of the biocontrol fungus, *Verticillium lecanii* (Zimm.) Viegas. *J. Plant Diseases Prot.* 103: 436-439.

Rao, M.S., Parvatha Reddy, P. and Mohandas, S. 1996. Effect of integration of *Calotropis procera* leaf and *Glomus fasciculatum* on the management of *Meloidogyne incognita* infesting tomato. *Nematol. Medit.* 24: 59-61.

Rao, M.S., Parvatha Reddy, P. and Nagesh, M. 1997a. Integrated management of *Meloidogyne incognita* on okra by castor cake suspension and *Paecilomyces lilacinus*. *Nematol. Medit.* 25: 17-19.

Rao, M.S., Parvatha Reddy, P. and Nagesh, M. 1997b. Effective use of neem cake extract for the management of root-knot nematodes infecting okra (*Abelmoschus esculentus*). *Pest Management in Hortil. Ecosystems* 3: 95-99.

Rao, M.S., Parvatha Reddy, P. and Nagesh, M. 1997c. Integration of *Paecilomyces lilacinus* with neem leaf suspension for the management of root-knot nematodes on eggplant. *Nematol. Medit.* 25: 249-252.

Rao, M.S., Parvatha Reddy, P. and Nagesh, M. 1997d. Management of root-knot nematode, *Meloidogyne incognita* on tomato by integration of *Trichoderma harzianum* with neem cake. *J. Pl. Dis. Prot.* 104: 423-425.

Rao, M.S., Parvatha Reddy, P. and Nagesh, M. 1998a. Integrated management of *Meloidogyne incognita* on tomato using *Verticillium chlamydosporium* and *Pasteuria penetrans*. *Pest Management in Hortil. Ecosystems* 4: 32-35.

Rao, M.S., Parvatha Reddy, P. and Nagesh, M. 1998c. Evaluation of plant-based formulations of *Trichoderma harzianum* for the management of *Meloidogyne incognita* on eggplant. *Nematol. Medit.* 26: 59-62.

Rao, M.S., Parvatha Reddy, P. and Nagesh, M. 1998d. Evaluation of neem-based formulation of *Paecilomyces lilacinus* for the management of root-knot nematodes on eggplant. *Pest Managmt. in Hortil. Ecosystems* 4: 116-119

Rao, M.S., Parvatha Reddy, P., Somasekhar, N. and Nagesh, M. 1997g. Management of root-knot nematodes, *Meloidogyne incognita* in tomato nursery by integration of endomycorrhiza, *Glomus fasciculatum* with castor cake. *Pest Managmt. in Hortil. Ecosystems* 3: 31-35.

Rao, M.S., Parvatha Reddy, P. and Sukhada, M. 1998f. Biointensive management of *Meloidogyne incognita* on eggplant, by integrating *Paeciloymces lilacinus* and *Glomus mosseae*. *Nematol. Medit.* 26: 217-219.

Rao, M.S., Parvatha Reddy, P., Sukhada, M., Nagesh, M. and Pankaj, S. 1998b. Management of root-knot nematodes on eggplant by integrating endomycorrhiza (*Glomus fasciculatum*) and castor (*Ricinus communis*) cake. *Nematol. Medit.* 26: 221-223.

Rao, M.S., Ramachandran, N., Sowmya, D.S. *et al.* 2009. Biological control of nematode induced disease complex in certain vegetable crops. *Int. Conf. on Hort.*, Bangalore.

Rao, M.S. and Shylaja, M. 2004. Role of *Pseudomonas fluorescens* (Migula) in induction of systemic resistance (ISR) and managing *Rotylenchulus reniformis* (Linford and Oliveira) on carrot. *Pest Management in Horticultural Ecosystems* 10: 87-93.

Rao, M.S., Tewari, R.P. and Pandey, M. 1992. Occurrence of mushroom nematodes in South India. *Mushroom Res.* 1: 135–136.

Rao, V.R. and Singh, D.B. 1978. Chemical control of root-knot nematodes in okra with the aid of nematicides. *Indian Journal of Nematology* 35: 151-154.

Rao, V.R. and Tikoo, S.K. 1973. Resistance of tomato cultivars to the root-knot nematode, *Meloidogyne incognita* (Kofoid and White, 1919) Chitwood, 1949. *Current Science* 44: 282-283.

Ravichandra, N.G., Krishnappa, K. and Setty, K.G.H. 1990. Interaction of *Meloidogyne javanica* and race 1, race 2, race 3 of *Meloidogyne incognita* with *Pseudomonas solanacearum* on a few brinjal lines. *Indian J. Nematol.* 20:138-147.

Ravichandra, N.G. and Reddy, B.M.R. 2008. Efficacy of *Pasteuria penetrans* in the management of *Meloidogyne incognita* infecting tomato. *Indian J. Nematol.* 38: 172-175.

Reddy, D.D.R. 1985. Analysis of crop losses in tomato due to *Meloidogyne incognita. Indian Journal of Nematology* 15: 55-59.

Reddy, D.D.R. and Seshadri, A.R. 1972. Studies on some systemic nematicides. II. Further studies on the action of thionazin and aldicarb on *Meloidogyne incognita* and *Rotylenchulus reniformis. Indian Journal of Nematology* 2: 182-190.

Reddy, D.D.R. and Seshadri, A.R.1975. Elimination of root-knot nematode infestation from tomato seedlings by chemical bare-root dips or soil application. *Indian Journal of Nematology* 5: 170-175.

Reddy, M.V., Raju, T.N., Sharma, S.B. *et al.* 1993. *Handbook of Pigeonpea Diseases*. Information Bulletin No. 42, International Crops Research Institute for the Semi-Arid Tropics, Patancheru, Telangana, India, 64 pp.

Reen, R.A., Thompson, J.P., Clewett, T.G., *et al.* 2014. Yield response in chickpea cultivars and wheat following crop rotations affecting population densities of *Pratylenchus thornei* and arbuscular mycorrhizal fungi. *Crop Pest. Sci.* 65: 428–441.

Renjhen, P.L. 1973. *History of Plant Quarantine and its functioning in India.* Minist. of Agri. Directorate of Pl. Prot. Quarantine and Storage, Govt. of India, Faridabad.

Richards, R.A. 2004. Physiological traits used in the breeding of new cultivars for water-scarce environments. In: R.A. Fischer, N. Turner, J. Angus, L. McIntyre, M. Robertson, A. Borrell and D. Lloyd (eds.) *New Directions for a Diverse Planet, Proc. 4th Int. Crop Sci. Cong.*, Brisbane.

Rivard, C.L., Sydorovych, O., O'Connell, S. *et al.* 2010. An economic analysis of two grafted tomato transplant production systems in the United States. *Horticulture Technology* 20: 794-803.

Robinson, A.F., Inserra, R.N., Caswell-Chen, E.P. *et al.* 1997. *Rotylenchulus* species: Identification, distribution, host ranges, and crop plant resistance. *Nematropica* 27: 127-180.

Rodriguez-Kabana, R. and King, P.S. 1977. Nematicidal activity of the fungicide ethazole. *J. Nematol.* 9: 203-206.

Routaray, B.N. and Sahoo, H. 1985. Integrated control of root-knot nematode, *Meloidogyne incognita* with neem cake and granular nematicides on tomato. *Indian Journal of Nematology* 15: 261.

Roy, K., Mukhopadhyay, A.K. and Pramanik, A. 2008. Pathogenicity of *Rotylenchulus reniformis* Linford & Oliveira, 1940 in brinjal, *Solanum melongena* L. *Ann. Plant Prot. Sci.* 16(1): 179-181.

Saikia, J, Borah, A and Bora, B.C. 2007. Efficacy of organic amendments in combination with the nematicide for management of *Meloidogyne incognita* in brinjal. *Indian J. Nematol.* 37(1): 179–181.

Samiyappan, R. 2003. Biological control of fungal and nematode complex diseases by plant growth promoting rhizobacteria (PGPR). *Winter School on Biological Control of Plant Parasitic Nematodes*, Tamil Nadu Agri. Univ., Coimbatore, pp. 69-77.

Sano, Z. 2002. Nematode management strategies in East Asian countries. *Nematology* 4: 129-130.

Savitha, R.M. and Lingaraju, S. 1996. Interaction of root-knot nematode and wilt bacterium on brinjal. *Karnataka J. Agri. Sci.* 9: 657-660.

Saxena, R. and Reddy, D.D.R. 1987. Crop losses in pigeon pea and mung bean by pigeon pea cyst nematode, *Heterodera cajani*. *Indian J. Nematol.* 17: 91–94.

Saxena, S.K., Alam, M.M. and Khan, A.M. 1974. Chemotherapeutic control of nematodes with Vydate oxamyl, VC-13 and Dazomet on chilli plants. *Indian Journal of Nematology* 4: 235-238.

Scoville, S. 2013. Implications of nanotechnology safety of sensors on homeland security industries. In *Nanotechnology Safety*, pp. 175–194. Elsevier B.V. https://doi.org/ 10.1016/B978-0-444-59438-9.00013-8.

Sebastian, S. and Gupta, P. 1995. Population dynamics of *Pratylenchus thornei* in infested fields at Allahabad. *Indian J. Mycol. Plant Pathol.* 25: 270–271.

Sen, K. and Dasgupta, M.K. 1981. Effect of rice hull ash on root-knot nematodes in tomato. *Tropical Pest Management* 27: 518-519.

Servin, A.D. and White, J.C. 2016. Nanotechnology in agriculture: Next steps for understanding engineered nanoparticle exposure and risk. *NanoImpact* 1: 9–12.

Seth, A. 1984. *Studies on The Incidence, Pathogenicity, Biology and Control of* Aphelenchoides *Species Associated with Mushroom*. Ph. D. thesis, Dept. of Entomol. and Apiculture, Dr. Y.S. Paramar Univ. of Hort. & Forestry, Solan, India.

Seth, A. and Sharma, N.K. 1986. Five new species of genus *Aphelenchoides* (Nematoda: Aphelenchida) infesting mushroom in North India. *Indian J. Nematol.* 16: 205-215.

Sethi, C.L., Nath, R.P., Mathur, V.K. and Ahuja, S. 1972. Interceptions of plant parasitic nematodes from imported seed/plant materials. *Indian Journal of Nematology* 2: 89-93.

Shahzad, S. and Gaffer, A. 1989. Use of *Paecilomyces lilacinus* in the control of root rot and root-knot complex in okra and mung bean. *Pakistan J. Nematol.* 7: 47-53.

Shanthi, A. 2010. Management of root-knot nematode, *Meloidogyne incognita* in tomato using bioinoculants. *National Conf. on Innovations in Nematol. Res. for Agri. Sustainability – Challenges and A Roadmap Ahead.* Tamil Nadu Agri. Univ., Coimbatore, p. 93.

Shantharam, S. 2015. Bt brinjal ban a costly mistake. *Mint* 11 Mar 2015.

Shanthi, A. 2016. Management of root-knot nematode, *Meloidogyne incognita* on ash gourd (*Benincasa hispida*) and pumpkin (*Cucurbita pepo*) using bio-control agents. *Annals of Plant Protection Sciences* 24(2): 399-401.

Sharma, R. and Swarup, G. 1988. *Pathology of Cyst Nematodes*. Malhotra Publishing house, New Delhi, 88 pp.

Sharma, A. and Trivedi, P.C. 1994. Interaction of root-knot nematode with VAM on *Abelmoschus esculentus* and their effect on fruit yield. In T. Singh and P.C. Trivedi (eds.) *Vistas in Seed Biology*, Vol. II, pp. 61-81, Printwell, Jaipur.

Sharma, H.K., Pankaj and Mishra, S.D. 2005. Polyethylene mulching in the management of plant parasitic nematodes. *Indian Journal of Nematology* 35: 82-85.

Sharma, H.K., Pankaj, Lal, J. and Singh, S. 2007. Combined effect of inorganic supplements with *Trichoderma viride* and *Kalisena* against root-knot nematode. *Indian J. Nematol.* 37: 192-196.

Sharma, G.L. 1989. Estimated losses due to root-knot nematodes, *Meloidogyne incognita* and *M. javanica* in pea crop. *International Nematology Network Newsletter* 6(1): 28-29.

Sharma, H.K. and Gill, J.S. 1992. Prevalence of races in root-knot nematode, *Meloidogyne incognita* in India. *Indian J. Nematol.* 22: 43-49.

Sharma, J.K., Singh, I. and Chhabra, H.K. 1980. Observations on the influence of *Meloidogyne incognita* and *Rhizoctonia bataticola* on okra. *Indian Journal of Nematology* 10: 148-151.

Sharma, K.P., Sharma, G.C., Thakur, N. and Kumari, S. 2015. Incidence of phytoparasitic nematodes in vegetable crops grown under protected cultivation in Himachal Pradesh. *Indian Journal of Nematology* 45(1): 7-11.

Sharma, N.K. and Khanna, A.S. 1992. Nematode pests of mushroom. In D.S. Bhatti and R.K. Walia (eds.) *Nematode Pests of Crops*, pp. 267–274. CBS Publishers and Distributors, Delhi, 381 pp.

Sharma, N.K. and Sethi, C.L. 1975. Leghaemoglobin content of cowpea nodules as influenced by *Meloidogyne incognita* and *Heterodera cajani. Indian Journal of Nematology* 5: 113-114.

Sharma, N.K. and Sethi, C.L. 1976a. Reaction of certain cowpea varieties to *Meloidogyne* incognita and *Heterodera cajani. Indian Journal of Nematology* 6: 99-102.

Sharma, N.K. and Sethi, C.L. 1976b. Interrelationship between *Meloidogyne incognita, Heterodera cajani* and *Rhizobium* sp. on cowpea (*Vigna sinensis* (L.) Savi). *Indian Journal of Nematology* 6: 117-123.

Sharma, N.K., Thapa, C.D. and Kaur, D. 1984. Estimation of loss in mushroom yield due to myceliophagous nematode, *Aphelenchoides sacchari. Indian J. Nematol.* 14: 121-124.

Sharma, N.K., Thapa, C.D. and Kaur, D. 1985. Pathogenicity and identity of *Ditylenchus myceliophagus* "Indian population" infesting *Agaricus bisporus* (Lange) Sing. in India. *Indian J. Nematol.* 15: 233-234

Sharma, N.K., Thapa, C.D. and Nath, A. 1981. Pathogenicity and identity of myceliophagous nematode infesting *Agaricus bisporus* (Lange) Sing. in Himachal Pradesh (India). *Indian J. Nematol.* 11: 230-231.

Sharma, R., Kaur, S., Dhillon, N.K. and Pathak, M. 2019. Identification of resistance in cultivated and wild bitter gourd against root-knot nematode, *Meloidogyne incognita. Indian Phytopathology* 72: 203–208.

Sharma, S.B. 1993. Pearly root of pigeon peas caused by *Heterodera cajani. Indian Journal of Nematology* 21: 169.

Sharma, S.B., Ali, S.S., Upadhyay, K.D. and Ahmed, F, 1996. Potential nematode constraints of pigeon pea in Uttar Pradesh in northern India. *Afro-Asian Journal of Nematology* 6(2): 151-155.

Sharma, S.B., Jain, K.C. and Lingaraju, S. 2000. Tolerance to reniform nematode (*Rotylenchulus reniformis*) race A in pigeon pea (*Cajanus cajan*) genotypes. *Ann. Appl. Biol.* 18: 197-203.

Sharma, S.B. and Nene, Y.L. 1985. Additions to host range of pigeon pea cyst nematode, *Heterodera cajani. International Pigeonpea Newsletter* 4: 42.

Sharma, S.B., Sharma, H.K. and Pankaj. 2002. Nematode problems in India. In D. Prasad and S.N. Puri, (eds.) *Crop Pest and Disease Management: Challenges for the Millennium*, pp. 267-275. Jyoti Publishers, New Delhi, India.

Sharma, S.B., Smith, D.H. and McDonald, D. 1992. Nematode constraints of chickpea and pigeonpea production in the semiarid tropics. *Plant Dis.* 76: 868–874.

Sharon, A., Balakrishnamurthy, G. and Jansirani, P. 2012. Influence of fertigation treatments on growth and yield of curry leaf (*Murraya koenigii* Spreng.) during off season. *Crop Research* 44: 461-465.

Sharon, M., Choudhary, A.K. and Kumar, R. 2010. Nanotechnology in agricultural diseases and food safety. *J. Phytol.* 2: 83–92.

Shaukat, S.S., Sayed, M., Khan, A. *et al.* 2010. Effect of different treatments on population of nematodes and yield of Guar (*Cyamopsis tetragonoloba* L.). *Pakistan Journal of Agricultural Research* 23: 143-146.

Sheela, M.S. and Nisha, M.S. 2004. Impact of biological control agents for the management of root-knot nematode in brinjal. *National Symp. on Paradigms in Nematol. Res. for Biodynamic Farming*, Univ. of Agri. Sci., Bangalore, pp. 63-64.

Sheela, M.S., Nisha, M.S. and Mohandas, C. 2002. Impact of trap cropping and crop rotation for the management of root-knot nematode in vegetables. *Proc. of Int. Conf. on Vegetables*, Bangalore, pp. 344 -346.

Shendge, A.G., Mhase, N.L., Landge, S.A. and Kadu, R.V. 2010. Integrated management of root-knot nematode, *Meloidogyne incognita* infesting okra [*Abelmoschus esculentus* (L.) Moench] *International Journal of Plant Protection* 3(2): 369-371.

Shetty, K.D. and Reddy, D.D.R. 1985. Resistance in *Solanum* species to root-knot nematode, *Meloidogyne incognita. Indian Journal of Nematology* 15: 230.

Shobha, G., Kantharaju, V. and Gururaj, S. 2016. Management of root-knot nematode, *M. incognita* in ridge gourd by using bio-agents, botanicals and chemicals. *Current Nematology* 27(1): 29–38.

Shukla, V.N. and Swarup, G. 1970. Interrelationships of the root-knot nematode, *Meloidogyne incognita* (Kofoid and White, 1919) Chitwood, 1949 and *Sclerotium rolfsii* Sacc. on the emergence of tomato seedlings. *Bulletin of Indian Phytopathological Society* 6: 52-54.

Siddiqi, M.A. and Alam, M.M. 1989a. Seed treatment with azadirachtin for the control of the stunt nematode attacking cabbage and cauliflower. *Ann. Appl. Biol.* 113: 4-5.

Siddiqi, M.A. and Alam, M.M. 1989b. Control of stunt nematode by bare-root dip in leaf extracts of margosa and Persian lilac. *Pakistan J. Nematol.* 7: 33-38.

Siddiqi, M.A. and Saxena, S.K. 1987a. Effect of interculture of margosa and Persian lilac with tomato and eggplant on root-knot and reniform nematodes. *Int. Nematol. Network Newsl.* 4(2): 5-8.

Siddiqi, M.A. and Saxena, S.K. 1987b. Studies on the control of the stunt nematode, *Tylenchorhynchus brassicae* by interculture of margosa and Persian lilac. *Int. Nematol. Network Newsl.* 4(4): 27-29.

Siddiqui, M.H., Al-Whaibi, M.H., Faisal, M. and Al Sahli, A.A. 2014. Nano-silicon dioxide mitigates the adverse effects of salt stress on *Cucurbita pepo* L. *Environ. Toxicol. Chem.* 33(11): 2429-2437.

Siddiqui, Z.A., Khan, A.M. and Saxena, S.K. 1972. Host range and varietal resistance of certain crucifers against *Tylenchorhynchus brassicae. Indian Phytopath.* 25: 275-281.

Siddiqui, Z.A., Khan, A.M. and Saxena, S.K. 1973. Studies on *Tylenchorhynchus brassicae* II - Effect of temperature and moisture on multiplication of nematode. *Indian J. Entomol.* 26: 139-147.

Siddiqi, Z.A., Khan, A.M. and Saxena, S.K. 1976. Nematode population and yield of certain vegetables as influenced by oil-cake amendments. *Indian Journal of Nematology* 6: 179-182.

Siddiqui, Z.A. and Mahmood, I. 1995a. Biological control of *Heterodera cajani* and *Fusarium udum* by *Bacillus subtilis*, *Bradyrhizobium japonicum* and *Glomus fasciculatum* on pigeon pea. *Fundamental and Applied Nematology* 18(6): 559-566.

Siddiqui, Z.A. and Mahmood, I. 1995b. Some observations on the management of the wilt disease complex of pigeon pea by treatment with a vesicular arbuscular fungus and biocontrol agents for nematodes. *Bioresource Technology* 54(3): 227-230.

Siddiqui, Z.A. and Mahmood, I. 1996. Biological control of *Heterodera cajani* and *Fusarium udum* on pigeon pea by *Glomus mosseae, Trichoderma harzianum*, and *Verticillium chlamydosporium. Israel Journal of Plant Sciences* 44(1): 49-56.

Sidhu, J. 2021. Alternative nematicides for root-knot nematodes in melons. *Progressive Crop Cosultant*, January 2021 Issue.

Sikora, R.A. 1992. Management of the antagonistic potential in agricultural ecosystems for the biological control of plant parasitic nematodes. *Annual Review of Phytopathology* 30: 245–270.

Sikora, R.A. and Fernandez, E. 2005. Nematode parasites. In: M. Luc, R.A. Sikora and J. Bridge (eds.) *Plant Parasitic Nematodes in Subtropical and Tropical Agriculture*, pp. 319- 392. CABI publishing, Wallingford, U.K.

Sikora, R.A., Greco, N. and Veloso Silva, J.F. 2005. Nematode parasites of food legumes. In: M. Luc, R.A. Sikora, and J. Bridge (eds.) *Plant Parasitic Nematodes in Subtropical and Tropical Agriculture*, pp. 259-318. CABI Publishing, Wallingford, UK.

Sikora, R.A., Molendijk, L.P.G. and Desaeger, J. 2022. Integrated nematode management and crop health: Future challenges and opportunities. In R.A. Sikora, J. Desaeger and L.P.G. Molendijk (eds.) *Integrated Nematode Management: State-of-The-Art and Visions for The Future*, pp. 3-10. CAB International. Wallingford, UK.

Sikora, R.A., Pocasangre, L., Zum Felde, A. *et al.* 2008. Mutualistic endophytic fungi and in-planta suppressiveness to plant parasitic nematodes. *Biological Control* 46: 15–23.

Sikora, R.A. and Roberts, P. 2018. Management practices: An overview of integrated nematode management technologies. In: R.A. Sikora, D. Coyne, J. Hallmann and P. Timper (eds.) *Plant Parasitic Nematodes in Subtropical and Tropical Agriculture*, pp. 795–838. CAB International, Wallingford, UK.

Singh, A.U. and Prasad, D. 2016. Evaluation of mung bean lines/germplasm for resistance to root-knot nematodes, *Meloidogyne incognita*. *J. Natur. Resource Devel.* 11: 41–43.

Singh, D. 1991. *Studies on the Effect of* Tagetes *species on* Meloidogyne javanica *Infectivity in Brinjal* (Solanum melongena). Ph. D. thesis, Haryana Agri. Univ., Hisar, 83 pp.

Singh, D.B. and Parvatha Reddy, P. 1981a. Influence of *Meloidogyne incognita* infestation on *Rhizobium nodule* formation in French bean. *Nematologia Mediterranea* 9: 1-5.

Singh, D.B. and Parvatha Reddy, P. 1981b. Note on chemical control of root-knot nematodes infesting French bean. *Indian Journal of Agricultural Science* 51: 534-535.

Singh, D.B. and Parvatha Reddy, P. 1982a. Cultivars of cowpea resistant to root-knot nematode, *Meloidogyne incognita* in India. *Tropical Pest Management* 28: 185-186.

Singh, D.B. and Parvatha Reddy, P. 1982b. Chemical control of *Meloidogyne incognita* infecting cowpea. *Indian Journal of Nematology* 12: 196-197.

Singh, D.B. and Parvatha Reddy, P. 1985. Nature of resistance to *Meloidogyne incognita* in cowpea, *Vigna unguiculata*. *Nematologia Mediterranea* 13: 127-132.

Singh, D.B., Parvatha Reddy, P. and Sharma, S.R. 1981b. Effect of the root-knot nematode, *Meloidogyne incognita* on *Fusarium* wilt of French beans. *Indian J. Nematol.* 11: 84-85.

Singh, D.B., Parvatha Reddy, P., Rao, V.R. and Rajendran, R. 1981. Reaction of some varieties and selections of French bean to *Meloidogyne incognita. Indian Journal of Nematology* 11: 81-83.

Singh, D.B., Parvatha Reddy, P., Rao, V.R. and Rajendran, R. 1981a. Cultivars of French bean resistant to root-knot nematode, *Meloidogyne incognita,* in India. *Trop. Pest Mangmt.* 27: 29-31.

Singh, D.B., Rao, V.R. and Parvatha Reddy, P. 1978. Evaluation of nematicides for the control of root-knot nematodes on brinjal. *Indian Journal of Nematology 8*: 64-66.

Singh, R. and Kumar, U. 2013. Assessment of nematode distribution and yield losses in vegetable crops of western Uttar Pradesh in India. *International Journal of Science and Research* 4(5): 2812-2816.

Singh, R.S. and Sitaramaiah, K. 1966. Incidence of root-knot of okra and tomato in oil cake amended soil. *Plant Disease Reporter* 50: 668-672.

Singh, R.S. and Sitaramaiah, K. 1971. Control of root-knot through organic and inorganic amendments of soil: Effect of saw dust and inorganic nitrogen. *Indian Journal of Nematology* 1: 80-84.

Singh, R.V., Midha, S.K. and Kumar, V. 2003. Major nematode pests of economically important crops in India and their management. In P.C. Trivedi (ed.) *Advances in Nematology*, pp. 233-256. Scientific Publishers (India), Jodhpur.

Singh, S. 2019. Integrated approach for management of root-knot nematode (*Meloidogyne incognita*) in bitter gourd (*Momordica charantia*). *Indian Journal of Agricultural Sciences* 89(9): 1439–1444.

Singh, S.P., Pant, V., Khan, A.M. and Saxena, S.K. 1984. Efficacy of certain nematicides and oil cakes for the control of stunt nematode, *Tylenchorhynchus brassicae* and parasitic fungi in the rhizosphere of cauliflower. *Indian Phytopathology* 37: 647-651.

Singh, T. and Patel, B.A. 2015. Management of root-knot nematode (*Meloidogyne incognita*) in bottle gourd using different botanicals in pots. *J. Parasit. Dis.* 39(3): 441–445.

Singh, T., Patel, A.D., Prajapati, A. and Patel, B.A. 2019. Biomanagement of *Meloidogyne* species in bitter gourd. *Int. J. Curr. Microbiol. App. Sci.* 8(10): 546-553.

Singh, B., Ali, S.S., Naimuddin and Askary, T.H. 2004. Combined effect of *Fusarium udum* and *Meloidogyne javanica* on wilt resistant accessions of pigeon pea. *Ann. Pl. Prot. Sci.* 12: 130133.

Singh, S., Dhawan, S.C. and Goswami, B.K. 2007. Histopathology of cowpea roots co-infected with root-knot nematode, *Meloidogyne incognita* and wilt fungus *Fusarium oxysporum. Indian J. Nematol.* 37: 156-160.

Širca, S., Stare, B.G. and Pleško, I.M. 2007. *Xiphinema rivesi* from Slovania transmit tobacco ringspot virus and tomato ringspot virus to cucumber bait plants. *Plant Dis.* 91(6):770.

Sitaramaiah, K. and Naidu, H. 2003. Plant parasitic nematodes-management - opportunities and challenges. In P.C. Trivedi (ed.) *Nematode Management in Plants*, pp. 57-80. Scientific Publishers (India), Jodhpur.

Sitaramaiah. K. and R. A. Sikora 1981. Influence of the endomycorrhizal fungus *Glomus mosseae* on *Rotylenchulus reniformis* penetration and development on bush bean, cucumber and muskmelon. *Med. Fac. Landbouww Rijksuniv Gent*. 46: 695 - 702.

Siti, E., Cohn, E., Katan, J. and Mordechai, M. 1982. Control of *Ditylenchus dipsaci* in garlic by bulb and soil treatments. *Phytoparasitica* 10: 93-100.

Sivakumar, C.V. 1998. Bacterial antagonists for suppression of plant parasitic nematodes. In S.P. Singh and S.S. Hussaini (eds.) *Biological Suppression of Plant Diseases, Phytoparasitic Nematodes and Weeds*, pp. 128-137. Project Directorate of Biological Control, Bangalore.

Sivakumar, C.V., Kuppuswamy, S. and Meerazainuddin, M. 1973. Evaluation of carbofuran seed treatment for the control of root-knot on okra. *Indian J. Nematol.* 3: 71-72.

Sivakumar, C.V. and Meerazainuddin, M. 1974. Influence of N, P and K on the reniform nematode, *Rotylenchulus reniformis* and its effect on the yield of okra. *Indian Journal of Nematology* 4: 243-244.

Sivakumar, C.V., Palanisamy, S. and Naganathan, T.G. 1976. Persistence of nematicidal activity in seed treatment of okra with carbofuran and aldicarb sulfone. *Indian Journal of Nematology* 6: 106-108.

Smittle, D.A., Dickens, W.L. and Stansell, J.R. 1994. Irrigation regimes affect yield and water use by bell pepper. *J. Amer. Soc. Hort. Sci.* 119:936–939.

Somasekhara, Y.M. and Ravichandra, N.G. 2010. Evaluation of bio-pesticides for the management of *Meloidogyne incognita* in carrot. *National Conf. on Innovations in Nematological Research for Agricultural Sustainability – Challenges and A Roadmap Ahead.* Tamil Nadu Agri. Univ., Coimbatore, p. 73.

Son, S.H., Khan, Z., Kim, S.G. and Kim, Y.H. 2009. Plant growth-promoting rhizobacteria, *Paenibacillus polymyxa* and *Paenibacillus lentimorbus* suppress disease complex caused by root-knot nematode and Fusarium wilt fungus. *J. Appl. Microbiol.* 107: 524-532.

Sonowal, B., Mahanta, B., Borah, A. and Dutta, P. 2020. Efficacy of *Glomus fasciculatum*, org-trichojal, vermicompost and carbofuran 3G in the management of *Meloidogyne incognita* on ivy gourd. *Int. J. Curr. Microbiol. App. Sci.* 9(5): 936-943.

Sowmya D.S. and Rao M.S., 2011. Biotechnological approaches for bio-management of disease complex in carrot (*Daucus carota* L.) by using fluorescent pseudomonads and *Paecilomyces lilacinus*. *National Nematology Symposium on "Nematodes: A Challenge Under Changing Climate and Agricultural Practices"*, 16-18 November, 2011, Thiruvananthapuram, Kerala, India, pp. 143-144.

Sowmya, D.S., Rao, M.S., Gopalakrishnan, C. and Ramachandran, N. 2010. Biomanagement of *Meloidogyne incognita* and *Erwinia carotovora* s. sp. *carotovora* infecting carrot (*Dacus carota* L.). *National Conf. on Innovations in Nematological Research for Agricultural Sustainability – Challenges and A Roadmap Ahead,* Tamil Nadu Agri. Univ., Coimbatore, p. 97.

Sowmya, D.S., Rao, M.S., Manoj Kumar, R. *et al.* 2012. Bio-management of *Meloidogyne incognita* and *Erwinia carotovora* in carrot (*Daucus carota* L.) using *Pseudomonas putida* and *Paecilomyces lilacinus*. *Nematol. Medit.* 40: 189-194

Sreegayathri, E., Karthikeyan, G., Rajendran, L. *et al.* 2018. Effect of root-knot nematode (*Meloidogyne incognita*) infestation on severity of wilt (*Fusarium solani*) in bitter gourd and its management. *Madras Agric. J.* 105(10-12): 459-468.

Srinivasan, A. 1974. *Studies on Root Lesion Nematode, Pratylenchus delattrei.* M. Sc. thesis, Tamil Nadu Agricultural University, Coimbatore.

Srivastava, S. S. and Swarup, G. 1986. Biomanagement of root-knot nematode (*Meloidogyne incognita*) through the predatory fungus, *Arthrobotrys conoides. National Conference on Plant Parasitic Nematodes of India - Problems and Progress*, Indian Agri. Res. Inst., New Delhi, p. 85.

Steyn, W.P., Daneel, M.S. and Slabbert, M.M. 2013. Evaluation of *Amaranthus* species for their host suitability to the root-knot nematodes, *Meloidogyne incognita* race 2 and *Meloidogyne javanica* in South Africa. *Acta Hortic.* 1007: 403-407.

Stirling, G.R., Berthelsen, J.E., Garside, A.L. and James, A.T. 2006. The reaction of soybean and other legume crops to root-knot nematodes (*Meloidogyne* spp.), and implications for growing these crops in rotation with sugarcane. *Australas. Plant Pathol.* 35: 707–714.

Su, X.L. and Li, Y. 2004, Quantum dot bio-labelling coupled with immunomagnetic separation for detection of *Escherichia coli* 0157:H7.2004. *Anal. Chem.* 76: 4806-4810.

Sud, U.C., Varaprasad, K.S., Seshadri, A.R. and Kher, K. 1984. Relationships between initial densities of Rotylenchulus reniformis and damage to cotton with a fit to Seinhorst's curve. *Indian J. Nematol.* 14: 148-151.

Sujata, Patil, J.A., Kumari, B. *et al.* 2022. Management of root-knot nematode, *Meloidogyne incognita* infecting okra using botanicals. *The Pharma Innovation Journal* 11(5): 2242-2244.

Sumita, D. and Das, K.K. 1990. Studies on avoidable losses in mung bean (*Vigna radiata*) due to root-knot nematode, *Meloidogyne javanica* and its control under field conditions. *Indian Journal of Nematology* 20: 148-151.

Sumita, K. and Das, D. 2014. Management of root-knot nematode, *Meloidogyne incognita* on green gram through bioagents. *International Journal of Plant, Animal and Environmental Sciences* 4(4): 287-289.

Sumner, D.R. and Johnson, A.W. 1973. Effect of root-knot nematodes on Fusarium wilt of watermelon. *Phytopathology* 63: 857–861.

Swamy, K.R.M. and Sadashiva, A.T. 2004. *Vegetable Varieties/Hybrids of IIHR.* Tech. Bull. 2, Indian Inst. of Hort. Res., Bangalore, 30 pp.

Talavera, M., and Valor, H. 2000. Influence of the previous crop on the anhydrobiotic ability of *Pratylenchus thornei* and *Merlinius brevidens*. *Nematol. Medit.* 28: 77-81.

Talukdar, R. 1993. *Management of* Meloidogyne incognita *(Kofoid and White, 1919) Chitwood 1949 on okra* (Abelmoschus esculentus (*L.*) *Moench*). M. Sc. (Ag) thesis, Assam Agri. Univ., Jorhat.

Tamilselvi, N.A. and Pugalendhi, L. 2015. Agronomic evaluation of bitter gourd grafts for yield and quality. *Bioscan* 10(3): 1331-1334.

Tamilselvi, N.A., Pugalendhi, L. and Sivakumar, M. 2016. Defence responses of cucurbitaceous rootstocks and bitter gourd scions against root-knot nematode *Meloidogyne incognita* Kofoid and White. *Vegetos* 29: 4. doi: 10.5958/2229-4473. 2016.00111.7.

Tamilselvi, N.A., Pugalendhi, L. and Sivakumar, M. 2017. Response of cucurbitaceous rootstocks and bitter gourd scions to root-knot nematode *Meloidogyne incognita* Kofoid and White. *International Journal of Horticulture and Floriculture* 5(1): 247-251.

Tanda, A.S. and Atwal, A.S. 1988. Effect of sesame intercropping against the root-knot nematode (*Meloidogyne incognita*). *Nematologica* 34: 484-492.

Tanimola, A.A. and Akarekor, C. 2014. Management of root-knot nematode (*Meloidogyne incognita*) on okra (*Abelmoschus esculentus* (L.) Moench) using carbofuran and some animal manures. *World Journal of Agricultural Sciences* 10(4): 185-193.

Tarafdar, J.C. 2015. *Nanoparticle Production, Characterization and its Application to Horticultural Crops*. ICAR- Central Arid Zone Research Institute, Jodhpur, Rajasthan, pp. 222–229.

Tarafdar, J.C., Raliya, R., Mahawar, H. and Rathore, I. 2014. Development of zinc nano-fertilizer to enhance crop production in pearl millet (*Pennisetum americanum*). *Agric. Res.* 3: 257–262.

Tempta, N. 2017. *Management of Root-Knot Nematode* Meloidogyne incognita *(Kofoid & White, 1919) Chitwood, 1949 in Cucumber,* Cucumis sativus *L.* M. Sc. Thesis, Dr. Y S Parmar Univ. of Hort. and Forestry, Solan, Himachal Pradesh, 70 pp.

Thakar, N.A. and Patel, C.C. 1984. Screening of few cowpea varieties against reniform nematode, *Rotylenchulus reniformis* Linford and Oliveira, 1940. *Indian J. Nematol.* 14: 204.

Thakar, N.A. and Patel, C.C. 1985. Control of root-knot nematode, *Meloidogyne javanica* on tomato by chemical bare-root dips. *Indian Journal of Nematology* 15: 110-111.

Thakur, R. 2016. *Management of Root-Knot Nematode* (Meloidogyne incognita *Kofoid & White*) *in Chilli* (Capsicum annuum *L.*). M. Sc. Thesis, Dr. Y.S. Parmar Univ. Hort. Forestry, Solan, Himachal Pradesh, 60 pp.

Thangaraju, D. 1983. Distribution of potato cyst nematodes in Kodaikanal Hills, Madurai district, Tamil Nadu. *Indian Journal of Nematology* 13: 222-223.

Thapa, C.D., Jandaik, C.L. and Sharma, N.K 1987. Comparative resistance in some cultivated edible fungi against two mushroom nematodes occurring in India. *Proc. Int. Conf. on Sci. Cultivation Technol. of Edible Fungi,* pp. 169-172. Regional Research Laboratory, Jammu and Kashmir, India.

Thapa, C.D., Sharma, N.K and Nath, A. 1981. Mushroom nematodes and their source of infestation in mushroom beds in India. *Mushroom Sci.* 11: 443-448.

Thirumalachar, M.J. 1951. Root-knot nematode on potato tubers in Shimla. *Curr. Sci.* 20: 104.

Thompson, J., Greco, N., Eastwood, R. *et al.* 2000. Integrated control of nematodes of cool season food legumes. In R. Knight (ed.) *Linking Research and Marketing Opportunities for Pulses in the 21st Century*, 3rd edn., pp. 491–506. Kluwer Academic Publishers, Dordrecht.

Trimbake, D.K. and Mail, N.S. 2021. Management of root knot nematodes, *Meloidogyne incognita* using bioagents and compost in cucumber, ridge gourd, bitter gourd and bottle gourd cultivars. *Annals of R.S.C.B.* 25(6): 13512–13516.

Tripathi, P.K. and Singh, C.S. 2006. Effect of some compatible biocontrol agents along with mustard cake and furadan on *Meloidogyne incognita* infecting tomato plant. *Indian J. Nematol.* 36: 309-311.

Trivedi, P.C. and Tiagi, A. 1984. Effect of some plants on *Meloidogyne incognita* population in chilli soil. *Indian J. Bot. Sci.* 63: 416-419.

Turner, N.C. 2004. Agronomic options for improving rainfall-use efficiency of crops in dryland farming systems. *J. Exp. Bot.* 55: 2413–2425.

Tzortzakakis, E.A. and Gowen, S.R. 1994. Evaluation of *Pasteuria penetrans* alone and in combination with oxamyl, plant resistance and solarization for control of *Meloidogyne* spp. on vegetables grown in greenhouses in Crete. *Crop Prot.* 13: 455–462.

Ugade, S.R., Ayare, B.L., Thorat, T.N. and Thokal, R.T. 2014. Effect of irrigation and fertigation levels on yield and nutrient uptake of brinjal (*Solanum melongena* L.). *International Journal of Agricultural Engineering* 7(1): 74- 80.

Umamaheswari, R., Somasekhar, N. and Joseph, T.A. 2010. Evaluation of biocontrol agents on potato cyst nematodes, *Globodera rostochiensis* and *G. pallida* infesting potato. *National Conf. on Innovations in Nematol. Res. for Agri. Sustainability – Challenges and A Roadmap Ahead.* Tamil Nadu Agri. Univ., Coimbatore, p. 86.

Umamaheswari1, R., Somasekhar, N., Manorama, K. and Joseph, T.A. 2012. Eco-friendly management of potato cyst nematodes in the Nilgiris of Tamil Nadu. *Potato J.* 39 (2): 185-190.

Upadhyay, K.D. and Dwivedi, K. 1987. Analysis of crop losses in pea and gram due to *Meloidogyne incognita. Int. Nematol. Network Newsl.* 4(4): 6-7.

USDA NIFA, 2015. *Description of Commercial Cucurbit Rootstocks as of February 5, 2015.* USDA NIFA.

Vadhera, I., Sheila, B.N. and Bhatt, J. 1995. Interaction between reniform nematode (*R. reniformis*) and *Fusarium solani* causing root rot of French bean. *Indian J. Agric. Sci.* 65: 774-777.

Vadivelu, R.S. and Rajendran, G. 1986. Damage potential of the root-knot nematode, *Meloidogyne arenaria* to onion. *Int. Nematol. Network Newsl.* 4(2): 5-8.

Vallotton, R. 1980. Le nematode a kyste *Heterodera carotae*, un nuveau ravageur de la carrotte en Suisse romande. *Le maraicher* 43(11/12): 258-261.

Vance, M.E., Kuiken, T., Vejerano, E.P. *et al.* 2015. Nanotechnology in the real world: Redeveloping the nanomaterial consumer products inventory. *Beilst. J. Nanotechnol.* 6: 1769–1780.

Varaprasad, K.S., Sharma, S.B. and Loknathan, T.R. 1997. Nematode constraints to pigeon pea and chickpea in Vidarbha region of Maharashtra in India. *Int. J. Nematol.* 7: 152–157.

Varma, M.K., Sharma, H.C. and Pathak, V.N. 1978. Evaluation of systemic granular nematicides against *Tylenchorhynchus* spp. parasitic on cabbage. *Indian Journal of Nematology* 8: 59-60.

Varshney, V.P., Khan, A.M. and Saxena, S.K. 1987. Interaction between *Meloidogyne incognita, Rhizoctonia solani* and *Rhizobium* species on cowpea. *Int. Nematol. Network Newsl.* 4(3): 11-14.

Vasu, D. and Reddy, M.S. 2013. Effect of fertigation on yield, quality, nutrient uptake, fertilizer and water use efficiency in cabbage (*Brassica oleracea*). *Agropedology* 23: 106-112.

Vats, R. and Dalal, M.R. 1997. Interaction between *R. reniformis* and *Fusarium oxysporum* f. sp. *pisi* on pea. *Ann. Biol., Ludhiana* 13: 239-242.

Veere Gowda, R., Singh, T.H., Nagesh, M. and Rao, M.S. 2001. Evaluation of carrot (*Dacus carota L.*) genotypes for resistance to root-knot nematode (*Meloidogyne incognita*). *Proc. Second National Symp. Integrated Pest Mangmt. Horti. Crops: New Molecules, Biopesticides and Environment*, Bangalore, p. 155.

Venkatesan, M. and Patel, B.A. 2013. Management of root-knot nematodes in bitter gourd using organic amendments, nematicides and bioagents. *Annals of Plant Protection Sciences* 21(1): 218-220.

Verma, K.K., Goel, S.R. and Paruthi, I.J. 2010. Studies on biomanagement of root-knot and root-rot complex in cowpea (*Vigna unguiculata*) by fungal antagonists. *National Conf. on Innovations in Nematological Research for Agricultural Sustainability – Challenges and A Roadmap Ahead*, Tamil Nadu Agri. Univ., Coimbatore, p. 83.

Verma, A.C. 2001. N.D. University of Agriculture and Technology, Kumarganj, Faizabad, Uttar Pradesh. In S.C. Dhawan *et al.* (eds.) *Indian Nematology-Progress and Perspectives*. Divn. of Nematol., Indian Agri. Res. Inst., New Delhi, pp. 121-125.

Verma, A.C. and Anwar, A. 1992. *First Annual Technical Report ICAR Ad-hoc Scheme Investigation on Root-Knot Nematode on Pointed Gourd*. Department of Nematology, NDAUAT, Faizabad, U.P., India.

Verma, A.C., Singh, H.K. and Khan, N. 2005. Management of root-knot nematode, *Meloidogyne incognita* through antagonistic approaches in pointed gourd. *Indian J. Nematol.* 35: 78-79.

Vetrivelkalai, P. and Subramanian, S. 2006. Plant parasitic nematode populations as affected by onion based coping sequences. *Indian J. Nematol.* 36: 10-14.

Viaene, N.M. and Abawe, G.S. 2000. *Hirsutella rhossiliensis* and *Verticillium chlamydosporium* as biocontrol of root-knot nematode *Meloidogyne hapla* on lettuce. *J. Nematol.* 32: 85-100.

Vicente, N.E., Sance, L.A. and Acosta, N. 1991. Effect of granular nematicides and the fungus *Paecilomyces lilacinus* in nematode control in watermelon. *J. Agri. Univ. Puerto Rico* 75: 307-309.

Vovlasa, Lucarellib, G., Sasanellia, N. *et al.* 2008. Pathogenicity and host-parasite relationships of the root-knot nematode *Meloidogyne incognita* on celery. *Plant Pathology* 57: 981–987.

Walia, R.K. and Chakraborty, P.K. (eds.). 2018. *Nematodes Problems of Crops in India*. ICAR-All India Coordinated Research Project on Nematodes in Agriculture, New Delhi, 402 pp.

Walia, K.K. and Gupta, D.C. 1986a. Antagonism between *Heterodera cajani* and *Rhizoctonia solani* on cowpea (*Vigna unguiculata*). *Indian J. Nematol.* 16:41-43.

Walia, K.K. and Gupta, D.C. 1986b. Effect of the fungus *Rhizoctonia solani* on the population development of pigeon-pea cyst nematode, *Heterodera cajani* on cowpea (*Vigna unguiculata*). *Indian J. Nematol.* 16: 131-132.

Walia, R.K., Bansal, R.K. and Bhatti, D.S. 1990. A new bacterial parasite (*Pasteuria* sp.), isolated from pigeonpea cyst nematode. *Proc. of Seminar on Bioagents in Nematode Management*, Indian Agri. Res. Inst., New Delhi, p. 57.

Walia, R.K., Bansal, R.K. and Bhatti, D.S. 1991. Effect of *Paecilomyces lilacinus* application time and method in controlling *Meloidogyne javanica* on okra. *Nematologia Mediterranea* 19: 247-149.

Walters, S.A., Wehner, T.C. and Barker, K.R. 1993. Root-knot nematode resistance in cucumber and horned cucumber. *HortScience* 28: 151–154.

Wang, K. 2019. *Reniform nematode (*Rotylenchulus reniformis*) Linford and Oliveira (Nematoda: Tylenchida: Tylenchoidea: Hoplolaimidea: Rotylenchulinae)*. University of Florida publication Series no. EENY-210 (IN367), p. 4.

Wang, K.H. 2010. Improving marigold cover cropping for nematode management. *The Food Provider* June-July-Aug 2010 Issue.

Wang, W.N., Tarafdar, J.C. and Biswas, P. 2013. Nanoparticle synthesis and delivery by an aerosol route for watermelon plant foliar uptake. *J. Nanopart. Res.* 15(1): 1417.

Wani, A.H. and Bhat, M.Y. 2012. Control of root-knot nematode, *Meloidogyne incognita* by urea coated with Nimin or other natural oils on mung, *Vigna radiata* (L.) R. Wilczek. *J Biopestic.* 5(Suppl.): 255–258.

Waykule, R.K., Palande, P.R. and Waykule, P.K. 2023. Assessment of avoidable yield loss due to root knot nematode, *Meloidogyne incognita* infesting ridge gourd, *Luffa acutangula* (L.) Roxb. *Int. J. Curr. Microbiol. App. Sci.* 12(1): 209-214.

Wepuhkhulu, M., Kimenju, J., Anyango, B. *et al.*, 2011. Effect of soil fertility management practices and *Bacillus subtilis* on plant pasitic nematodes associated with common bean, *Phaseolus vulgaris*. *Trop. Subtrop. Agroecosyt.* 13(1): 27-34.

Whitehead, A.G. 1998. *Plant Nematode Control.* CAB International

Whitehead, A.G., Dunning, R.A. and Cooke, D.A. 1971. Docking disorder and root ectoparasitic nematodes of sugar beet. *Report of Rothamsted Experimental Station for 1970*, part II, pp. 219-236.

Wilcken, S.R.S., Garcia, M.J.M. and Silva, N. 2005. Resistência de alface do tipo americana a *Meloidogyne incognita* raça 2. *Nematologia Brasileira* 29(2): 267-271.

Wyatt, J.E., Fassuliotis, G., Hoffman, J.C. and Deakin, J.R. 1983. "Nemasnap" snap bean. *Hort Science* 18: 776.

Yadav, P.K., Baheti, B.L. and Rathore, B.S. 2009. Pathogenicity of root-knot nematode, *Meloidogyne incognita* on cluster bean (*Cyamopsis tetragonoloba*). *Journal of Mycology and Plant Pathology* 39: 172-173.

Zakaria, H.M., Kassab, A.S., Shamseldean, M.M. *et al.* 2013. Controlling the root-knot nematode, *Meloidogyne incognita* in cucumber plants using some soil bioagents and some amendments under simulated field conditions. *Annals of Agricultural Sciences* 58(1): 77-82.

Zaki, F.A. 1998. Biological control of *Meloidogyne javanica* in tomato by *Paecilomyces lilacinus* and castor leaves. *Indian J. Nematol.* 28: 132-139.

Zaki, F.A. and Bhatti, D.S. 1986a. Effect of pigeon pea cyst nematode, *Heterodera cajani* Koshy, 1967 on macro and micronutrients of pigeon pea and moth. Indian Journal of Nematology 16(1): 103–105.

Zaki, F.A. and Bhatti, D.S. 1986b. Control of pigeon-pea cyst nematode, *Heterodera cajani* Koshy, 1967 by chemical seed treatment. *Indian Journal of Nematology* 16(1): 106-108.

Zaki, F.A. and Bhatti, D.S. 1991. Effect of castor *Ricinus communis* and the biocontrol fungus, *Paecilomyces lilacinus* on *Meloidogyne javanica*. *Nematologica* 36: 114-122.

Zamioudis, C. and Pieterse, C.M.J. 2012. Modulation of host immunity by beneficial microbes. *Molecular Plant–Microbe Interactions* 25: 139–150.

Zhang, M., Gao, B., Chen, J. *et al.* 2014. Slow-release fertilizer encapsulated by graphene oxide films. *Chem. Eng. J.* 255: 107–113.

Zhang, S., Gun, X. and Wang, Y. 2006. Effect of bur cucumber (*Sicyos angulatus* L) as rootstock on growth physiology and stress resistance of cucumber plants (in Chinese). *Acta Hort. Sinica* 33: 1231–1236.

Zuckermann, B.E., Dicklow, M.B. and Ascosta, N. 1993. A strain of *Bacillus thuringiensis* for the control of plant parasitic nematodes. *Biocont. Sci. Technol.* 3: 41-46.

FSC
www.fsc.org
MIX
Papier | Fördert gute Waldnutzung
FSC® C083411

Kolibri 360 GmbH
Ferdinand-Jühlke-Straße 7
99095 Erfurt, Deutschland
produktsicherheit@kolibri360.de